OCEAN
SOLUTIONS
EARTH
SOLUTIONS

Edited by
DAWN J. WRIGHT

Foreword by David G. Gallo

Esri Press
REDLANDS|CALIFORNIA

Contents

Dawn J. Wright, ed.; 2015; *Ocean Solutions, Earth Solutions*; http://dx.doi.org/10.17128/9781589483651

Foreword

We are living on an ocean planet. In fact, more than 70% of the earth's surface is covered by ocean water. The ocean is vast, deep, and in places represents some of the most remote, hostile, and dynamic environments on Earth. To date, we have explored less than 10% of the world beneath the sea, yet in our voyage of discovery, we find the most dramatic and dynamic topography on the planet; we find thriving ecosystems where we were sure there could be no life at all; and perhaps most important, we find keys and clues to the evolution of Earth itself. Although we are a long way from understanding the relationship between humanity and the sea, one important headline has emerged: a healthy global ocean is critical to human health. Any review of recent scientific literature, anecdotal evidence from those that live on or make their living on the sea, and environmental evidence all lead to one conclusion: the ocean is in deep trouble, and the world is facing a near-term crisis.

We used to think that the ocean was self-healing, too big to fail, and that there were no limits to its bounty. We now know those statements to be untrue. Over time, human activity has slowly but surely changed the chemistry and temperature of the sea. In addition, we have managed to collapse the stock of some of the world's greatest fisheries, and it's unclear how bad the situation really is. Through the growth of civilization, we are putting increasing environmental pressure on ocean ecosystems. No matter where on Earth we live, what we put on our fields, backyards, golf courses, and city streets eventually makes its way to the streams, rivers, estuaries, and the ocean; and in many cases, the results are catastrophic. Dead zones are becoming more widespread and long lived than ever. Today, more than 30% of the world's seven billion people live within the coastal zone, and 15 of the world's most populated cities are built either on the ocean or on estuaries. Population and coastal growth are continuing to rise, and along with that comes the risk from coastal storms, hurricanes, typhoons, and tsunamis. There is a desperate need to better manage our coastal zone.

If there is any "good news," it's that "now we know." We know that the ocean provides us with the greatest portion of the air we breathe, the food we eat, and the water we drink. We know that environmental pressure on the sea due to human activity is more, not less. We also know that regardless of where we live on planet Earth, we have an impact on the ocean. Conversely, regardless of where we live on Earth, the ocean has an impact on our everyday lives.

Despite knowing these important things, there is a knowledge gap that needs to be filled. The ocean is complex, dynamic, and ever changing. It's clear that we need to create policy to better manage the ocean, but to be effective these policies need to be based on fact as much as emotion. The ocean has a strong emotional pull, but if we are not careful in our actions, we will suffer from unintended consequences and ultimately we will love this planet to death.

The past several decades have seen rapid growth in our ability to explore and understand the world beneath the sea. Observation is the cornerstone of science, and where the ocean is concerned, observation requires sophisticated technology. The development and deployment of new platforms (satellites, ships, submarines, remotely operated vehicles, autonomous undersea vehicles, towed systems, buoys, drifters, gliders, and more) and new sensors (sonars; cameras; physical, chemical, and geophysical sensors) allow us to "see" the seafloor and waters above with unprecedented clarity, accuracy, and precision. This data, combined with airborne and satellite information, is for the first

Dawn J. Wright, ed.; 2015; *Ocean Solutions, Earth Solutions*; http://dx.doi.org/10.17128/9781589483651

time providing an effective means of sampling the ocean at discrete locations as well as conducting long-term observations over entire ocean basins. All this data combined, complemented by new GIS analytical and modeling technologies and techniques, has provided humanity with a new view of planet Earth. In so doing, it is becoming easier to recognize and observe the relationship between humanity and the sea. Herein lies the topic of this book.

To understand the ocean, we need to understand what's "out there." What are the components? How are they arranged? How do they change in time and space? The ability to recognize and observe relationships, patterns, and trends is the true gift of maps and the trilogy of GIS.

The ability to collect oceanographic data, especially multisensor/multiscale data, has heralded the beginning of a new generation in ocean sciences. Combining these sensors and platforms with the ability to visualize and share information through GIS has led to an exponential growth of confidence in our ability to describe the ocean world. *So the publication of this book comes at a critical time. For the very first time, we can balance the term "ocean crisis" with the term "ocean solutions." For the first time, we can share data and information with a global audience in a manner that allows them to "participate."*

The challenge of understanding the relationship between humanity and the sea remains daunting. The goal of implementing policy to best ensure that the ocean is managed wisely is even more difficult. Nevertheless, these things are necessary and critical to a healthy ocean, a thriving planet, and human survival. We can perhaps draw from the wisdom that the longest journey begins with the first small steps forward. We now have the tools, and we can let the journey begin.

—David G. Gallo
Director of Special Projects, Woods Hole Oceanographic Institution
4-time TED speaker
CNN analyst

Acknowledgments

I thank all the contributors to this book for their enthusiasm and skill in authoring the chapters and preparing and contributing the supplemental resources. In addition, the book benefited greatly from the external peer review process afforded to each of the chapters. Reviewers were chosen for their domain science and GIS technical expertise, specific knowledge about the material in a particular chapter, and ability to be fair and insightful in their assessments. The reviews were conducted to ensure standards of objectivity, clarity, responsibility, and overall scientific quality rivaling that of many peer-reviewed journals. I wish to thank the following individuals for taking the time to provide such thorough, candid, and constructive reviews:

Becky Allee, National Oceanic and Atmospheric Administration (NOAA) Gulf Coast Services Center

Christine Baier, NOAA Alaska Fisheries Science Center

Rob Braun, South Florida Water Management District

Jerry Davis, San Francisco State University

Moe Doucet, Quality Positioning Services (QPS)

Annette Dougherty, NOAA Alaska Fisheries Science Center

Sally Duncan, Oregon State University

Lainie Edwards, Florida Department of Environmental Protection

Edward Game, The Nature Conservancy

Felimon Gayanilo, Texas A&M University, Corpus Christi

Jim Graham, Humboldt State University

Ellen Hines, San Francisco State University

Alejandro Iglesias Campos, UNESCO Intergovernmental Oceanographic Commission

Carissa Klein, University of Queensland, Australia

Rick Lathrop, Rutgers University

Yuanjie Li, NOAA National Oceanographic Data Center

Daniel Martin, NOAA Office for Coastal Management

Ann Matarese, NOAA Alaska Fisheries Science Center

Emilio Mayorga, University of Washington

Carrie McDougall, NOAA Office of Education

John McLaughlin, NOAA Office of Education

Nazila Merati, ClipCard

Ivonne Ortiz, NOAA Alaska Fisheries Science Center

Daniel Palacios, Oregon State University

Jaynya Richards, Esri

Emily Shumchenia, E&C Enviroscape

Melissa Stevens, The Nature Conservancy

David Stoms, California Energy Commission

Robert Sullivan, Argonne National Laboratory

Ben Waltenberger, Channel Islands National Marine Sanctuary
Christine White, Esri
Curt Whitmire, NOAA Northwest Fisheries Science Center
Olga Wilhelmi, National Center for Atmospheric Research
Brooke Wikgren, New England Aquarium

These reviewers were not asked to endorse any of the findings, conclusions, or recommendations in the chapters, nor do these necessarily reflect the views of Esri or Esri Press.

Finally, I would like to express deep appreciation for the team at Esri Press. They offered invaluable advice and provided excellent editorial oversight and support. I am especially grateful to Esri President Jack Dangermond for his continued vision and support of a healthy ocean.

Dawn J. Wright
Esri Chief Scientist

Introduction

Dawn J. Wright

The ocean dominates the surface of the earth and greatly affects our daily lives. It regulates Earth's climate, plays a critical role in the hydrologic cycle, sustains a large portion of Earth's biodiversity, supplies food and mineral resources, constitutes an important medium of national defense, provides an inexpensive means of transportation, is the final destination of many waste products, is a major location of human recreation, and inspires our aesthetic nature. However, it also carries with it the threat of deadly tsunamis and hurricanes, industrial accidents, and outbreaks of waterborne pathogens. (National Academy of Sciences National Research Council 2011)

Our ability to measure change in the ocean is increasing, not only because of improved measuring devices and scientific techniques, but also because new GIS technology is aiding us in better understanding this dynamic environment. The domain of GIS has progressed from applications that merely collect and display data to complex simulation and modeling and the development of new research methods and concepts.

GIS has traditionally provided effective technological solutions to the integration, visualization, and analysis of heterogeneous, georeferenced data *on land*. As early as the late 1980s, Esri began to make significant progress in the application of *ocean* GIS solutions to nautical charting, commercial shipping, oil spill response, and defense and intelligence. Recent years have seen GIS increasingly used in ocean science, conservation, and resource management communities, in large part because of the Esri Ocean GIS initiative launched in 2012 (Wright 2012). Many challenges remain, especially regarding inconsistencies in ocean data models, formats, standards, tools, services, and terminology.

Toward this end, Esri held a one-time-only Oceans Summit at its headquarters in Redlands, California, in November 2012 (Wright 2013). Intermediate-to-advanced ocean GIS analysts and developers attended this invitation-only, high-level strategic workshop. Their goal was to help Esri move forward in its approaches to ocean-centric software, associated data formats, tools, workflows, and computing platforms. The success of that summit led to the first annual, all-comers Esri Ocean GIS Forum, held in November 2013, again in Redlands (Pratt 2013). Presentations at the forum covered a wide range of topics, including ocean exploration and science, coastal management and marine spatial planning, coastal resilience and conservation, hydrographic surveying, commercial ship tracking, and closing technological gaps in multidimensional data handling and analysis.

Attendees lauded these earnest efforts to apply GIS to the ocean (the open ocean as well as the nearshore or coast) and resolve a range of technological gaps. Going forward, they also understand that the health and sustainability of the planet still lies in the balance. With water covering 71% of the planet's surface, it's clear that the ocean is critical to our lives, our energy, and our economy. This vital role is increasingly evident in the latest warnings from the UN Intergovernmental Panel on Climate Change (2013), the Third US National Climate Assessment (Melillo et al. 2014), the

Dawn J. Wright, ed.; 2015; *Ocean Solutions, Earth Solutions*; http://dx.doi.org/10.17128/9781589483651

launch of the US White House Climate Data Initiative, and the search for Malaysia Airlines flight MH 370 (ongoing at the time of this writing). The nongovernment sector has also recognized the need for action. The Global Partnership for Oceans, XPRIZE Foundation (and its focus on ocean health), Mission Blue Alliance, and Schmidt Ocean Research Institute are just a few of the new organizations that have emerged to focus on ocean sustainability.

Human-caused pressures have put the ocean in a state of deep crisis. *And if the ocean is in crisis, the earth is in crisis.* At this pivotal juncture in history, we need good, digestible science to underpin solutions for

- protecting the ocean and ensuring our safety,
- managing and mitigating conflict among multiple and simultaneous uses of the ocean,
- geodesigning the ocean, and
- discovering and exploring a part of our planet that remains less understood than the moon, Mars, and Venus.

Ocean Solutions, Earth Solutions is about use-inspired science and realistic solutions for the ocean and thus the earth. With chapters drawn from among the best science presented at the 2013 Esri Ocean GIS Forum, this book seeks to put that science into the hands of government decision-makers and ocean/coastal science researchers, state and local coastal zone managers, and ocean/coastal GIS practitioners. It also seeks to preserve good scholarship, including the emerging scholarship of students working on theses and dissertations. The editor and a small cadre of experts conducted a standard academic peer review of all chapters. The book encourages GIS best practices. Toward this end, it features an extensive digital supplement, including datasets with accompanying digital object identifiers (DOIs, in keeping with data publication trends; e.g., The Royal Society 2012; Leadbetter et al. 2013; Parsons and Fox 2013), geoprocessing workflows, GIS tools packaged as desktop extensions or web services, mobile apps, Python scripts, and story maps based on the Esri Story Map app. Digital content for this book, described under "Supplemental Resources" at the end of pertinent chapters, can be accessed on the Esri Press "Book Resources" webpage at esripress.esri.com/bookresources. Then, in the list of Esri Press books, click *Ocean Solutions, Earth Solutions*. On the *Ocean Solutions, Earth Solutions* resource page, click a chapter link to access that webpage and the links to the digital content for that chapter. URLs and QR codes are also provided in the book. Some websites may require Adobe Flash Player.

The solutions developed in these chapters lend credence to the hope that we can fix the problems of the ocean before it is too late—to build what Knowlton (2014) and others have tagged as a new #OceanOptimism hashtag on Twitter. In a similar vein, the Mission Blue initiative of the Sylvia Earle Alliance has established "hope spots" (Mission Blue 2014). Indeed, the efforts of scientists, resource managers, coastal engineers, and local communities have shown many promising results around the world. As Knowlton urges, we need to trumpet and seed these solutions.

The chapters of *Ocean Solutions, Earth Solutions* are divided into four main themes:

1. Server/cloud GIS
2. Coastal and marine spatial planning
3. Analytical and mapping tools
4. Visualization

During the 1980s and '90s, GIS architectures were largely file-oriented, with Esri datasets structured as coverages and later shapefiles. In the late 1990s, geographic datasets were placed into relational databases to allow for multiuser access. The web has evolved more recently to embrace a service model powered by "on premise" servers and/or the cloud, extending the old desktop data models with layers and web maps that travel *with* the data.

Chapter 1 represents the critical role that **server/cloud GIS** now plays in fisheries oceanography, particularly as local desktop resources are less able to handle the huge streams of data from ocean observatories, the complexity of numerical models, and the massive output files these models generate. The chapter describes the on-premise, server-based Ecosystems Fisheries Oceanography Coordinated Investigations (EcoFOCI) Data Access and Analysis Tool (EcoDAAT) and the cloud-based LarvaMap, both developed for data integration and particle modeling to support fisheries oceanography and fisheries management. And because server/cloud GIS has become so important for data discovery and collaborative geodesign as part of coastal and marine spatial planning, it segues to the next theme of **coastal and marine spatial planning**.

Chapter 2 focuses on best practices for decision support guided by the popular Marxan approach when used in concert with GIS to inform multiobjective ocean conservation and spatial planning. Chapter 3 describes coastal and marine spatial planning programs under way in Martin County, Florida, and provides specific exemplars for data consolidation and organization, a workable data model and geodatabase schema, and GIS applications for managing artificial reefs and beaches.

The next theme highlights chapters presenting specific **analytical and mapping tools**. Chapter 4 introduces tools for use in conjunction with the Coastal and Marine Ecological Classification Standard (CMECS), the first consistent US national classification for mapping and integrating ecosystem observations. The NatureServe observation toolkit and the CMECS crosswalk tool of the National Oceanic and Atmospheric Administration (NOAA) are examples of effective solutions for standardizing the collection of data in the field and for uploading, validating, aggregating, and integrating that data with existing archives. Chapter 5 provides important guidelines and a desktop solution for accessing, mapping, and interpreting the output of numerical models used for projecting the effects induced by changes in atmospheric greenhouse gases. It describes the methodology used to extract the relevant ocean variables and considers the consequences of making decisions based on climate change scenarios. Policy makers and resource managers alike need such insights for climate adaptation and mitigation plans. A method and an accompanying ArcGIS tool for calculating a pollutant exposure index (PEI) for the Southern California Bight are the subject of chapter 6, which provides Python scripts for geoprocessing the exposure data and producing georeferenced PEI rasters that may be used with other spatial data to examine relative pollution risk for any area of interest within the mapped region.

Chapter 7 emphasizes mapping tools, in this case a map viewer to facilitate the management and monitoring of the Archipelago de Cabrera National Park in the Balearic Islands of Spain. The chapter includes useful guidelines for designing an intuitive interface with the Adobe Flash Integrated Development Environment along with the ArcGIS Viewer for Flex and creating Open Geospatial Consortium Web Map Services. Chapter 8 describes design principles and development of both a web app and a mobile app for mapping marine mammal distributions, drawing in large part from data contributed by citizen scientists. Chapter 9 discusses the benefits of providing multiple computer pathways within the powerful web tool of the West Coast Ocean Data Portal to better enable search and connection to the services and datasets published by members of this network. These pathways include Catalog Services for the Web, the Esri Geoportal representational state transfer (REST) application programming interface (API), the OpenSearch API, and the Apache Solr REST API.

Chapter 10 analyzes land-cover data from the East End Marine Park of Saint Croix, US Virgin Islands, applying a Landscape Development Intensity Index for adjacent watersheds along with an analysis of benthic habitat data, both from benthic habitat maps and in-water surveys. The resulting maps are a powerful tool to understand the connections between actions on land and impacts at sea, specifically the impacts of land-based sources of pollution on benthic habitats and species composition on fringing coral reefs. Chapter 11 employs a suite of both desktop and ArcGIS Online mapping and spatial analysis tools to develop a collaborative proposal to the Pacific Fishery Management Council. The proposal, representing nearly a year of discussion and the building of trust among commercial fishermen, conservationists, and local government, is for the design of spatial modifications to the current Pacific Coast groundfish essential fish habitat (EFH) conservation areas. The development of an Esri Story Map app, in particular, helped this process. Chapter 12 focuses on story maps again, presenting the Esri Story Map app as one of many tools that the Aquarium of the Pacific uses to engage the public in lifelong learning experiences. The chapter provides best-practice examples in employing these tools to create a rich and informal science education experience that inspires learners to take action to protect and conserve.

Chapter 13 begins the next major theme on practical solutions for improved **visualization** of the ocean. It reviews principles for integrating GIS and landscape visualization software to produce immersive and interactive four-dimensional models of the ocean. Planners and scientists provide feedback in a case study that tests these approaches within a marine protected area. Chapter 14 describes advances in visualization methods based on the Worldwide Telescope virtual globe and applies these methods to the management and analysis of data from marine microbial ecology and physical transport processes in the ocean. The discussion has implications for how to more effectively analyze and understand new harvests of large complex datasets in three- and four-dimensional software environments that combine virtual globes with web GIS. Chapter 15 presents a framework adopted by the State of Oregon for managing the visual landscape of the territorial sea. The framework is part of the state's effort to plan for marine renewable energy development. We could have included this chapter in the coastal and marine spatial management section of the book but put it here because of the emphasis on a visual resource scenic quality inventory of publicly accessible viewpoints of the territorial sea, the adoption of visual class standards, and the modeling of viewsheds and viewshed class values. As such, the use of visualization tools described in this chapter may provide an important exemplar for other coastal states seeking to bring public participation and community values into the spatial framework for decision-making, thereby helping communities define and conceptualize management, sustainability, and resilience.

The final study, chapter 16, brings the discussion full circle back to server/cloud GIS while drawing also on the theme of visualization. The chapter presents workflows using ArcGIS 10.2 GeoEvent Extension for Server as well as a series of Python scripts to map in near real time the observations of wave and profile gliders monitoring the shallow water depths of the Gulf of Mexico. Researchers will apply this process for visualizing future observations from these platforms, which are critical for understanding and reducing ocean acidification and oxygen-poor conditions in the Gulf.

May this book not only inspire #OceanOptimism but drive real action with GIS. Such action needs a continued infusion of ideas and best practices. Hence, we envision *Ocean Solutions, Earth Solutions* as the first in a series of research monographs, based on future Esri Ocean GIS Forum events as they continue to evolve.

References

Intergovernmental Panel on Climate Change. 2013. "Summary for Policy Makers." In *Climate Change 2013: The Physical Science Basis.* Contribution of Working Group I to the Fifth Assessment Report of the Intergovernmental Panel on Climate Change, edited by T. F. Stocker, D. Qin, G.-K. Plattner, M. Tignor, S. K. Allen, J. Boschung, A. Nauels, Y. Xia, V. Bex, and P. M. Midgley. Cambridge: Cambridge University Press.

Knowlton, N. 2014. "Why Do We Have Trouble Talking about Success in Ocean Conservation?" http://po.st/IyKMYe. Last accessed August 17, 2014.

Leadbetter, A., L. Raymond, C. Chandler, L. Pikula, P. Pissierssens, and E. R. Urban. 2013. *Ocean Data Publication Cookbook.* IOC Manuals and Guides No. 64. Oostende, Belgium: UNESCO IOC International Oceanographic Data and Information Exchange. http://bit.ly/1lcHa6I.

Melillo, J. M., T. C. Richmond, and G. W. Yohe, eds. 2014. *Climate Change Impacts in the United States: The Third National Climate Assessment.* Washington, DC: US Global Change Research Program. doi:10.7930/J0Z31WJ2.

Mission Blue. 2014. Hope Spots. http://mission-blue.org/hope-spots-new. Last accessed August 23, 2014.

National Academy of Sciences National Research Council. 2011. *An Ocean Infrastructure Strategy for US Ocean Research in 2030.* Washington, DC: National Academies Press.

Parsons, M. A., and P. A. Fox. 2013. "Is Data Publication the Right Metaphor?" *Data Science Journal* 12:WDS32-WDS46.

Pratt, M. 2013. "Saving the Blue Stuff: The First Ocean GIS Forum." *ArcUser* 17 (1): 58–63.

The Royal Society. 2012. *Science as an Open Enterprise: Open Data for Open Science.* The Royal Society Science Policy Centre Report 02/12. London: The Royal Society. https://royalsociety.org/policy/projects/science-public-enterprise.

Wright, D. J. 2012. *The Ocean GIS Initiative: Esri's Commitment to Understanding Our Oceans.* Esri White Paper/e-Book J10129. Redlands, CA: Esri.

Wright, D. 2013. "How Is GIS Meeting the Needs of Ocean (and Other) Sciences? Plus, Minus, Interesting." http://blogs.esri.com/esri/arcgis/?p=28054. Last accessed August 17, 2014.

CHAPTER 1

Cloudy with a Chance of Fish: ArcGIS for Server and Cloud-Based Fisheries Oceanography Applications

Tiffany C. Vance, Stephen Sontag, and Kyle Wilcox

Abstract

Wright et al. (2013) defined cyber-GIS as "GIS detached from the desktop and deployed on the web." This chapter describes two cyber-GIS applications—the Ecosystems Fisheries Oceanography Coordinated Investigations Data Access and Analysis Tool and LarvaMap—developed for data integration and particle modeling in support of fisheries oceanography research and fisheries management. Both were developed to make it easier for researchers to locate, link, analyze, and display in situ and model data. Because geospatial/temporal relationships are critical in understanding the early life history of commercially valuable fish species such as walleye pollock (*Gadus chalcogrammus*), integrating the generation of model results and in situ environmental data using GIS provides critically needed tools.

The Ecosystems Fisheries Oceanography Coordinated Investigations Data Access and Analysis Tool uses an ArcGIS for Server front end for selection, display, and analysis of data stored in an Oracle database. LarvaMap is a cloud-based tool for running particle dispersion models for fish and invertebrate larvae. It supports scenario testing and makes it easy for nonmodelers to configure and run models. The front end is a web-based map interface, and running the tracking model in a cloud resource allows scaling of computer resources to meet computational needs. Both applications provide basic display and analysis of data and can be integrated with desktop GIS for advanced analyses and two-, three-, and four-dimensional visualizations. The results from these projects can be translated to other types of particle tracking, other data management needs for environmental studies, and used as case studies of deployment of GIS tools to cyber-GIS computing resources.

Dawn J. Wright, ed.; 2015; *Ocean Solutions, Earth Solutions*; http://dx.doi.org/10.17128/9781589483651

Problem and Challenge

Knowledge of the complex relationships between the targeted fish species and its food sources, predators, and environment will aid the new emphasis on ecosystem-based approaches to fisheries management (Link 2002). Understanding these relationships can guide the management of a fishery and the setting of quotas, timing of fishing periods, and management of related fisheries or species that depend on the target species. Investigating these relationships requires a combination of field research and computer modeling. Field research gathers large amounts of disparate data that needs to be managed, integrated, analyzed, and visualized. Many times, the only common characteristic of this data is the fact that it can be geolocated. Spatial analyses and geospatial measures are critical in understanding this data. Models of ocean circulation, fish life histories, or complicated ecological interactions can require huge amounts of computing resources, generate massive output files, and be best understood using advanced visualization and analysis tools.

Providing the computational and analytical environment to support these studies is challenging. Local desktop resources are frequently insufficient. Tightly focused databases can make integrating information almost impossible, and visualization tools may be hard to use. New developments in cyber-GIS, specifically the expansion of ArcGIS for Server and increased access to cloud computing resources, have enabled us to develop and test a number of applications to address solutions for integrating disparate data for easy retrieval and analysis using ArcGIS for Server, either locally or in the cloud, and explore the possibilities of running models in the cloud and storing and displaying their results in a geospatially enabled application.

Introduction and Background

Parallel technological advances in computing and marine fisheries research have contributed greatly to truly implementing ecosystem approaches to fisheries management. Technologies first developed during World War II have been used both to improve sampling in marine systems and support modeling and geospatial analyses. Military surplus vessels formed the core of the oceanographic and fisheries research fleet in the 1950s and onward (Rainger 2000; Rozwadowski and Van Keuren 2004; Hamblin 2005); sound navigation and ranging (SONAR) techniques were adapted to become the core of fisheries hydroacoustics (Fornshell and Tesei 2013); and biological sampling became increasingly electronic (Dickey 2003; Wiebe and Benfield 2003). Computing resources evolved from large mainframe computers to minicomputers and personal computers, to computers on a chip embedded in deployed instruments, and finally back to large centralized, elastic computing resources in the cloud. Geospatial analysis has gone from the early days of GIS in SyMAP and the Canada Geographic Information System (Coppock and Rhind 1991) to ArcInfo to open-source GIS, cyber-GIS, and ArcGIS for Server providing GIS in the cloud.

GIS and other spatial analysis tools provide important tools to understand natural systems. The advent of cyber-GIS has made these tools even more powerful by removing constraints on the size and complexity of analyses through elastic computing resources and cloud-based storage for terabytes of data. Of course, what has been called the "grand challenges" remain, of data intensity, computing intensity, concurrent intensity, and spatiotemporal intensity (Yang et al. 2011). The data integration and complex system modeling tools we have created with the Ecosystems Fisheries Oceanography Coordinated Investigations (EcoFOCI) Data Access and Analysis Tool (EcoDAAT) and LarvaMap take advantage of early answers to these challenges.

Wright et al. (2013) have written of cyber-GIS as GIS removed from the desktop to centralized, usually cloud-based resources. Wang et al. (2013) describe cyber-GIS as "cyberinfrastructure-based GIS" and a subset of the broader realm of "spatial cyberinfrastructure" as defined in Wright and Wang (2011). Yang et al. (2013) have identified a number of disciplines in which cyber-GIS can support critical advances. These include topics related to fisheries such as climate science, ecology, environmental health, and disaster management. They call for advancements in system architectures, which could directly support modeling and visualization and are key to understanding the outputs of complicated ecosystem models; data storage, which would support both the storage of field data and outputs of models; and data and process colocation, which may be less critical because of developments such as the grid caching we developed for LarvaMap which is described later in this chapter (Yang et al. 2013).

Nyerges (2010) specifically addresses the application of cyber-GIS to marine systems by looking at uses of cyber-GIS for modeling coastal marine policy and governance. Although his focus is on human communities, many of the elements he cites are equally applicable to marine ecosystems that also have a "large number of interacting components," creating systems that are "computationally complex and intense" with spatiotemporal elements that "require tremendous amounts of data and computational support." EcoDAAT uses the cyber-GIS concepts of moving processing and data storage from the desktop to a centralized resource with a web-based thin client providing the user interface. The results can be used either for analyses via ArcGIS for Server or for local GIS or statistical analyses using ArcGIS or tools such as *R*. LarvaMap takes advantage of cloud-based computing resources by running its models on a cloud computing resource. The ability to launch additional model runs in the cloud provides easily scalable computer resources, and data can be stored either in the cloud or in local Open-Source Project for a Network Data Access Protocol (OPeNDAP)/Thematic Real-Time Environmental Distributed Data Services (THREDDS) servers. Both of these tools help meet the ecosystem-based management needs of integrating field data with models, integrating physical and biological data, and providing information on the early life history of commercially important species.

History and Status of the Walleye Pollock Fishery in Alaska

EcoDAAT and LarvaMap are first being applied to studies of the early life of walleye pollock (*Gadus chalcogrammus*), hereafter simply *pollock* (figure 1.1). Pollock are a groundfish found primarily in the north Pacific and Bering Sea that supports the largest commercial fishery in the United States by weight (Kendall et al. 1996; Alaska Fisheries Science Center 2013). The primary human uses of pollock are in making fish sticks and fish sandwiches, and as surimi to make imitation crab. Juvenile pollock are a critical food source for flatfish (Lang et al. 2000), and adult seabirds serve as food for Pacific cod, Pacific halibut, sea lions, and other marine mammals.

Figure 1.1. **Walleye pollock larvae, adults, the products made from pollock, and major predators of pollock.** By Ingrid Spies and other photographers from the Alaska Fisheries Science Center, NOAA Fisheries Service; additional photograph by Nazila Merati.

In the Gulf of Alaska, pollock start as eggs spawned in March and April. The eggs are spawned at a depth of 100–250 m and hatch in two to three weeks. Larvae occur at an average depth of 40 m and develop over the next few months. At this stage, they are still at the mercy of oceanic currents and transported as though they were inanimate particles. Once the larvae are about 60 days old, they develop rudimentary swimming skills. Though not able to swim long distances, they are able to swim in bursts to consume prey and avoid potential predators. After one hundred days, they are competent swimmers and can move up and down thorough the water column. At the age of three or four years, they are large enough to be commercially harvested and reach sexual maturity at age four or five. They can live for more than 20 years, but fish that old are unusual (AFSC 2010).

The Gulf of Alaska and Bering Sea pollock fishery started with a small trawl fishery in the 1950s, and catches increased rapidly in the 1970s and 1980s as the foreign fishing fleet targeted the population. The discovery of a spawning population of the fish in Shelikof Strait in 1981 led to the development of a roe fishery. Intense fishing in the international waters of the Donut Hole area of the western Bering Sea in the 1980s led to the collapse of that pollock population (Bailey 2013).

By 1988, changes in fisheries regulations excluded the foreign fleet from American waters and transformed the pollock fishery into a domestic US fishery with the primary consumer products being frozen filets and surimi. In the 1990s, the rise of ecosystem-based management led to consideration of related species in managing the fishery, and in 1999, the Aleutian Islands portion of the fishery was closed in response to declines in Stellar sea lion populations, with the thought that fishing was outcompeting sea lions for fish (National Research Council 2003). This fishery was reopened in 2005 (AFSC 2010). A complete history of the fishery, both domestic and international, can be found in Bailey (2013).

Management of the fishery depends on, among other things, an understanding of the early life history of pollock. Success in the early stages of life is one factor determining how many fish are eventually available to be harvested while preserving a sustainable population. The EcoFOCI program at the National Oceanic and Atmospheric Administration (NOAA) studies the factors that affect the variability in recruitment of pollock and other commercially harvested species in the Gulf of Alaska, north Pacific, and Bering Sea. Oceanographic and biological factors are studied using a combination of field and model studies. Field research includes physical oceanographic surveys and biological sampling to look at life stages from eggs to juveniles and post-age-1 fish. A particular emphasis of the sampling is studying pollock early life history. The zooplankton that support the fish populations are also sampled and described. The samples collected also support age and growth, genetic, and food habit studies. The field data supports an understanding of the biological and physical factors affecting recruitment. The data also provides initial conditions and validation for model studies.

Understanding the Early Life History of Pollock Using Individual-Based Models

One of the "grand challenges" in fisheries research is understanding the fluctuations in the size and structure of fish stocks. The classic work by Hjort (1914) described two hypotheses to explain the variability: The first, the Critical Period hypothesis, posits that the size and strength of a year class of fish is determined very early in life, when the fish have exhausted their yolk sack and started feeding, and is a function of the availability and quality of the plankton they are feeding on. The second, the Aberrant Drift hypothesis, examines the role of transport in taking fish larvae so far from place of birth that they are unable to return to support future populations. A variety of hypotheses have developed in the century since Hjort's work, including the Match-Mismatch hypothesis of Cushing (1974), which relates the timing of larvae hatching and bloom of plankton to feed them; the effect of vertical stratification in the Stable Ocean hypothesis of Lasker (1978); and a number of other hypotheses described and related in Houde (2008) and Hare (2014).

In studies of marine fish recruitment, early studies to test hypotheses were primarily descriptive and considered aspects such as the scale of distributions of plankton (Haury et al. 1978). More recently, advances in the collection of quasi-synoptic in situ and satellite data, the rise of ecosystem approaches to studying marine systems (Levin 2006; Vance and Doel 2010), and rapid and striking advances in computers and data storage (Gentleman 2002; Yang et al. 2011) have radically changed research on recruitment. Hydrodynamic models are now able to resolve details at the scales that affect recruitment (tens of kilometers), and they can be run for weeks to months of model time to cover the early life history of slowly developing species. Biological and physical processes can now be coupled and modeled, and advances in visualization tools have made visualizing and interpreting large model outputs easier (Megrey et al. 2002; Vance et al. 2006; Vance 2008).

One requirement these various hypotheses share is a need to know where fish larvae are located at various stages of life. This need has led to the development of techniques to model the dispersion of larvae based on hydrodynamic models of the ocean. At their simplest, these models distribute larvae as though they were inanimate particles drifting in ocean currents. More detailed models include larval behaviors such as the ability to swim, motion toward or away from light, and avoidance of warm or cool water (Hinckley et al. 1996; Hermann et al. 2001; Miller 2007; Hinckley et al. 2009). These are the simplest of what are called *individual-based models* (IBM), which are called *agent-based models* in other disciplines. An example is the DisMELS model (Dispersal Model for Early Life History Stages), which incorporates behavior in early life stages to study the dispersion of eggs and larvae (Stockhausen 2006). More complex models describe both larval distributions and the distributions of prey for the larvae. These tools, often called *nutrient-phytoplankton-zooplankton* (NPZ) models, describe spatial and temporal interactions (Franks 2002). Upper-trophic level models such as FEAST (Forage and Euphausiid Abundance in Space and Time) add predator and bioenergetic parameters (Aydin et al. 2010; Ortiz 2014).

Integrating and Analyzing Field Data using EcoDAAT

Web-based GIS applications for data discovery and marine planning encompass a wide range of tools. SeaSketch (http://mcclintock.msi.ucsb.edu/projects/seasketch) supports collaborative design but does not support data storage or analysis. SIMoN (http://sanctuarysimon.org/regional_sections/maps/mpaviewer/) and similar tools provide mapping of the location of data collection sites and links to project websites but do not allow for analysis of the data. GIS-based data discovery/download-only tools, such as many of those listed for ecosystem-based management (EBM) on the EBM Tools Network (http://www.ebmtools.org/?q=ebm_tools_taxonomy#), acquire and manage data but do not focus on geoanalysis. The strength of EcoDAAT is that it allows location and downloading of raw data as well as simple geoanalyses.

Because the field sampling for EcoFOCI work on pollock and other forage species is multidisciplinary, the datasets gathered are in disparate formats. The datasets for ichthyoplankton, zooplankton, and environmental measurements were originally stored in separate Microsoft Access databases. This made it hard to answer questions such as "What are the water temperatures where we find larger than average pollock larvae?" in a timely manner. Although the datasets were not enormous—the largest had about 250,000 records—network response time made it impractical to store the data in a single shared database on a network drive. Rather, it was necessary to have the complete databases on the desktop system of each user. This was a maintenance challenge and caused databases to get out of sync if updates failed or were delayed. The physical oceanographic datasets of EcoFOCI collaborators contain up to six million conductivity-temperature-depth (CTD) records, making it impossible to deploy them as part of the existing databases.

The project to create EcoDAAT had a number of goals: to integrate the databases so that cross-data-type queries were possible, move the integrated database to a larger and more robust database application, and make the data selection and display more geospatial by providing a map-based interface and ArcGIS for Server-based analytical tools. To meet these goals, an ArcGIS for Server/Flex front end connected to an Oracle database was chosen. The system layout is shown in figure 1.2. The datasets are stored in a single Oracle database designed to contain, and link, the various types of data being collected. The data cascade is started at sea, where a stand-alone Java/SQLite application called the Cruise Operations Database (MasterCOD) is used to input metadata and collection information about every operation. The information includes the latitude and longitude of the operation, time and date, type(s) of equipment deployed, types of sample(s) gathered, and any notes about the operations. These records are used to track samples as they are analyzed and also to geolocate and cross-reference operations and samples. These datasets are used to populate the Haul tables of the Oracle database with core information and also to populate tables in two associated applications, ZooPPSI and IchPPSI. These two Java/SQLite applications are used by researchers at the Polish Plankton Sorting and Identification Center to record identifications and measurements of zooplankton and ichthyoplankton in samples sent to them for processing. Data from completed ZooPPSI and IchPPSI tables is loaded into the EcoDAAT Oracle tables using a series of Standard Query Language (SQL) scripts. These datasets are used to populate sample and specimen tables and associated tables to hold species and diet information. Information on larger age-0 and older fish collected during trawling operations is stored in Microsoft Excel spreadsheets at sea. The datasets are loaded into the Oracle database in a similar process to the loading of ichthyoplankton data. Laboratory analyses of stomach contents and other diet and condition-related information are loaded from text files and spreadsheets. Physical oceanographic datasets, including CTD and Seacat measurements, are processed using Seabird programs, reformatted into comma-separated value (CSV) files, and loaded using SQL scripts. Both CTD and bottle datasets are stored and can be linked to sample data.

The EcoDAAT front end uses the Flex viewer available with ArcGIS for Server (figure 1.2). This was chosen for ease of implementation and enhanced functionality. The front end allows for data selection by type of data (e.g., ichthyoplankton or oceanographic data), geographic location using graphical or text-based searches, time of collection, and combined criteria. The interfaces for data selection are dynamic, and the available selection criteria vary based on the type of selection being made. Because some of the data in EcoDAAT is still undergoing quality assurance/quality control (QA/QC), the front end uses Lightweight Directory Access Protocol (LDAP) authentication for secure access. In the future, a tiered access scheme, in which data can be made fully public, shared with collaborators only, or limited to program scientists, will be implemented.

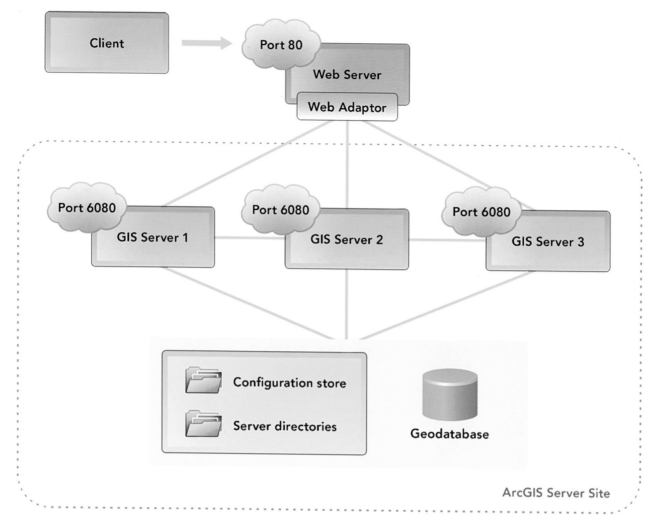

Figure 1.2. EcoDAAT system architecture showing integration of Flex, Oracle, and ArcGIS for Server. Esri.

The results of a search are available both in summary form on the map portion of the front end (with data locations highlighted and the standard GIS Identify tool implemented to query data) and as a table that can be output as a shapefile or .csv file for further analysis (figure 1.3). The results of a search for haul information can also be linked to searches for associated data to create complex queries.

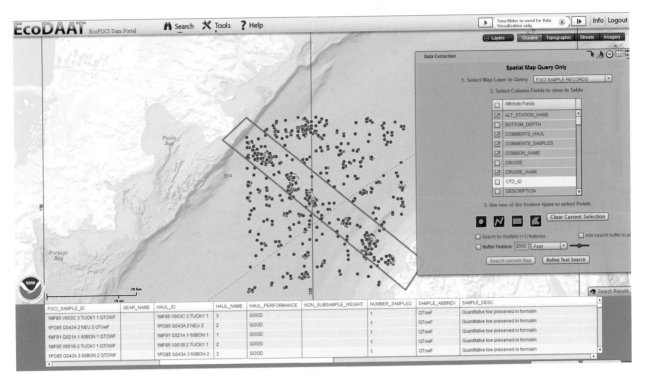

Figure 1.3. **Example of the map-based selection tool in the Ecosystems Data Access and Analysis tool (EcoDAAT) and a table of data.** By NOAA Alaska Fisheries Science Center; data sources: NOAA Alaska Fisheries Science Center, Esri, DeLorme, GEBCO, NOAA NGDC, National Geographic, HERE, Geonames.org, and other contributors.

Simple geoanalyses such as creating and plotting a bathymetric cross section are available (figure 1.4). These types of tools will be expanded in the future by taking advantage of the geoprocessing tools available in ArcGIS for Server.

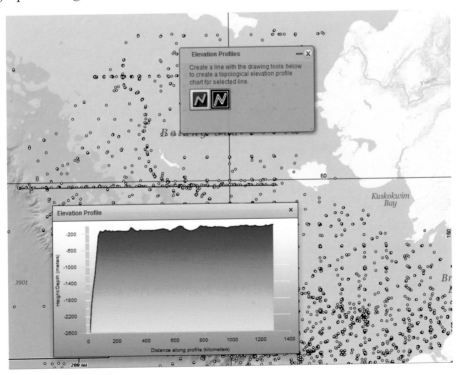

Figure 1.4. **Example of the cross section tool in EcoDAAT.** By NOAA Alaska Fisheries Science Center; data sources: NOAA Alaska Fisheries Science Center, Esri, DeLorme, GEBCO, NOAA NGDC, National Geographic, HERE, Geonames.org, and other contributors.

Tracking the Transport and Dispersion of Larvae Using LarvaMap

Particle-tracking models enable the study of the motion of particles in the atmosphere or the ocean, from oil globules from an oil spill to plumes of smoke or sediment to even the ash from a volcanic eruption. Many models run locally and require setting up complicated parameter files for a model run. Examples include Ichthyop for Lagrangian modeling of the Bay of Biscay and Mediterranean Sea
(Lett et al. 2008) and the Lagrangian Transport Model (LTRANS), which was originally created to model the dispersion of oyster larvae in Chesapeake Bay (http://northweb.hpl.umces.edu/LTRANS .htm). Model runs can take hours to days to complete, depending on the number of particles released, duration of the run, and power of the underlying computer(s). There are an increasing number of web-based particle-tracking models, including the HYbrid Single-Particle Lagrangian Integrated Trajectory (HYSPLIT) Volcanic Ash Model (Draxler and Rolph 2003), which allows users to model ash transport from a list of volcanoes using one of five common atmospheric models. Connie2 (http://www.csiro.au/connie2/) allows the user to set up and run particle-tracking models for the seas around Australia and the Mediterranean Sea. Connie2 allows for either passive/ inanimate particles or particles showing limited behaviors such as the ability to swim.

Particle tracking is a critical element of modeling the early life history of pollock and other commercially important species. The models can be as simple as one that tracks inanimate particles and as complex as ones that describe the interactions at many trophic levels in an ecosystem. Lagrangian models follow an individual through space and time by calculating a path line and tracking the motion of the particle. Eulerian models use positions on a fixed spatial grid and streamlines to describe transport through the location. LarvaMap is a cloud-hosted implementation of a Lagrangian particle-tracking model supporting a variety of behaviors in a target species. It is an individual-based model capable of using output from a variety of ocean circulation models to drive the dispersion and movement of particles. The front end is a series of web pages for setting up larval behaviors, choosing the circulation or hydrodynamic model, configuring the model run, and launching the run. The model itself is run on a cloud computing resource (currently the Amazon cloud), but it could also be configured to run on a local resource or private cloud. The results are visualized as a track line in the web client, and results are available as an Esri shapefile and network Common Data Format (netCDF) file for further analysis.

LarvaMap is a combination of four pieces of technology working together (figure 1.5), consisting of:

- A behavior library to create, catalog, and share larval behaviors: http://bit.ly/WFiZjK
- A fate and transport model written as a Python library: http://bit.ly/16dFk9x
- A model service implementing the fate and transport model in a cloud architecture
- A web client for interacting with the Model Service through a representational state transfer application programming interface (REST API): http://bit.ly/14aZinD

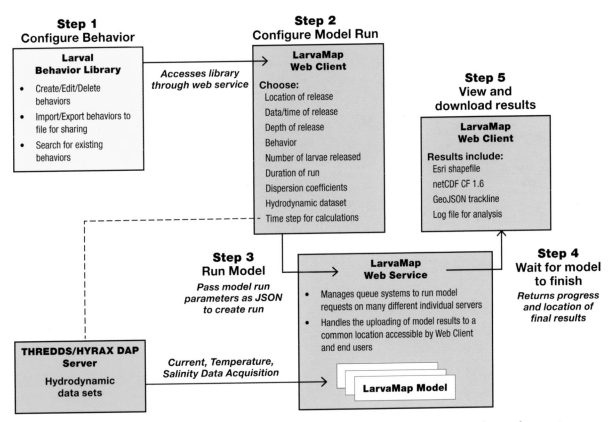

Figure 1.5. LarvaMap architecture showing the interconnections between the various elements and the process of creating a model run. RPS/ASA.

Setting Larval Behaviors with the Behavior Library

The first step in running LarvaMap is discovering or creating a set of larval behaviors for an organism or species of interest using the Larval Behavior Library (figure 1.6). The library allows the creation of new behavior files, searching for existing files, and sharing files. By allowing behaviors to be exchanged between researchers, modelers, and students, it supports creating behaviors as a social process, with the ability to see how others have defined behaviors, to run models using both your own and community behavior files, and to work collaboratively on building behaviors. With the ease of running the models, it is possible to quickly and easily compare the results for different behaviors.

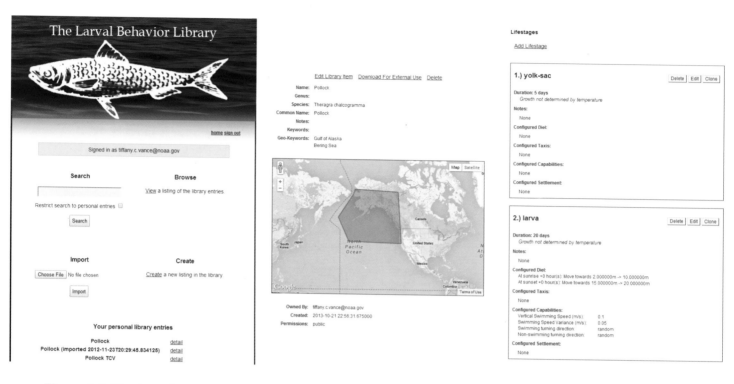

Figure 1.6. The interface to the Larval Behavior Library is used to define the behaviors of a specific species. By NOAA Alaska Fisheries Science Center; data sources: NOAA Alaska Fisheries Science Center, Esri, DeLorme, GEBCO, NOAA NGDC, National Geographic, HERE, Geonames.org, and other contributors.

Behaviors that can be defined include:

- The geographic range of the organism and any keywords to be used for regional searches
- Adding life stages (e.g., egg, larval, juvenile) with associated duration in days or by temperature range
- Adding behaviors to life stages, including capabilities such as swimming speed and vertical migration, reactions such as avoidance behaviors, and sensory reactions such as avoiding salinities or temperatures
- The timing of the transition out of the plankton stage and where the organism will end up—settling on the bottom for benthic species and where in the water column pelagic species will reside

Configuring the Model Run Using the Web Client: Setting Parameters and Choosing Circulation Models

Configuring the model requires setting the location, date, time, and depth of the release and specifying the number of particles to be released. The behavior model is specified, usually by species. Vertical and horizontal dispersion coefficients can be set, and the underlying hydrodynamic model is specified (figure 1.7). Finally, the time step for the model runs is specified. Model durations of months and years can be supported, but the time step for the model must be selected to be reasonably small to provide the level of detail needed without requiring the calculation of an unnecessary number of time steps.

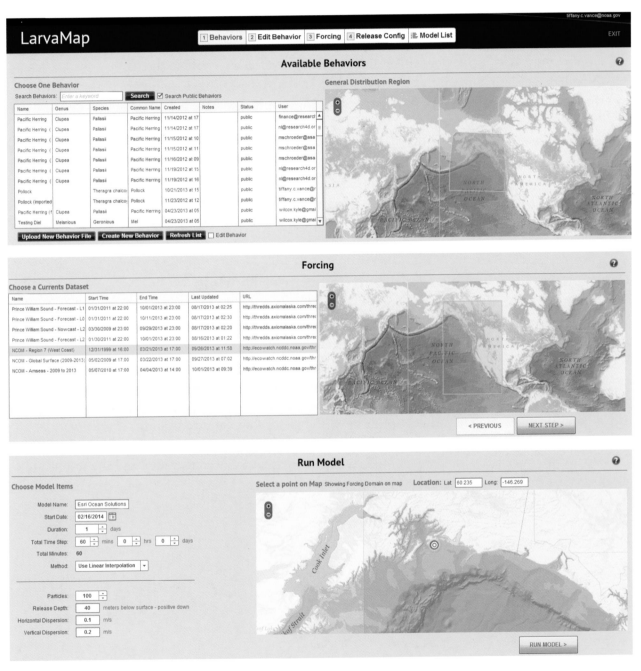

Figure 1.7. **The interface for configuring and running a LarvaMap model.** By NOAA Alaska Fisheries Science Center; data sources: NOAA Alaska Fisheries Science Center, Esri, DeLorme, GEBCO, NOAA NGDC, National Geographic, HERE, Geonames.org, and other contributors.

Hydrodynamic models include those provided via THREDDS/OPeNDAP server or available as a local file. OPeNDAP and file-based datasets can be added to the LarvaMap system as they are discovered. An administrator of the LarvaMap system can add a dataset through an interface, and datasets must be registered in LarvaMap by an administrator of the LarvaMap system before they can be used for particle forcing. Both local and OPeNDAP datasets can be registered in LarvaMap. A Python library called paegan (https://github.com /kwilcox/paegan) was developed as part of the LarvaMap project to facilitate consistent access to many formats of hydrodynamic model data. If the dataset opens in paegan as a CommonDataset object, it is usable in LarvaMap. Currently, paegan supports regular grids (delta y == delta x), curvilinear grids (sometimes called *i/j grids*), and static triangular meshes. Dynamically changing grids and nontriangular meshes are not yet supported. LarvaMap does not support uploading of any ocean circulation data files for internal storage. File-based datasets must be made accessible to all model runs in the LarvaMap architecture before being used. Installations of LarvaMap using the distributed file system GlusterFS can achieve this.

The Regional Ocean Modeling System (ROMS) and US Navy Coastal Ocean Model (NCOM) models have been used to look at transport of pollock and herring in the north Pacific. Lowry (2013) used LarvaMap linked to an ROMS model for Prince William Sound to model drift and recruitment of herring. The NCOM region 7 model has been used to look at transport of sea urchin larvae off the California coast. The NCOM AmSeas model has been implemented for use in the Gulf of Mexico.

The Heart of LarvaMap: Fate and Transport Models

During a model run, the transport process is broken into smaller fragments called *models*. Each model implements a *move* method that moves a particle based on environmental and behavioral properties. Each model has access to environmental conditions from the hydrodynamic model (u, v, w; temperature; and salinity) at the current location and time of the particle. In a few specific scenarios, environmental data for a particle's location and time may be unknown: particle has left the domain of the hydrodynamic dataset, or there is a data gap in the hydrodynamic dataset. Under these conditions, the last known environmental conditions the particle experienced are used in the model.

LarvaMap implements three main models. The models are run sequentially, and the results are checked against the shoreline and bathymetry at the end (figure 1.8). A typical time step in a model run looks like this:

- Run transport (moves particle).
- Check shoreline.
- Check bathymetry.
- Run dispersion (moves particle).
- Check shoreline.
- Check bathymetry.
- Run behavior (moves particle).
- Check shoreline.
- Check bathymetry.

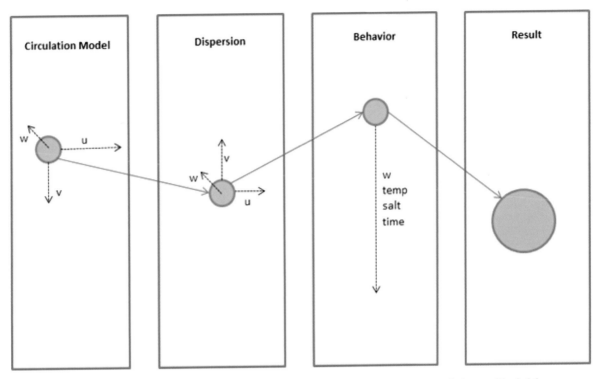

Figure 1.8. Elements of the transport process during a single time step in a model run. Variables *u*, *v*, and *w* are the circulation model velocity components. RPS/ASA.

Transport

The transport model for LarvaMap is a parallelized Lagrangian transport model using hydrodynamic circulation data stored as netCDF/OPeNDAP files. The model is written in Python and uses the paegan library, which provides a data model for ocean and meteorological data stored in netCDF files and the paegan-transport libraries for the model itself. It moves a particle based on the current velocity components (*u* [zonal or east–west], *v* [meridional or north–south], and *w* [vertical]) from the hydrodynamic model. The vertical component (*w*) is optional. If the transport model receives no *u*, *v*, or *w* data for the place and time of the particle, the last known conditions are used.

Dispersion

Each model run is configurable with a horizontal and vertical dispersion coefficient (in m/s). These coefficients are used to compute the *u*, *v*, and *w* dispersion.

Behavior

The behavior model is set up using a configuration from the Larval Behavior Library. A configuration consists of many life stages, each with its own behaviors. The behavior model is responsible for calling the correct life stage model based on such factors as a particle's development, age, and mass, and adding a "dead" life stage to every behavior configuration. This model does not actually move the particle, the life stage model does. It can be thought of as a *container* model.

Life Stage

The life stage model grows each particle based on time or as a function of temperature. It also calls the correct behavioral models associated with the particle's life stage. For example, if there are four diel behaviors configured for a single life stage, this life stage model determines which diel behavior to call based on the particle's time. The behavioral models described below work a bit differently from the transport and life stage models. They do not return a new location for the particle, they only return the u, v, and w vectors. The vectors are summed from all behaviors by this life stage model, and the particle is only moved once, after all behaviors are completed. If the life stage model receives no temperature or salinity data for the place and time of the particle, the last known conditions are used.

Life Stage: Diel

The diel model moves a particle up and down in the water column based on the sun's position in the sky. Each life stage of a larva has a configured static vertical swimming speed that is used here to determine how fast the particle can move to its desired depth range. If the particle's vertical swimming speed will put the particle beyond the desired depth range, the swimming speed is recomputed for a single model time step to land the particle in the middle of the desired depth range. If the desired depth range will not be reached, or the particle will end up inside the depth range, the vertical swimming speed configured in the behavior is used. A particle below desired depth will swim up, a particle below desired depth will swim down, and a particle at the desired depth will do nothing.

Life Stage: Settlement

The settlement model decides whether a particle should settle under its current conditions. As described in the Larval Behavior Library section, there are two types of settlement:
- Benthic: settles particle at 1 m above the bathymetry if bathymetry is within configured settlement depth range
- Pelagic: settles particle at its current depth if it is within configured depth range

The settlement model does not return u, v, or w vectors. It directly settles or moves the particle to the desired settlement location.

Life Stage: Dead

The dead model's purpose is to kill a particle that has not settled and has aged beyond its settlement life stage. It continues to track environmental conditions around the dead particle, and the particle will continue to be forced by the other models (transport/dispersion), but the dead model always returns 0 m/s u, v, and w.

Under the Hood: Circulation, Shoreline, and Bathymetric Data for Model Runs

Hydrodynamic Caching Strategy

One of the challenges in running a particle tracking model is that the particles are sparsely distributed. The transport model supports long-running models on the order of months and years, and there is actually no upper time limit imposed by the trajectory model. Although the circulation model data needs to cover a large area to support any possible particle locations, at any given time step the particles occupy a small subset of this total area. Downloading the entire circulation/hydrodynamic model data for each time step is prohibitive. But because LarvaMap allows for remote data access over the data access protocol (DAP), the entire hydrodynamic dataset for the area of interest never has to be downloaded and stored locally. The data can be accessed on the fly, and the model will download only the data it needs to run the model (figure 1.9). Data that is downloaded by the model is reused from particle to particle, so the same data is only accessed and downloaded once. Particles use the locally cached data in forcing algorithms.

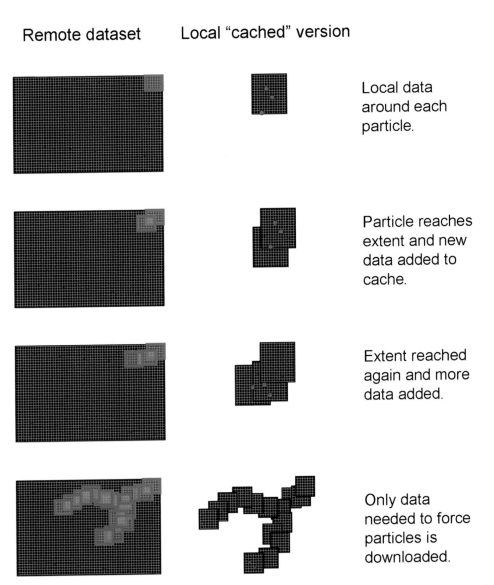

Figure 1.9. The data caching scheme used to reduce the size of downloads from the underlying circulation model. RPS/ASA.

Shoreline and Bathymetry

The fate and transport model uses land polygon objects to test for shoreline collisions. The land polygons can be stored in an Esri shapefile or accessed from an Open Geospatial Consortium (OGC) web feature service (WFS) server. Every particle loads the shoreline on creation and indexes the shoreline around its current location. Every time a particle tests its location against the shoreline file, it first makes sure it has the appropriate data from the source shoreline dataset to do its calculation (figure 1.10). If it does not, it accesses the raw shoreline data, whether it is a shapefile or WFS server, and pulls a chunk of the data around its current location. This acts as a caching mechanism so the particles are not reading from the shoreline data at every detection check. A similar process is used with bathymetric data to constrain particles to the water column.

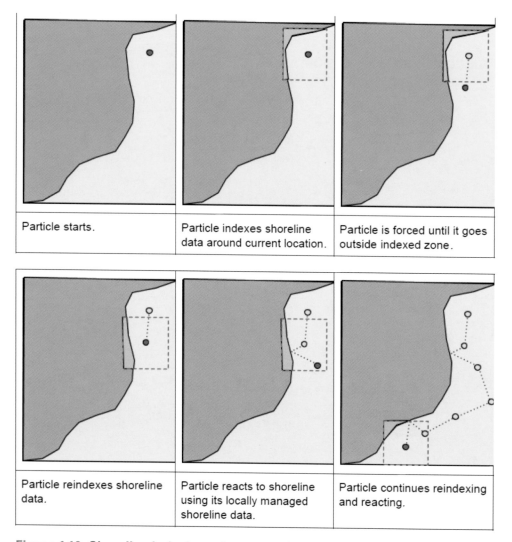

Figure 1.10. Shoreline indexing scheme used to locate the appropriate shoreline segments in the vicinity of a particle. Avoidance behavior when a particle nears a shoreline is shown in the last two parts of the figure. RPS/ASA.

The Model Service: Implementing LarvaMap in the Cloud

The model service queues model run requests received from the web client and distributes them within a network of "worker" servers. Worker servers can be running in many different places, including the Amazon cloud and local data centers. Each central processing unit (CPU) core on a worker server can force a single particle for the entire duration of its model. If one hundred worker cores are available, one hundred particles can be run in parallel. This can be a single model run with one hundred particles, or 10 model runs with 10 particles each. The model service will automatically create new worker servers when the load demands it. Exactly when to spawn a new worker server is determined by how many particles are in the queue waiting to be run. If the number of particles in the queue is three times more than the number of CPU cores available on all worker servers, another worker server is initialized. Worker servers are shut down as the number of particles in the queue decreases. When no particles are in the queue, there is a minimum of one worker server running. Cloud computing technology allows LarvaMap to scale the number of worker servers dynamically as they are needed. In the near future, the model service will also be able to distribute a single particle's model run onto multiple servers, which can reduce model run times to one-tenth of the current times.

The model service collects the results from all the worker servers and pieces them back together into the final results for each model run. It creates the final output formats and uploads the result files to a common location for the web client to access. Utilities are available to monitor the status of the model service currently running jobs and details about any failed jobs.

Outputs from Model Runs

Output from LarvaMap is available in a number of formats. A GeoJSON track line is created for display in the web client. A shapefile, a netCDF file, and a Hierarchical Data Format 5 (HDF5) file are produced for display and analysis in ArcGIS and other tools (figure 1.11). The output files contain the date, latitude, longitude, and depth of the particle; the temperature and salinity values at the point if they are available from the hydrodynamic model; and the u, v, and w speed components at the grid point. A log file is also produced for troubleshooting and system analysis. TopoJSON is being considered in the future to allow a web client to show the track lines for each individual particle instead of using the single centroid track line GeoJSON output format currently available. An experimental TopoJSON output format is available at https://github.com/kwilcox/geojson _examples/blob/master/particles/huge_particle_tracklines.topojson.

Post Processing and Product Generation: Gridded Products

LarvaMap contains some utilities for post-processing the netCDF output into gridded products. These are typically used for stochastically representing the probability of a Lagrangian element passing through a grid cell or, in the case of larvae, settlement occurring in a grid cell based on an ensemble of model runs. The gridded products can be output as GeoTIFFs for use in GIS applications such as ArcGIS ArcMap. This workflow is not automated as part of the model implemented in the cloud, but could be easily added to the cloud-based system.

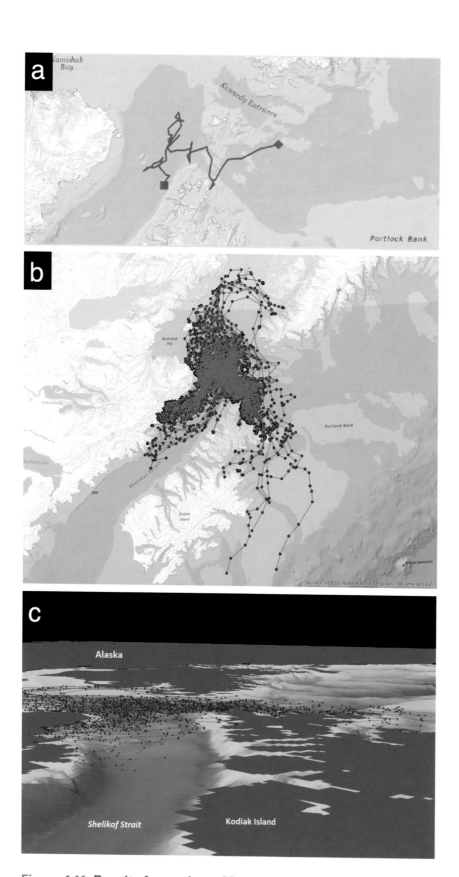

Figure 1.11. Results from a LarvaMap run shown in (a) the LarvaMap interface and the track lines shown in the (b) ArcMap and (c) ArcScene applications in ArcGIS. By NOAA Alaska Fisheries Science Center; data sources: NOAA Alaska Fisheries Science Center, Esri, DeLorme, GEBCO, NOAA NGDC, National Geographic, HERE, Geonames.org, and other contributors.

Role of Cloud and ArcGIS for Server, Future Needs

Creating EcoDAAT and LarvaMap has allowed us to explore practical applications of cyber-GIS advancements. Although EcoDAAT is not currently a cloud solution, the architecture of ArcGIS for Server means it could be redeployed using cloud storage and computation. The web client can use a variety of back-end architectures. LarvaMap is a more classic implementation of cyber-GIS. It uses cloud storage, optimized data transfer schemes, cloud-based computation, and local analysis of model results. In this way it requires solutions to Yang et al.'s (2011) challenges of data intensity, computing intensity, concurrent intensity, and spatiotemporal intensity.

Implementation of these tools has produced some challenges, and some unexpected opportunities. The EcoDAAT integration of data and ability to make data easily accessible has also raised challenges concerning data sharing and data use. Although the recent White House memorandum on open data (President Barack Obama 2013) calls for free and open distribution of data, this must be balanced with the need to make sure the data being distributed is fully quality controlled and the right of the researchers to their intellectual property is respected. Tiered access and other technical solutions supported by cyberinfrastructure will allow us to balance these needs.

Deploying LarvaMap raises challenges in procuring cloud resources for government use. Full deployment would require the security accreditation of cloud storage and computing and improved procedures for paying for cloud time. Most or all of these challenges could be solved by the creation of private clouds within the federal system or deploying LarvaMap in local high-performance computing resources.

One totally unexpected use of LarvaMap is as a teaching tool. Because of its ease of use and computational capabilities, it can be used by students to explore larval transport. A recent biological oceanography course in the joint Massachusetts Institute of Technology–Woods Hole Oceanographic Institution graduate program had the students using the tool to create and run scenarios for their final projects (L. Mulineaux, personal communication). This also provided us the benefit of extended stress testing of the system as the students created and ran numerous models.

These tools have proven that "GIS detached from the desktop and deployed on the web" can provide critical resources to support fisheries research and management. We plan to continue taking advantage of future cyber-GIS developments to improve and expand our capabilities.

Acknowledgments

This chapter was greatly improved by reviews from Christine Baier, Annette Dougherty, Ann Matarese, and Nazila Merati and an anonymous reviewer. This research was supported, in part, with funds from the NOAA High Performance Computing and Communications (HPCC) program and the National Marine Fisheries Service Ecosystems and Fisheries Oceanography Coordinated Investigations and funding to Kyle Wilcox from the Oil Spill Recovery Institute. This is contribution EcoFOCI-N819 to the NOAA North Pacific Climate Regimes and Ecosystem Productivity research program. The findings and conclusions in this paper are those of the authors and do not necessarily represent the views of the NOAA National Marine Fisheries Service. Reference to trade names does not imply endorsement by the NOAA National Marine Fisheries Service.

References

AFSC (Alaska Fisheries Science Center). 2010. Walleye Pollock. http://www.afsc.noaa.gov/Education/factsheets /10_Wpoll_FS.pdf. Last accessed February 15, 2014.

———. 2013. Walleye Pollock Research. http://www.afsc.noaa.gov/species/pollock.php. Last accessed February 11, 2014.

Aydin, K., N. Bond, E. N. Curchitser, M. Dalton, G. A. Gibson, K. Hedström, A. J. Hermann, E. Moffitt, J. Murphy, I. Ortiz, A. Punt, and M. Wan. 2010. "Integrating Data, Fieldwork, and Models into an Ecosystem-Level Forecasting Synthesis: The Forage-Euphausiid Abundance in Space and Time (FEAST) Model for the Bering Sea Integrated Research Program." *ICES CM 2010* L:21. http://info.ices.dk/products/CMdocs/CM-2010/L/L2110.pdf.

Bailey, K. M. 2013. *Billion-Dollar Fish: The Untold Story of Alaska Pollock*. Chicago: University of Chicago Press.

Coppock, J. T., and D. W. Rhind. 1991. "The History of GIS." In *Geographical Information Systems: Principles and Applications*, edited by D. J. Maguire, M. F. Goodchild, and D. Rhind, 21-43. Vol. 1. New York: Longman Scientific & Technical; New York: Wiley.

Cushing, D. H. 1974. "The Natural Regulation of Fish Populations." In *Sea Fisheries Research*, edited by F. R. Harden-Jones, 399–412. New York: John Wiley and Sons.

Dickey, T. D. 2003. "Emerging Ocean Observations for Interdisciplinary Data Assimilation Systems." *Journal of Marine Systems* 40/41: 5–48.

Draxler, R. R., and G. D. Rolph. 2003. HYSPLIT (HYbrid Single-Particle Lagrangian Integrated Trajectory) Model. http://www.arl.noaa.gov/HYSPLIT.php. Last accessed June 4, 2014.

Fornshell, J. A., and A. Tesei. 2013. "The Development of SONAR as a Tool in Marine Biological Research in the Twentieth Century." *International Journal of Oceanography* 3:1–9.

Franks, P. J. S. 2002. "NPZ Models of Plankton Dynamics: Their Construction, Coupling to Physics, and Application." *Journal of Oceanography* 58 (2): 379–87.

Gentleman, W. 2002. "A Chronology of Plankton Dynamics in silico: How Computer Models Have Been Used to Study Marine Ecosystems." *Hydrobiologia* 480 (1–3): 69–85.

Hamblin, J. D. 2005. *Oceanographers and the Cold War: Disciples of Marine Science*. Seattle: University of Washington Press.

Hare, J. A. 2014. "The Future of Fisheries Oceanography Lies in the Pursuit of Multiple Hypotheses." *ICES Journal of Marine Science*. doi:10.1093/icesjms/fsu018.

Haury, L. R., J. A. McGowan, and P. H. Wiebe. 1978. "Patterns and Processes in the Time-Space Scales of Plankton Distributions." In *Spatial Pattern in Plankton Communities*, edited by J. H. Steele, 277–327. New York: Plenum Press.

Hermann, A. J., S. Hinckley, B. A. Megrey, and J. M. Napp. 2001. "Applied and Theoretical Considerations for Constructing Spatially Explicit Individual-Based Models of Marine Larval Fish that Include Multiple Trophic Levels." *ICES Journal of Marine Science* 58: 1030–41.

Hinckley, S., A. J. Hermann, and B. A. Megrey. 1996. "Development of a Spatially Explicit, Individual-Based Model of Marine Fish Early Life History." *Marine Ecology Progress Series* 137:47–68.

Hinckley, S., J. M. Napp, A. J. Hermann, and C. Parada. 2009. "Simulation of Physically Mediated Variability in Prey Resources of a Larval Fish: A Three-Dimensional NPZ Model." *Fisheries Oceanography* 18 (4): 201–23.

Hjort, J. 1914. "Fluctuations in the Great Fisheries of Northern Europe in the Light of Biological Research." *Rapports et Proces-verbaux des Réunions. Conseil International pour l'Éxploration de la Mer* 20:1–228.

Houde, E. D. 2008. "Emerging from Hjort's Shadow." *Journal of Northwest Atlantic Fishery Science* 41:53–70.

Kendall, A. W., J. D. Schumacher, and S. Kim. 1996. "Walleye Pollock Recruitment in Shelikof Strait: Applied Fisheries Oceanography." *Fisheries Oceanography* 5:4–18.

Lang, G., R. Brodeur, J. Napp, and R. Schabetsberger. 2000. "Variation in Groundfish Predation on Juvenile Walleye Pollock Relative to Hydrographic Structure near the Pribilof Islands, Alaska." *ICES Journal of Marine Science* 57 (2): 265–71.

Lasker, R. 1978. "The Relation between Oceanographic Conditions, and Larval Anchovy Food in the California Current: Identification of Factors Contributing to Recruitment Failure." *Rapports et Proces-verbaux des Réunions. Conseil International pour l'Éxploration de la Mer* 173:212–30.

Lett, C., P. Verkey, C. Mullon, C. Parada, T. Brochier, P. Penven, and B. Blanke. 2008. "A Lagrangian Tool for Modelling Ichthyoplankton Dynamics." *Environmental Modelling & Software* 23:1210–14.

Levin, L. A. 2006. "Recent Progress in Understanding Larval Dispersal: New Directions and Digressions." *Integrative and Comparative Biology* 46 (3): 282–97.

Link, J. S. 2002. "What Does Ecosystem-Based Fisheries Management Mean?" *Fisheries* 27 (4): 18–21.

Lowry, N. 2013. *Larva Map: Final report to the Oil Spill Response Institute.* Seattle, WA: Research4D.

Megrey, B., S. Hinckley, and E. Dobbins. 2002. "Using Scientific Visualization Tools to Facilitate Analysis of Multi-dimensional data from a Spatially Explicit, Biophysical, Individual-Based Model of Marine Fish Early Life History." *ICES Journal of Marine Science* 59 (1): 203–15.

Miller, T. J. 2007. "Contribution of Individual-Based Coupled Physical-Biological Models to Understanding Recruitment in Marine Fish Populations." *Marine Ecology Progress Series* 347:127–38.

National Research Council. 2003. *The Decline of the Stellar Sea Lion in Alaskan Waters: Untangling Food Webs and Fishing Nets.* Washington, DC: The National Academies Press.

Nyerges, T. 2010. *Cyber-Enabled Platforms for Regional Ocean and Coastal Governance.* National Science Foundation TeraGrid Workshop on Cyber-GIS. Washington, DC: National Science Foundation. http://www.cigi.illinois .edu/cybergis/docs/Nyerges_Position_Paper.pdf.

Ortiz, I., K. Aydin, A. J. Hermann, and G. Gibson. In review, 2014. "Climate to Fisheries: A Vertically Integrated Model for the Eastern Bering Sea." *Deep Sea Research II.*

President Barack Obama. 2013. Memorandum on Open Data Policy–Managing Information as an Asset (May 9, 2013). http://www.whitehouse.gov/sites/default/files/omb/memoranda/2013/m-13-13.pdf. Last accessed June 4, 2014.

Rainger, R. 2000. "Patronage and Science: Roger Revelle, the US Navy, and Oceanography at the Scripps Institution." *Earth Sciences History* 19 (1): 58–89.

Rozwadowski, H. M., and D. K. Van Keuren. 2004. *The Machine in Neptune's Garden: Historical Perspectives on Technology and the Marine Environment.* Canton, NJ: Science History Publications.

Stockhausen, B. 2006. DisMELS: A Dispersal Model for Early Life History Stages. http://www.afsc.noaa.gov /quarterly/jas2006/divrptsREFM6.htm. Last accessed February 23, 2014.

Vance, T. C. 2008. "If You Build It, Will They Come? Evolution towards the Application of Multi-dimensional GIS to Fisheries-Oceanography." PhD diss. Corvallis, OR: Oregon State University.

Vance, T. C., and R. E. Doel. 2010. "Graphical Methods and Cold War Scientific Practice: The Stommel Diagram's Intriguing Journey from the Physical to the Biological Environmental Sciences." *Historical Studies in the Natural Sciences* 40 (1): 1–47.

Vance, T. C., B. A. Megrey, and C. W. Moore. 2006. "GeoModeler: Integration of a Nutrient-Phytoplankton-Zooplankton (NPZ) Model and an Individual-Based Model (IBM) with a Geographic Information System (GIS)." Paper presented at the ICES Annual Science Conference, Masstricht, Netherlands.

Wang, S., L. Anselin, B. Bhaduri, C. Crosby, M. F. Goodchild, Y. Liu, and T. L. Nyerges. 2013. "CyberGIS Software: A Synthetic Review and Integration Roadmap." *International Journal of Geographical Information Science* 27 (11): 2122–45.

Wiebe, P. H., and M. C. Benfield. 2003. "From the Hensen Net toward Four-Dimensional Biological Oceanography." *Progress in Oceanography* 56 (1): 7–136.

Wright, D. J., and S. Wang. 2011. "The Emergence of Spatial Cyberinfrastructure." In *Proceedings of the National Academy of Sciences* 108 (14): 5488–91.

Wright, D., S. Kopp, and C. Brown. 2013. "What Is Cyber GIS?" *Esri Insider.* http://blogs.esri.com/esri /arcgis/2013/10/01/what-is-cybergis/. Last accessed January 14, 2014.

Yang, C., M. Goodchild, Q. Huang, D. Nebert, R. Raskin, Y. Xu, M. Bambacus, and D. Fay. 2011. "Spatial Cloud Computing: How Can the Geospatial Sciences Use and Help Shape Cloud Computing?" *International Journal of Digital Earth* 4 (4): 305–29.

Yang, C., Y. Xu, and D. Nebert. 2013. "Redefining the Possibility of Digital Earth and Geosciences with Spatial Cloud Computing." *International Journal of Digital Earth* 6 (4): 297–312.

Supplemental Resources

URLs and QR codes are provided for the digital content that goes with this chapter.

Hyperlinks are also available on the Esri Press "Book Resources" webpage for chapter 1. Go to esripress.esri.com/bookresources. Then, in the list of Esri Press books, click *Ocean Solutions, Earth Solutions*. On the *Ocean Solutions, Earth Solutions* resource page, click the chapter 1 link to access that webpage and the hyperlinks listed there.

LarvaMap-Related Resources

- Larval Behavior Library to create, catalog, and share larval behaviors, at
  http://behavior.larvamap.axiomdatascience.com/

- LarvaMap Transport Model written as a Python library, at
  http://github.com/asascience-open/paegan-transport

- LarvaMap web client for interacting with a web service through a REST API, at
  http://services.asascience.com/mapapp/larvamap

- Paegan Common Access Library, at https://github.com/asascience-open/paegan

- Online archive of particle-tracking models, at
  http://www.nefsc.noaa.gov/drifter/particles.html

- OceanNOMADS for access to hydrodynamic models, at http://ecowatch.ncddc.noaa.gov

- Access to high-resolution operational hydrodynamic models, at
  http://www.opc.ncep.noaa.gov/newNCOM/NCOM_currents.shtml

EcoDAAT-Related Resources

- Ichthyoplankton Information System for information on the early life histories of
  fish species in the Bering Sea and Gulf of Alaska, at
  http://access.afsc.noaa.gov/ichthyo/index.php

- Survey data on the occurrence of adult pollock and other species, at
  http://www.afsc.noaa.gov/RACE/groundfish/survey_data/default.htm

- Mapping essential fish habitat for a variety of species, including pollock, at
  http://www.habitat.noaa.gov/protection/efh/efhmapper/index.html

CHAPTER 2

What GIS Experts and Policy Professionals Need to Know about Using Marxan in Multiobjective Planning Processes

Heather M. Coleman and Jeff A. Ardron

Abstract

Decision support tools are increasingly becoming part of marine conservation planning, and a body of case studies and good practices is growing around their usage. Nevertheless, misconceptions persist among managers and stakeholders about what these tools can provide, and among technical experts, who question how the tools may offer an advantage in planning processes. This chapter provides a summary of these issues with a focus on challenges and problems faced in real-world marine conservation planning. The popular decision support tool Marxan, designed to inform multiobjective spatial conservation and resource-use planning decisions, provides a good example of decision support tool applications and lessons over two decades of use. Marxan helps select networks of areas that represent target amounts of features at minimal "cost." Many challenging decisions and considerations are inherent in solving this problem, including setting targets, determining appropriate socioeconomic values, and interpreting and communicating results. In the end, however, handling relationships with stakeholders and managers in a planning process can make or break acceptance of the tool and the use of its results.

Dawn J. Wright, ed.; 2015; *Ocean Solutions, Earth Solutions*; http://dx.doi.org/10.17128/9781589483651

Introduction

Marine protected areas (MPAs) have become an increasingly applied tool yet remain a controversial issue in spatial resource management. In part, this is because of the lack of a systematic process and clear definition and implementation of planning criteria, as well as heavy emphasis on MPA placement to avoid conflict with existing human uses (Devillers et al. 2014). However, specific elements have recently been identified to explain the success, failure, and indifference of many MPAs to biodiversity and resource enhancement (Edgar et al. 2014). Just as these common elements should aid development of effective MPA networks, recognizing themes of successful multiobjective tool use should help guide the process of meeting MPA network criteria. Although this chapter focuses largely on experiences gained through marine conservation planning, we believe that much of its content is also applicable to marine/maritime spatial planning (MSP) and multiobjective resource planning in a broader sense.

In order to meet a range of conservation and other planning objectives, a systematic approach has taken the place of the ad hoc, site-by-site approaches of the past that resulted in overrepresentation of certain habitats or ecosystems at the expense of others (Pressey et al. 1993; Simberloff 1998; Possingham et al. 2000). The process of systematic conservation planning (SCP) was developed to describe the location, design, and management of areas that comprehensively represent regional biodiversity and targeted resources (Margules and Pressey 2000; Margules and Sarkar 2007; Moilanen et al. 2009). A clear set of objectives, incorporation of the best available information, inclusion of stakeholders, transparent analysis and reporting, and iterative decision-making all characterize good SCP practices. However, the full promise of SCP has remained elusive because of emerging complexities such as proper accounting of the costs associated with proposed solutions and appropriate use of available decision support tools (McDonald 2009; Ardron et al. 2010).

The Role of Spatial Tools in a Systematic Planning Process

Decision support tools (DSTs) are designed to benefit planning processes by assisting with complicated problems that are beyond human intuition or conventional approaches. Conservation-focused DSTs are software programs that (1) guide decisions intended to promote protection of biodiversity, ecological structure and function, ecosystem services, and/or scenery; and (2) at minimum, identify either sets of complementary sites needed to achieve quantitative targets for biodiversity features or the complementary contribution that individual sites make to conservation within a region (Sarkar et al. 2006).

In recent years, DSTs that help guide ecosystem-based management, integrated ocean management, adaptive management, MSP, and SCP processes have proliferated. Many of them are free or inexpensive to use, but a commonly associated problem is a lack of dedicated maintenance and/or adequate user support, often making the tools unreliable and difficult to use (Curtice et al. 2012). In some cases, proper usage and good practices are documented (Watts et al. 2009; Ardron et al. 2010; Fulton et al. 2011; Guerry et al. 2012). However, there are few published examples that compare functions and features of these tools (cf. Sarkar et al. 2006; Coleman et al. 2011; Rozum and Carr 2013). Rarer still is discussion of how DSTs have been effectively used in the broader context of SCP or MSP processes. Many of the most informative published studies present results from multiobjective analyses and tend to feature the spatial planning tool Marxan (e.g., Chan et al. 2006; Klein, Chan, et al. 2008; Klein, Steinback, et al. 2008; Ban, Picard, et al. 2009; Christensen et al. 2009; Grantham et al. 2013). Given the broad experience that has been built in terms of planning contexts and study locations, this chapter focuses on Marxan; however, lessons discussed here can often apply to the broader use of multiobjective planning tools in general (e.g., Zonation from Moilanen 2014, ConsNet from Ciarleglio et al. 2009, and C-Plan from Watts et al. 2014).

With more than nine thousand downloads from users in at least 140 countries over the past four years alone (M. Watts, personal communication, June 2014), Marxan has become the most popular tool of its kind. It was developed to address the "minimum-set problem" of protecting at least a set amount of conservation features for the minimum cost (McDonnell et al. 2002). Over the past 15 years, experience in implementing Marxan results has shown that part of its success comes from the ability to help practitioners adhere to the stages of SCP (box 2.1), which promotes a comprehensive, flexible, complementary, repeatable, and efficient process (Possingham et al. 2010). For example, using Marxan to plan a network of protected areas requires users to set explicit targets for species and habitat inclusion (step 4 of the SCP process), a potentially politically charged consideration that too easily can be deferred or neglected in planning situations without tools that require this difficult, yet essential, discussion to occur (Lieberknecht et al. 2010).

Box 2.1. Stages of SCP

Systematic conservation planning can be described in 11 key stages, based on guidelines presented by Pressey and Bottrill (2009) and including feedback between steps. This set is meant to be indicative and should not be considered definitive (for example, Possingham et al. [2010] propose an eight-stage approach):

1. **Scope and budget the planning process** to make decisions about boundaries, the planning team, budget, necessary additional funds, and how each step in the process will be addressed, if at all.

2. **Identify and involve stakeholders** (those who will influence, be affected by, or implement actions) from the start of a planning process to encourage information exchange, enable collaborative decision-making, foster buy-in, and increase accountability. Different groups of stakeholders will need to be involved in different ways in specific stages of planning.

3. **Describe the context for conservation areas** in terms of social, economic, and political factors. Identify the types of threats to natural features that can be mitigated by spatial planning and the broad constraints and opportunities around conservation actions.

4. **Identify clear goals and objectives** to articulate priorities for protection and restoration/enhancement of biodiversity (e.g., representation, persistence) and socioeconomic interests (e.g., ecosystem services, livelihoods).

5. **Collect spatially explicit data on socioeconomic variables and threats,** including human uses, tenure, extractive uses, costs of conservation, constraints and opportunities to which planners can respond, and predictions about the expansion of threatening processes.

6. **Collect spatially explicit data on biodiversity and other natural features,** including representation (e.g., vegetation types), focal species, habitat types, biodiversity proxies, ecological processes, and ecosystem services.

7. **Establish targets** by interpreting goals/objectives to quantitatively specify how much of each feature to protect within the network, and establish design principles to influence the geographic configuration (including size, shape, number, and connectivity of sites) of the network.

8. **Review current achievement of objectives** and identify network gaps (often through remote data and field surveys) to determine the extent to which targets are already met.

9. **Select new conservation areas** (often with stakeholders) from multiple location and configuration options that complement existing protected areas for cohesive networks that meet targets and design criteria. Factors influencing decisions usually include costs, constraints, and opportunities for effective conservation.

10. **Implement conservation action** by making decisions on fine-scale boundaries, appropriate management measures, institutional arrangements, implementation sequencing priority, and other site-specific considerations to ensure that selected areas are given the most feasible and appropriate management and that areas are prioritized for action when resources are limited.

11. **Maintain and monitor** the protected area network to evaluate whether management is effectively preserving ecological integrity and long-term persistence in the context of original goals and objectives.

Strengths and Limitations in the Use of Decision Support Tools such as Marxan

When used correctly, planning tools such as Marxan and its expansion, Marxan with Zones, save users time and resources, act as a guide for moving from data to decision-making, build on past work rather than repeating it, reduce the need for human expertise, help explore a wider range of alternatives and trade-offs, document decisions about inputs and parameters, and help integrate planning across diverse sectors (Curtice et al. 2012; Possingham et al. 2010). However, these benefits cannot be realized if DSTs are used without the proper data, as detailed below.

The use of tools such as Marxan is seldom a quick process, and we caution that it comes with an initially steep learning curve. Furthermore, using Marxan is not recommended in regions where data is inconsistent (patchy) and/or scarce, as the tool will simply highlight the few areas that contain the most data, and cannot take into consideration features for which there is no data. This uneven and/or inadequate coverage of a planning region may miss important ecological features or socioeconomic information that could dramatically influence site selection results. In such cases, it could be beneficial to incorporate facilitated and structured discussion, which for "simpler problems" (i.e., those with few data layers and associated targets and costs) can often lead to similar results after less time and technical effort spent (Ban, Hansen, et al., 2009; Ardron et al. 2014). In such situations, if planners decide to also use a DST, the tool should play a secondary role in the decision-making process.

In many data-poor situations, it is possible to address some deficiencies, especially at coarse scales, with surrogates. For example, if species occurrence data is unavailable, climate profiles and other environmental parameters may be appropriate surrogates to develop predictive habitat models. Models can be updated or later replaced in the analysis if and when more data becomes available (e.g., Sarkar et al. 2005). Likewise, socioeconomic information such as fishing effort could be approximated by more easily collected data, such as boat counts (Weeks et al. 2010). Also, surrogates such as sea surface temperature variation can be used to inform the selection of suitable future habitat to protect a range of management options considering climate change (Makino et al. 2014).

In general, more data is often available than may be first thought. Various institutions often hold a number of different datasets describing an area at global, regional, and national scales. Taking stock of these various data sources is one of the first and often most time-consuming tasks in SCP, and is necessary regardless of whether a DST is ultimately used or not.

How Marxan Can Fit into a Systematic Planning Process

Marxan is freely available, spatially explicit software hosted by the University of Queensland, Australia, to support spatial planning. Commonly, practitioners use Marxan analyses to protect at least a certain percentage (termed a *target*) of each representative habitat type, endangered species, climatic zone, and so forth (the *features* of an analysis). The site selection process involves (1) collecting planning units containing at least the targeted amount of features, while (2) minimizing the sum of selected area perimeters, and (3) avoiding socioeconomically important areas (termed *costs*, for which high values make a planning unit less efficient to include) where possible. Marxan allows flexibility in aspects of an analysis such as the number and types of features included, the target set (or not set) for each feature, whether any planning units have a preexisting status (e.g., necessarily included or excluded from all network configurations), and the cost of including each planning unit in the reserve network. Variable emphasis can also be placed on the importance of meeting specific targets and on clumping sites together versus creating a dispersed network (more information available in Ardron et al. 2010). For example, figure 2.1 shows the powerful effect that calibration can have on results through the boundary length modifier (BLM) setting (PacMARA 2012).

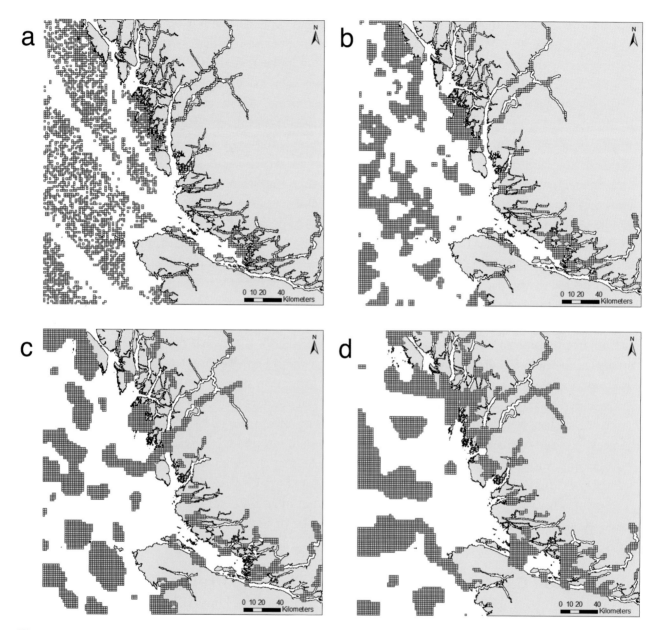

Figure 2.1. **Results of a demonstration Marxan analysis used for Pacific Marine Analysis and Research Association training courses describing hypothetical planning along the central coast of British Columbia, Canada. Purple shading indicates areas selected as part of a complementary network for each scenario. With all other parameters held constant, (a) shows no weighting of boundary length (BLM = 0), while (b), (c), and (d) show increasingly strong influence of total boundary length minimization (BLM = 0.1, 1, and 10, respectively) on network configuration.** Data source: British Columbia Marine Conservation Analysis Project Team. 2011. *Marine Atlas of Pacific Canada: A Product of the British Columbia Marine Conservation Analysis.* Available at www.bcmca.ca.

Although initially conceived in the context of marine conservation planning, Marxan is also applicable to freshwater, terrestrial, or a combination of systems (Tallis et al. 2008; Ball et al. 2009; Beger et al. 2010; Hazlitt et al. 2010; Makino, Beger, et al. 2013; Klein, Jupiter, and Possingham 2014; Klein, Jupiter, Watts, et al. 2014). Marxan has also been used in a variety of ways other than developing recommendations for new conservation sites (for example, considering existing networks [Stewart et al. 2003; Fuller et al. 2010], examining the efficiency of ad hoc planning processes [Klein, Chan, et al. 2008; Mills et al. 2012], and planning for persistence in the face of climatic change [Game et al. 2008; Makino et al. 2014]). Other examples of innovative Marxan uses include optimizing a measure of ecosystem services as well as biodiversity (Chan et al. 2006; Luck et al. 2012), prioritizing a network of fishing sites to maintain viability of the industry (Ban and Vincent 2009), and using conservation priority as the cost for industrial site selection and vice versa to inform potential conflict analysis (PacMARA 2012).

In perhaps the best-known case worldwide of MPA creation involving Marxan (e.g., Ecology Centre of the University of Queensland 2009), the Australian government cosponsored an analysis with stakeholder engagement to inform protected area designation in the Great Barrier Reef (GBR). Substantial stakeholder involvement in planning promoted impartiality in the recommendations Marxan returned, and resulting maps comprised an effective platform for communication and compromise when used iteratively to zone the GBR (Lewis et al. 2003; Davis 2004; Pattison et al. 2004). However, by systematically avoiding areas of high human "cost," the GBR zoning plan has recently come under criticism (Devillers et al. 2014). This emerging controversy highlights the tendency of DSTs to simply "do as they are told," and the need to address larger questions concerning the objectives and trade-offs that arise in planning processes.

Marxan with Zones has since been developed to more efficiently allocate zones for multiple objectives and has been used to inform planning in many regions globally (e.g., Makino, Klein, et al., 2013). Marxan with Zones is based on the same software as Marxan (akin to layers of Marxan nested within one another) but benefits from more data (i.e., zone-specific datasets) to operate at its full potential. There are also more variables to control and more explicit judgments required when creating Marxan with Zones scenarios, which complicates usage but allows more flexibility and range of options than by using Marxan alone. The contribution of zones to different targets, the costs of implementing different zones in different locations, and the interactions between zones are all considered with this advanced tool (Watts et al. 2009). Marxan with Zones makes it possible to create a range of conservation schemes from "no-take" to allowing only certain types of extractive activities, depending on the designation (Klein et al. 2009).

Both programs help users evaluate how well each scenario meets conservation and socioeconomic objectives, thereby facilitating exploration of trade-offs. However, Marxan with Zones has the distinct advantage of allowing what would be treated as *cost layers* in Marxan (e.g., a specific fishery) to become separate features within the analysis with specific targets set for each associated zone (e.g., multiple use, specific industry access, indigenous access). This functionality allows users to avoid a major shortcoming of Marxan, which is requiring that all cost data be combined into a single value for each planning unit. However, the greater power of Marxan with Zones leads to fundamental questions regarding the efficacy of different management treatments (e.g., differing levels of restrictions) in different zones, or *zone effectiveness* (Makino, Klein, et al., 2013), and their effects on biodiversity, which can be difficult to predict.

Lessons Learned

Further recommendations on the use of Marxan, including the topics briefly mentioned in this section, have been covered in detail in the *Marxan Good Practices Handbook* (Ardron et al. 2010), as well as in two workshop reports on the topic (PacMARA 2008; BCMCA and PacMARA 2010). Following, we expand upon considerations that have emerged as central in the course of technical and managerial DST trainings through our involvement in the Pacific Marine Analysis and Research Association (PacMARA). PacMARA is a charitable organization based in British Columbia, Canada, dedicated to building capacity in marine and coastal planning internationally. The organization provides facilitation, training, analysis, and support for the ocean planning community, both within and external to planning processes, through workshops, courses, symposia, and spatial planning research case studies. In addition to our own work, we have learned a great deal from participants while leading hands-on and conceptual capacity building courses throughout the world (PacMARA has held 40-plus courses and workshops for 800-plus participants from 50-plus countries).

In the following sections, we focus on insights that we believe to be relevant to practitioners wishing to understand how spatial tools and their outputs fit into the larger context of systematic planning. These are, for the most part, personal observations; however, while many of these "tips" are Marxan-specific and unpublished, they are often similar to good practices that have been highlighted in the marine planning and GIS literature more generally (e.g., Mitchell 2001; Ehler and Douvere 2009). In addition, there are good practices in cartography and the display of quantitative information that deserve close attention; however, these are amply covered in the existing literature (e.g., Monmonier 1996; Tufte 2001) and are not addressed further here.

Asking the Right Question

For a spatial planning tool to appropriately prioritize areas for protection, practitioners must be cautious to ensure they are asking the right question of the tool. Game et al. (2013) elaborate on this issue with several valuable insights, and here we highlight a few practical considerations. Marxan and Marxan with Zones can be used in a seemingly endless variety of ways (Watts et al. 2009; Martin et al. 2010); however, the value of their outputs will depend in large part on whether they are addressing the legal and policy requirements of a particular audience and planning process. For example, although displaying results for an MPA network that protects 20% of a region can have heuristic value, the usefulness in implementation will be limited if governing bodies are interested in only meeting lesser targets (e.g., the CBD Aichi Target 11 of 10%) (Convention on Biological Diversity 2011). Likewise, study areas that mismatch or fall outside the jurisdiction of a given planning process, even if they make more "ecological sense," will have little value in terms of the planning process's mandate.

Common Misconceptions

The original and rewritten Marxan manuals explain the algorithm and its objective function (Ball and Possingham 2000; Game and Grantham 2008). However, the explanation is necessarily technical with some mathematical notation.

Marxan Is a 'Black Box'

The black-box argument often arises when stakeholders are wary of a planning process, and hence, any analyses associated with it. In our view, this is not about understanding the math, simulated annealing, or other aspects of Marxan's objective function; rather, it is mostly a question of building trust. If stakeholders are given an overview of what Marxan does, and can see that their inputs have been incorporated and actually affect Marxan's outputs, a great deal of the mistrust dissipates, and concerns about how Marxan works are minimized.

Marxan Is a Model

Marxan is, in fact, an optimization algorithm. However, this technical difference is often lost on a nontechnical audience. For a politician, the term *model* can indicate that the results are not real, based largely on assumptions, and are therefore not applicable. There is probably no easy way to completely correct this misperception; however, emphasizing that Marxan simply suggests a number of good solutions based on user-defined planning targets can steer discussion away from notions that it constructs artificial realities or otherwise "knows" things about the system.

The 'Best Solution' Is the Ultimate Planning Answer

Marxan terminology does not always translate readily into planning process discussions, and the term *best solution* is a case in point: it is only meant to indicate the solution with the best overall score (which is based to a large degree on the efficiency of the network). However, identifying the lowest score in no way suggests that the other favorable, but slightly less efficient, solutions are not worthy of consideration. Indeed, given that one of the great advantages of the tool is its ability to produce a variety of good solutions, it would be counterproductive to limit discussion to just one. Additionally, many other considerations that are not (or cannot) be quantified into data layers will affect planning choices. Marxan's solutions should be seen as suggested starting points and topics for discussion and engagement rather than end points.

Selection Frequency Is the Same as Ecological Irreplaceability

Selection frequency vis-à-vis ecological irreplaceability is another unfortunate misunderstanding of terminology that arose because of the use of the word *irreplaceability* in an early version of the Marxan manual, which then crept into the early Marxan literature. How often a given area is selected in Marxan solutions can be a good indication of how "useful" it is in a variety of scenarios (Ardron 2008). However, this may or may not mean that it is irreplaceable (Fischer et al. 2010). This misunderstanding is discussed in more detail in the *Marxan Good Practices Handbook* (Wilson et al. 2010), but we have listed it here because misuse of the term irreplaceability is still pervasive.

Marxan Is Only Used for Identifying Marine Protected Areas

Although recommending MPA network locations has been a common application of the Marxan tool, it is also frequently used in terrestrial and freshwater settings. In any case, Marxan can also be used to identify areas for extractive or other types of uses beyond conservation (e.g., Ban and Vincent 2009), as nothing about the tool inherently predisposes it to MPA site selection.

Including Socioeconomic Considerations Detracts from a 'Pure' Ecological Analysis

The misconception over including socioeconomic considerations is based on the legitimate fear that these considerations can often dilute or trade off conservation values in planning processes. However, in the case of a Marxan analysis, adding socioeconomic interests adds relevant constraints and increases the practicality of solutions, among other benefits (Ban and Klein 2009). If some ecological targets are no longer being met after the addition of human use or interest data, Marxan parameters can be adjusted until they are met again, according to procedures described in Game and Grantham (2008).

Various 'Costs' Can Be Added Together

A shortcoming of Marxan is that it has just one cost function, and hence constraints that are measured differently (e.g., the values of various fisheries as well as other human activities) will need to be reconciled. There are various strategies for dealing with this issue (Ban and Klein 2009) (for example, translating the different constraints into a common *currency*, such as dollars or jobs). However, adding different measures together as though they were the same, like "apples and oranges," is inappropriate (Cameron et al. 2010). Although Marxan with Zones, unlike Marxan, allows for multiple cost layer inclusion, the costs are simply added internally, and so the issue largely still remains (Watts et al. 2009). Although this has led to a further misconception that the problem has been solved, using Marxan with Zones can indeed alleviate the issue of multiple costs if some of the costs are available as input layers and can be associated with zoning targets. (Costs are discussed again later in this chapter in the context of stakeholder relations.)

A Marxan Analysis Can Be Completed in the Time It Takes to Learn How to Use the Tool

Grossly underestimating the time required for a Marxan analysis is perhaps the most common misconception of all (e.g., a typically expected time frame might be a few weeks or perhaps a few months). However, this misconception has a grain of truth: learning Marxan should not take more than a few weeks, and less if one takes a course. What consumes the months (and often years) of time in an analysis is mostly data related: acquisition, preparation, quality assurance, collation into single layers, and finally conversion into input files (Martin et al. 2010). Another often significant drain on time in an inclusive process is securing stakeholder and/or scientific agreement on targets (BCMCA and PacMARA 2010; Ban et al. 2013).

Working with Stakeholders and Their Information

One of the most challenging components of conservation planning processes is working with a diverse range of stakeholders. In the authors' experiences, most conflicts in conservation planning processes stem from stakeholder engagement, especially concerning socioeconomic issues. The acceptance of such tools as Marxan will hinge to a large degree on stakeholder acceptance of a planning process more generally. Cultural context is critical in this regard, as the levels of engagement and power expected from resource users, indigenous groups, industry associations, nongovernmental organizations (NGOs), and the like depend largely on the historical and cultural context of where a process takes place. For example, stakeholders in Germany have, to date, been minimally engaged in marine planning (Von Nordheim et al. 2006) and arguably expect less than those in the United Kingdom, which has seen much greater engagement (e.g., Lieberknecht et al. 2013). Further, stakeholder expectations can vary greatly across different groups (Maguire et al. 2011).

Being clear from the beginning of a process regarding the scope of planning, mandate(s), goals and objectives, what is and is not "on the table" for discussion, process limitations, and how stakeholders can expect to be engaged can avoid misunderstandings and mismatched expectations as progress is made. If stakeholders arrive later in a process timeline, it will likely be necessary to revisit some of the basics, including the agreed-on terms of reference, and again outline the benefits of participation. As the appropriate level of inclusivity can be time consuming, it is clearly preferable to involve all the necessary stakeholders from the start (Ban et al. 2013).

Incorporation of human-use data and values into Marxan analyses has substantially evolved over its history, from largely nonexistent (e.g., cost simply as an area of selected planning units) to single measure to more integrated options such as combined costs, measures of naturalness, cumulative impacts, and habitat suitability (Ban and Klein 2009). Classifying, mapping, and integrating human uses and values, including economically and culturally outstanding areas, into DSTs helps keep the process systematic by representing ecological and socioeconomic contributions as partners in conservation site prioritization. However, such data characterizations are imperfect and will still require robust stakeholder discussion to examine nuances and corrections.

Marxan is limited in a number of cost assumptions, including that cost values are uniform through time and human activities affect all features similarly. Practitioners should be sensitive to the fact that stakeholders may adamantly oppose the combination of socioeconomic datasets in an analysis, which leaves the option of running separate Marxan scenarios for each different cost, or potentially using the more sophisticated (but more complicated) Marxan with Zones instead (e.g., BCMCA and PacMARA 2010). Whereas including costs in an analysis is usually complicated, it has become good practice to use more than the simple area of selected planning units (e.g., Ban, Picard, et al. 2009). Figure 2.2 shows the influence that different cost inputs can have on Marxan site selection results.

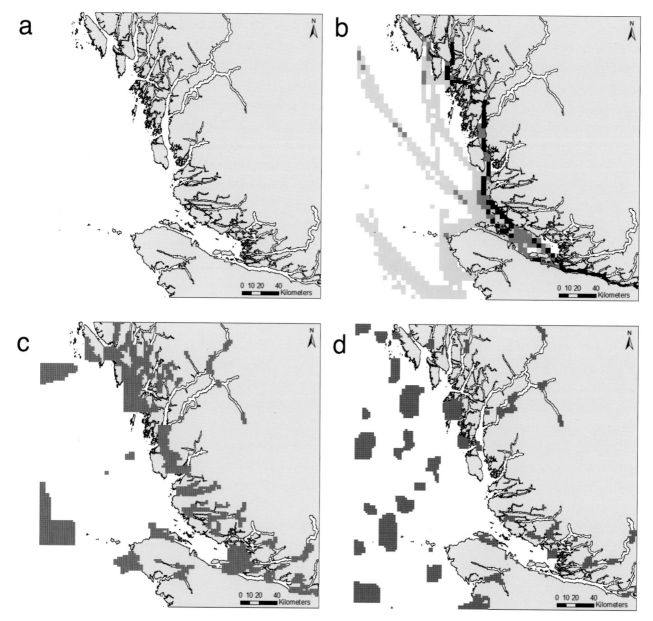

Figure 2.2. Marxan site selection results as a function of different costs with all other parameters held constant: (a) area only as cost (i.e., all planning units have a uniform value); (b) vessel density shown as the cost layer (in which darker planning units were more frequently selected; (c) selection results considering area only as cost (red planning units); and (d) selection results considering vessel density as cost (blue planning units) in an example used for Pacific Marine Analysis and Research Association training courses describing hypothetical planning along the central coast of British Columbia, Canada. Data source: British Columbia Marine Conservation Analysis Project Team. *2011. Marine Atlas of Pacific Canada: A Product of the British Columbia Marine Conservation Analysis.* Available at www.bcmca.ca.

The key to understanding the incorporation of socioeconomic data into a Marxan analysis is recognizing that whenever multiple human uses are combined, Marxan cannot provide equity among those uses. As discussed, adding radically different costs together should be avoided, such as social values (e.g., spiritual, history, identity) and economic information (e.g., profits, jobs).

A strategy to successfully engage stakeholders in DST use is to keep an analysis simple initially and add complexity iteratively as the discussion progresses, when new data is available, when new stakeholders join the process, and/or when results are reviewed and returned to the project team (BCMCA and PacMARA 2010). Reiteration of analyses, the interactive involvement of stakeholders (e.g., through the use of interactive tools such as Zonae Cogito [Segan et al. 2011; Ecology Centre of the University of Queensland 2012] or SeaSketch [McClintock 2012]), and the openness of practitioners to creative uses of Marxan and criticism that can improve previous analyses should all help stakeholders and decision-makers view the systematic process as legitimate.

Also, initial simplicity helps analysts assess whether the program is working correctly and whether core datasets appear to be accurate, even if first results are unrealistic (with the caveat that future iterations contain sufficient data to provide meaningful and balanced solutions). Using a range of feature targets (and potentially also clumping together factors, penalties, and costs) emphasizes the flexibility of Marxan and its solutions, as well as that of the process itself. It has been observed that communicating initial and midterm outputs and soliciting feedback may be as important to building stakeholder relationships, and hence to the planning process, as the actual Marxan results themselves (BCMCA and PacMARA 2010).

Ultimately, successful stakeholder engagement in SCP depends more on a clear understanding of the context and trusting the planning process than on the specifics of the tool being used. However, it remains important that participants hear an explanation of why Marxan is being employed. Common reasons for this include (1) the tool's ability to explore more potential solutions than could be comprehended manually, (2) finding unforeseen good solutions, (3) comprehensively including stakeholder uses and values, and (4) helping systematically explore and potentially dissolve controversy through trade-off options. Marxan also provides a framework for stakeholders to trust that their data is being transparently considered alongside ecological information, which often supports postreserve implementation success. However, it is also important to temper expectations for DSTs to address contentious issues and resolve conflicts. In reality, Marxan can help users change the focus to exploring trade-offs and informing conflict, but it cannot resolve differences in values or world views (PacMARA 2008).

Presentation and Interpretation of Results

For each set of targets and costs specified (known as a *scenario*), each run of Marxan will produce a different solution (e.g., figure 2.3a–b), providing a shapefile of the solution and summary tables indicating how well each run met the targets and its overall score. The score in Marxan is a sum of the perimeters of each set of planning units included in the solution, the cost of each unit included, and the penalty for missing any targets that were specified in the solution. Planning unit configurations with lower scores are considered more favorable. To obtain an accurate overall picture of the intended solution space, several runs (preferably, one hundred or more) are recommended for each scenario. With multiple runs, the software will produce a selection frequency output (e.g., figure 2.3c–d) that overlays all solutions to indicate which planning units are chosen most often. These compiled, or *summed*, solutions can highlight places that frequently recur, thereby suggesting that these places are likely to be important under a variety of planning scenarios.

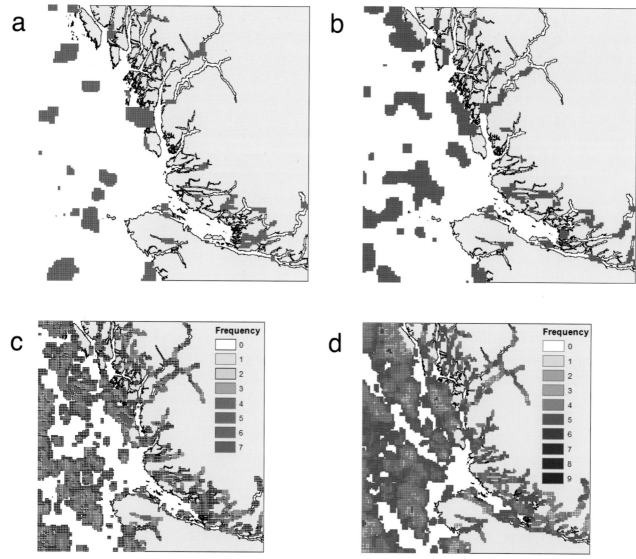

Figure 2.3. The best solutions resulting from an analysis with (a) red planning units, targets of 10%; (b) blue planning units, targets of 20%; the corresponding selection frequencies for (c), in which darker red indicates more frequent selection, targets of 10%; and (d), in which darker blue indicates more frequent selection, targets of 20%. All as used for Pacific Marine Analysis and Research Association training courses describing hypothetical planning along the central coast of British Columbia, Canada. Data source: British Columbia Marine Conservation Analysis Project Team. 2011. *Marine Atlas of Pacific Canada: A Product of the British Columbia Marine Conservation Analysis.* Available at www.bcmca.ca.

Maps of *selection frequency/summed solution* (the two terms have the same meaning) are often appropriate in discussions with planners and stakeholders because they identify areas that are chosen most often under conditions of the analysis with a range of targets, costs, penalties, and so forth. Selection frequency maps should be accompanied by the maps of two or three different good solutions to show possible spatial configurations that meet parameters and conditions of the analysis (e.g., figure 2.4). This practice helps avoid an audience confusing selection frequency with a network of complementary sites that combine to meet feature targets. It can be helpful to use cluster analysis to highlight the range of solutions available (Airame 2005; Fischer et al. 2010).

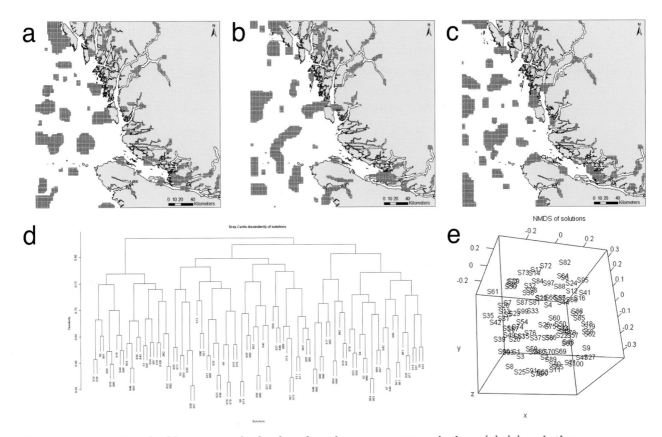

Figure 2.4. Results of a Marxan analysis showing three separate solutions (a)–(c), solutions indicated by purple shaded planning units, resulting from a single scenario used for Pacific Marine Analysis and Research Association training courses to describe hypothetical planning along the central coast of British Columbia, Canada. Panel (d) shows a dendrogram, and (e) shows a three-dimensional, nonmetric multidimensional scaling representation of relatedness between one hundred different solutions. Data source: British Columbia Marine Conservation Analysis Project Team. 2011. *Marine Atlas of Pacific Canada: A Product of the British Columbia Marine Conservation Analysis.* Available at www.bcmca.ca.

Selection frequency maps using dotted lines or blurred boundaries, and presented with logical and intuitive color gradients, can help dissipate fears that a single result is a complete and ready answer to the planning problems at hand. In smaller print, including information such as the target values, number of features, and parameters used (including BLM and species penalty factor) is good practice, even though the majority of stakeholders will not read such information (see Ardron et al. 2010 for Marxan terminology explanations). More important, maps should represent different scenarios, each of which should have a descriptive title in plain language (e.g., "Scenario 3: Moderate conservation targets combined with values from commercial fisheries"). When presenting initial results, maps should be designed to reflect their draft status and be easy to interpret. Whereas a large watermark indicating "draft" is de rigueur, such labeling will not necessarily convince all stakeholders, underlining the need to build trust before any results are released and for careful communication throughout the process (Nicolson et al. 2010). For example, the MPA planning process in Gwaii Haanas, British Columbia, saw a reduction from the desired 30% protection as no-take to a final result of just 3% of the designated marine area. This disparity was largely because of problems with communication and presentation of Marxan results as well as late stakeholder engagement (Sloan 2014).

Effectively presenting the outputs of a Marxan analysis can be a substantial challenge to technical teams that may lack the nonbias necessary to communicate with those unfamiliar with software tools. For example, economic activities such as fisheries are typically treated as "costs" in the Marxan lexicon (but see Ban and Vincent 2009). However, stakeholders may object to such terminology as not reflecting their true economic and social values. Replacing the word *cost* with *value* when communicating Marxan results can greatly ease such communication potholes (BCMCA and PacMARA 2010).

Arranging a third party that is skilled in communication to present results can be beneficial to encourage focus on the key messages and avoid distracting details that may not be relevant to the audience. This communication strategy applies to public presentations as well as internal, in which it is important for senior management to appreciate, in general terms, the value and limitations of Marxan results as they relate to the agency mandate. However, ensuring the presence of competent technical personnel is always good practice to confirm that results are being interpreted correctly, and to answer detailed questions that may arise.

In general, starting communication with project team members and stakeholders early in the process can help avoid problems later. Presenting midterm results can allow for the early identification of errors and omissions. That said, practitioners are advised to be prudent about sharing initial and midterm results with groups with which trust may be an issue. Maps can easily be misinterpreted and spread beyond the meeting room without accompanying explanations and caveats. Therefore, it can be advisable to select a subset of "friendly" partners and stakeholders to preview early results, using projected images rather than paper maps.

As with all technical work, it is also good practice to avoid jargon, and limit the technical details to those necessary to describe an analysis in its broadest terms. Focusing on key messages will mean narrowing the results that are presented, even if many more scenarios were actually completed. At the same time, showing multiple options for any one scenario reinforces the message that there is flexibility in the number of good solutions to a planning problem (see figure 2.4). This practice also helps reinforce to stakeholders that the role of an analyst is to provide results that inform decisions, not make the planning decisions themselves.

Presenting visual results that clearly show how stakeholder values and concerns have been accounted for will speak more loudly than any number of theoretical discussions on the process of simulated annealing (McClintock 2012) and other technical details. However, stakeholders and other planning process participants should clearly understand what Marxan is designed to do: illustrate multiple options of spatially protecting targeted features for the least cost.

Anticipating conflicts and presenting them up front, either through a conflict analysis applied to Marxan results (e.g., comparing competing cost scenarios) or simply by listing issues emerging through dialog with critics, will help dispel the misunderstanding that planning tools circumvent public dialog. Rather, the emphasis should be on their use as a conflict resolution or minimization tool. That some conflicts cannot be resolved through the use of Marxan (or similar tools) should be acknowledged and will help keep discussions realistic in expectations of what the tool can achieve. All messages should ultimately relate back to the agreed-on objectives of the process and how the software helps address them.

Discussion

We suggest that a multiobjective planning process should consist of merging many of the successful elements of ecosystem-based management, integrated ocean management, adaptive management, MSP, and SCP (Ardron 2010). Such approaches have demonstrated the benefit of stating process objectives explicitly, recognizing that weaknesses in an analysis often stem from issues in the problem development stage. Not only does this practice facilitate decision-making within Marxan, such as setting appropriate targets, but it also helps illustrate the policy, political, and social framework within which conservation and resource management decisions are made.

Parties involved in a comprehensive spatial planning process can learn from systematic planning experiences about the need to bound the scope of analysis, specifically around what the pertinent legislation can and cannot do (e.g., Kirlin et al. 2013). Although the final implemented result often differs significantly from Marxan outputs, practitioners nonetheless emphasize the value of the tool in reaching agreement (e.g., Horsman et al. 2011). Additionally, the inclusive multiobjective nature of systematic tools can help avoid "end runs" where a party negotiates bilaterally with a regulator outside the process (PacMARA 2008).

Although conservation planning DSTs have advanced considerably in recent years, there is still room for growth and improvement. For example, persistent challenges faced in marine spatial and systematic conservation planning include (1) significant data requirements of many tools that are unfulfilled because of missing or poor-quality information; (2) stakeholder isolation because of the complexity of optimization tools; (3) overwhelmed managers and planners, stemming from the quantity and/or lack of straightforward interoperability of available DSTs; (4) availability of capacity building and trainings for managers; and (5) lack of ground-truthing of optimal model outputs (McClintock et al. 2013).

Conclusion

The appropriate role for Marxan, as with other DSTs, is to support decision-making. These tools can suggest efficient sets of areas for conservation or other types of management. However, they will not provide final answers or solve difficult value-laden decisions. Using planning tools does not diminish the importance of human input and collaboration nor the need to manage prejudice, politics, territory, mistrust, or technophobia. DST-generated options will inevitably need to be revised to yield a final plan that considers the full range of political, socioeconomic, and practical factors.

Although the use of Marxan as a DST can facilitate stakeholder engagement, and encourage a systematic approach, it is far from a guarantee of fruitful participation and acceptance of the planning process. For that, building trust through political and social channels is necessary. Practitioners will need to be ready to cautiously interpret, to both their colleagues and superiors and to the public, factors such as outputs in the context of data limitations, parameters used, functioning of the tool itself, and how all these pieces relate to the planning process objectives. Nevertheless, appropriate use of DSTs can provide an effective, efficient, and viable means to link science with policy, while negotiating the minefield of legal, scientific, and stakeholder requirements, leading to policy options that are otherwise unlikely to be identified.

Acknowledgments

We thank Hugh Possingham, Matt Watts, and Ian Ball for developing, maintaining, and supporting Marxan software, and for codeveloping training materials and protocols. We also thank Norma Serra-Sogas, PacMARA trainers, and participants of numerous courses and workshops for their insights and help in improving PacMARA trainings and providing contributions to this text; as well as two anonymous reviewers for providing excellent suggestions to improve the content and clarity of this chapter. Finally, we thank Esri for organizing and executing the 2013 Esri Ocean GIS Forum and this book.

References

Airame, S. 2005. "Channel Islands National Marine Sanctuary: Advancing the Science and Policy of Marine Protected Areas." In *Place Matters: Geospatial Tools for Marine Science, Conservation, and Management in the Pacific Northwest*, edited by D. Wright and A. J. Scholz, 92–133. Corvallis, OR: Oregon State University Press.

Ardron, J. A. 2008. "Modelling Marine Protected Areas (MPAs) in British Columbia." In *Proceedings of the Fourth World Fisheries Congress: Reconciling Fisheries with Conservation*. American Fisheries Society Symposium 49, Vancouver, British Columbia, May 2–4, 2004. http://web.fisheries.org/proofs/wfc/ardron.pdf.

———. 2010. "Marine Planning: Tragedy of the Acronyms." *Marine Ecosystems and Management* 4 (2): 6. Revised 2012 as a blog. http://openchannels.org/blog/jeffardron/marine-planning-tragedy-acronyms.

Ardron, J. A., M. R. Clark, A. J. Penny, T. F. Hourigan, A. A. Rowden, P. K. Dunstan, L. E. Watling, T. M. Shank, D. M. Tracey, M. R. Dunn, and S. J. Parker. 2014. "A Systematic Approach towards the Identification and Protection of Vulnerable Marine Ecosystems." *Marine Policy* 49:146–54.

Ardron, J. A., H. P. Possingham, and C. J. Klein, eds. 2010. *Marxan Good Practices Handbook*, version 2. Victoria, BC: PacMARA. http://pacmara.org/wp-content/uploads/2010/07/Marxan-Good-Practices-Handbook-v2-2010.pdf.

Ball, I. R., and H. P. Possingham. 2000. *Marxan (V1.8.2): Marine Reserve Design Using Spatially Explicit Annealing: A Manual*. http://www.uq.edu.au/marxan/docs/marxan_manual_1_8_2.pdf. Last accessed June 21, 2014.

Ball, I. R., H. P. Possingham, and M. E. Watts. 2009. "Marxan and Relatives: Software for Spatial Conservation Prioritization." In *Spatial Conservation Prioritization Quantitative Methods and Computational Tools*, edited by A. Moilanen, K. A. Wilson, and H. P. Possingham, 185–95. Oxford, UK: Oxford University Press.

Ban, N. C., K. M. Bodtker, D. Nicolson, C. K. Robb, K. Royale, and C. Short. 2013. "Setting the Stage for Marine Spatial Planning: Ecological and Social Data Collation and Analyses in Canada's Pacific Waters." *Marine Policy* 39:11–20.

Ban, N.C., G. J. Hansen, M. Jones, and A. C.J. Vincent. 2009. Systematic Marine Conservation Planning in Data-Poor Regions: Socioeconomic Data Is Essential. *Marine Policy* 33 (5): 794–800.

Ban, N. C., and C. J. Klein. 2009. "Spatial Socioeconomic Data as a Cost in Systematic Marine Conservation Planning." *Conservation Letters* 2 (5): 206–15.

Ban, N. C., C. R. Picard, and A. C. J. Vincent. 2009. "Comparing and Integrating Community-Based and Science-Based Approaches to Prioritizing Marine Areas for Protection." *Conservation Biology* 23 (4): 899–910.

Ban, N. C., and A. C. J. Vincent. 2009. "Beyond Marine Reserves: Exploring the Approach of Selecting Areas Where Fishing Is Permitted, Rather Than Prohibited." *PLoS ONE* 4 (7): e6258.

BCMCA and PacMARA (British Columbia Marine Conservation Analysis and Pacific Marine Analysis and Research Association). 2010. *Marxan Workshop Proceedings Report*. Vancouver, BC: PacMARA. http://pacmara.org/bcmcamarxan-good-practices-workshops. Last accessed May 2014.

Beger, M., H. S. Grantham, R. L. Pressey, K. A. Wilson, E. L. Peterson, D. Dorfman, P. J. Mumby, R. Lourival, D. R. Brumbaugh, and H. P. Possingham. 2010. "Conservation Planning for Connectivity across Marine, Freshwater, and Terrestrial Realms." *Biological Conservation* 143 (3): 565–75.

Cameron, S. E., C. J. Klein, R. Canessa, and L. Geselbracht. 2010. "Addressing Socioeconomic Objectives." In *Marxan Good Practices Handbook,* version 2, edited by J. A. Ardron, H. P. Possingham, and C. J. Klein, 49–56. Victoria, BC: PacMARA. http://pacmara.org/wp-content/uploads/2010/07/Marxan-Good-Practices -Handbook-v2-2010.pdf.

Chan, K. M. A., M. R. Shaw, D. R. Cameron, E. C. Underwood, and G. C. Daily. 2006. "Conservation Planning for Ecosystem Services." *PLoS Biology* 4 (11): e379.

Christensen, V., Z. Ferdaña, and J. Steenbeek. 2009. "Spatial Optimization of Protected Area Placement Incorporating Ecological, Social, and Economical Criteria." *Ecological Modelling* 220 (19): 2583–93.

Ciarleglio, M., J. W. Barnes, and S. Sarkar. 2009. "ConsNet: New Software for the Selection of Conservation Area Networks with Spatial and Multi-criteria Analyses." *Ecography* 32 (2): 205–9. http://uts.cc.utexas .edu/~consbio/Cons/consnet_home.html.

Coleman, H. M., M. Foley, E. Prahler, M. Armsby, and G. Shillinger. 2011. *Decision Guide: Selecting Decision Support Tools for Marine Spatial Planning.* http://pacmara.org/wp-content/uploads/2013/03/cos_pacmara_msp_guide .pdf. Last accessed June 21, 2014.

Convention on Biological Diversity. 2011. Aichi Biodiversity Targets. http://www.cbd.int/sp/targets. Last accessed January 17, 2015.

Curtice, C., D. C. Dunn, J. J. Roberts, S. D. Carr, and P. N. Halpin. 2012. "Why Ecosystem-Based Management May Fail without Changes to Tool Development and Financing." *BioScience* 62 (5): 508–15.

Davis, J. B. 2004. "Using Computer Software to Design Marine Reserve Networks: Planners Discuss Their Use of Marxan." *MPA News* 6:1–3.

Devillers, R., R. L. Pressey, A. Grech, J. N. Kittinger, G. J. Edgar, T. Ward, and R. Watson. 2014. "Reinventing Residual Reserves in the Sea: Are We Favoring Ease of Establishment over Need for Protection?" *Aquatic Conservation: Marine and Freshwater Ecosystems.* doi:10.1002/aqc.2445.

Ecology Centre of the University of Queensland. 2009. *Scientific Principles for Design of Marine Protected Areas in Australia: A Guidance Statement.* http://ecology.uq.edu.au/mpa-principals-guidance-statement. Last accessed June 21, 2014.

———. 2012. Marxan R&D. http://www.uq.edu.au/marxan/latest-r-d. Last accessed June 21, 2014.

Edgar, G. J., R. D. Stuart-Smith, T. J. Willis, S. Kininmonth, S. C. Baker, S. Banks, and N. S. Barrett. 2014. "Global Conservation Outcomes Depend on Marine Protected Areas with Five Key Features." *Nature* 506:216–20.

Ehler, C., and F. Douvere. 2009. *Marine Spatial Planning: A Step-by-step Approach toward Ecosystem-Based Management.* Intergovernmental Oceanographic Commission and Man and the Biosphere Programme. IOC Manual and Guides No. 53, ICAM Dossier No. 6. Paris: UNESCO.

Fischer, D.T., H. M. Alidina, C. Steinback, P. I. Ramirez de Arellano, Z. Ferdaña, and C. J. Klein. 2010. "Ensuring Robust Analysis." In *Marxan Good Practices Handbook,* version 2, edited by J. A. Ardron, H. P. Possingham, and C. J. Klein, 75–96. Victoria, BC: PacMARA. http://pacmara.org/wp-content/uploads/2010/07/Marxan -Good-Practices-Handbook-v2-2010.pdf.

Fuller, R. A., E. McDonald-Madden, K. A. Wilson, J. Carwardine, H. S. Grantham, J. E. Watson, C. J. Klein, D. C. Green, and H. P. Possingham. 2010. "Replacing Underperforming Protected Areas Achieves Better Conservation Outcomes." *Nature* 466 (7304): 365–67.

Fulton, E. A., J. S. Link, I. C. Kaplan, M. Savina-Rolland, P. Johnson, C. Ainsworth, P. Horne, R. Gorton, R. J. Gamble, A. D. M. Smith, and D. C. Smith. 2011. "Lessons in Modelling and Management of Marine Ecosystems: The Atlantis Experience." *Fish and Fisheries* 12 (2): 171–88.

Game, E. T., and H. S. Grantham. 2008. *Marxan User Manual: For Marxan Version 1.8.10.* Vancouver, BC: University of Queensland and PacMARA. http://www.uq.edu.au/marxan/docs/Marxan_User_Manual_2008.pdf.

Game, E. T., P. Kareiva, and H. P. Possingham. 2013. "Six Common Mistakes in Conservation Priority Setting." *Conservation Biology* 27:480–85.

Game, E. T., M. Watts, S. Wooldridge, and H. P. Possingham. 2008. "Planning for Persistence in Marine Reserves: A Question of Catastrophic Importance." *Ecological Applications* 18:670–80.

Grantham, H. S., V. N. Agostini, J. Wilson, S. Mangubhai, N. Hidayat, A. Muljadi, C. Rotinsulu, M. Mongdong, M. W. Beck, and H. P. Possingham. 2013. "A Comparison of Zoning Analyses to Inform the Planning of a Marine Protected Area Network in Raja Ampat, Indonesia." *Marine Policy* 38:184–94.

Guerry, A. D., M. H. Ruckelshaus, K. K. Arkema, J. R. Bernhardt, G. Guannel, C. K. Kim, and M. Marsik. 2012. "Modeling Benefits from Nature: Using Ecosystem Services to Inform Coastal and Marine Spatial Planning." *International Journal of Biodiversity Science, Ecosystem Services & Management* 8: 107–21.

Hazlitt, S. L., T. G. Martin, L. Sampson, and P. Arcese. 2010. "The Effects of Including Marine Ecological Values in Terrestrial Reserve Planning for a Forest-Nesting Seabird." *Biological Conservation* 143 (5): 1299–1303.

Horsman, T. L., A. Serdynska, K. C. T. Zwanenburg, and N. L. Shackell. 2011. *Report on the Marine Protected Area Network Analysis for the Maritimes Region of Canada*. Report Number 2917. Dartmouth, Nova Scotia: Department of Fisheries and Oceans.

Kirlin, J., M. Caldwell, M. Gleason, M. Weber, J. Ugoretz, and E. Fox. 2013. "California's Marine Life Protection Act Initiative: Supporting Implementation of Legislation Establishing a Statewide Network of Marine Protected Areas." *Ocean & Coastal Management* 74:3–13.

Klein, C. J., A. Chan, L. Kircher, A. J. Cundiff, N. Gardner, Y. Hrovat, A. Scholz, B. E. Kendall, and S. Airame. 2008. "Striking a Balance between Biodiversity Conservation and Socioeconomic Viability in the Design of Marine Protected Areas." *Conservation Biology* 22 (3): 691–700.

Klein, C. J., S. D. Jupiter, and H. P. Possingham. 2014. "Setting Conservation Priorities in Fiji: Decision Science versus Additive Scoring Systems." *Marine Policy* 48:204–5.

Klein, C. J., S. D. Jupiter, M. Watts, and H. P. Possingham. 2014. "Evaluating the Influence of Candidate Terrestrial Protected Areas on Coral Reef Condition in Fiji." *Marine Policy* 44:360–65.

Klein, C. J., C. Steinback, A. J. Scholz, and H. P. Possingham. 2008. "Effectiveness of Marine Reserve Networks in Representing Biodiversity and Minimizing Impact to Fishermen: A Comparison of Two Approaches Used in California." *Conservation Letters* 1 (1): 44–51.

Klein, C. J., C. Steinback, M. Watts, A. J. Scholz, and H. P. Possingham. 2009. "Spatial Marine Zoning for Fisheries and Conservation." *Frontiers in Ecology and the Environment*. doi:10.1890/090047.

Lewis, A., S. Slegers, D. Lowe, L. Muller, L. Fernandes, and J. Day. 2003. "Use of Spatial Analysis and GIS Techniques to Re-zone the Great Barrier Reef Marine Park." *Coastal GIS*, 431–51.

Lieberknecht, L., J. A. Ardron, R. Wells, N. C. Ban, M. Lötter, J. L. Gerhartz, and D. J. Nicolson. 2010. "Addressing Ecological Objectives through the Setting of Targets." In *Marxan Good Practices Handbook*, version 2, edited by J. A. Ardron, H. P. Possingham, and C. J. Klein, 24–38. Victoria, BC: PacMARA. http://pacmara.org/wp-content/uploads/2010/07/Marxan-Good-Practices-Handbook-v2-2010.pdf.

Lieberknecht, L. M, W. Qiu, and P. J. S. Jones. 2013. *Celtic Sea Case Study Governance Analysis: Finding Sanctuary and England's Marine Conservation Zone Process*. http://www.homepages.ucl.ac.uk/~ucfwpej/pdf/FindingSanctuaryGovernanceAnalysisFull.pdf. Full report last accessed June 21, 2014. http://www.homepages.ucl.ac.uk/~ucfwpej/pdf/FindingSanctuaryGovernanceAnalysisSummary.pdf. Summary report last accessed June 21, 2014.

Luck, G. W., K. M. A. Chan, and C. J. Klein. 2012. "Identifying Spatial Priorities for Protecting Ecosystem Services." *F1000Research*, 1–17.

Maguire, B., J. Potts, and S. Fletcher. 2011. "Who, When, and How? Marine Planning Stakeholder Involvement Preferences: A Case Study of the Solent, United Kingdom." *Marine Pollution Bulletin* 62 (11): 2288–92.

Makino, A., M. Beger, C. J. Klein, S. D. Jupiter, and H. P. Possingham. 2013. "Integrated Planning for Land–Sea Ecosystem Connectivity to Protect Coral Reefs." *Biological Conservation* 165:35–42.

Makino, A., C. J. Klein, M. Beger, S. D. Jupiter, and H. P. Possingham. 2013. "Incorporating Conservation Zone Effectiveness for Protecting Biodiversity in Marine Planning." *PLoS ONE* 8 (11): e78986.

Makino, A., H. Yamano, M. Beger, C. J. Klein, Y. Yara, and H. P. Possingham. 2014. "Spatio-Temporal Marine Conservation Planning to Support High-Latitude Coral Range Expansion under Climate Change." *Diversity and Distributions*, 1–13.

Margules, C. R., and R. L. Pressey. 2000. "Systematic Conservation Planning." *Nature* 405:243–53.

Margules, C. R., and S. Sarkar. 2007. *Systematic Conservation Planning*. Cambridge: Cambridge University Press.

Martin, T. G., J. L. Smith, K. Royle, and F. Huettmann. 2010. "Is Marxan the Right Tool?" In *Marxan Good Practices Handbook*, version 2, edited by J. A. Ardron, H. P. Possingham, and C. J. Klein, 12–17. Victoria, BC: PacMARA. http://pacmara.org/wp-content/uploads/2010/07/Marxan-Good-Practices-Handbook-v2-2010.pdf. Last accessed May 2014.

McClintock, W. 2012. McClintock Lab: SeaSketch. http://mcclintock.msi.ucsb.edu/projects/seasketch. Last accessed June 21, 2014.

McClintock, W., F. Kershaw, and A. Chassanite. 2013. "Planning Networks including MSP: Innovating Tools and Methods." In *Proceedings of the Third International Marine Protected Areas Conference (IMPAC 3)*, Marseilles, France, October 27–30. http://www.impac3.org.

McDonald, R. I. 2009. "The Promise and Pitfalls of Systematic Conservation Planning." In *Proceedings of the National Academy of Sciences* 106 (36): 15101–2.

McDonnell, M. D., H. P. Possingham, I. R. Ball, and E. A. Cousins. 2002. "Mathematical Methods for Spatially Cohesive Reserve Design." *Environmental Modeling and Assessment* 7:107–14.

Mills, M., V. M. Adams, R. L. Pressey, N. C. Ban, and S. D. Jupiter. 2012. "Where Do National and Local Conservation Actions Meet? Simulating the Expansion of Ad Hoc and Systematic Approaches to Conservation into the Future in Fiji." *Conservation Letters* 55:387–98.

Mitchell, A. 2001. *The Esri Guide to GIS Analysis*, vol. 1. Redlands, CA: Esri Press.

Moilanen, A. 2014. Zonation, C-BIG Conservation Biology Informatics Group. http://cbig.it.helsinki.fi/software/zonation/. Last accessed June 21, 2014.

Moilanen, A., K. A. Wilson, and H. P. Possingham, eds. 2009. *Spatial Conservation Prioritization: Quantitative Methods and Computational Tools*. Oxford, UK: Oxford University Press.

Monmonier, M. 1996. *How to Lie with Maps*, 2nd ed. Chicago: University of Chicago Press.

Nicolson, D. J., M. Lötter, L. Lieberknecht, R. Wells, J. L. Gerhartz, N. C. Ban, and J. A. Ardron. 2010. "Interpreting and Communicating Outputs." In *Marxan Good Practices Handbook*, version 2, edited by J. A. Ardron, H. P. Possingham, and C. J. Klein, 97–109. Victoria, BC: PacMARA. http://pacmara.org/wp-content/uploads/2010/07/Marxan-Good-Practices-Handbook-v2-2010.pdf.

PacMARA (Pacific Marine Analysis and Research Association). 2008. *Summary Report: Addressing the Practicalities and Problems of Incorporating Conservation Planning Tools into Decision-Making*. Vancouver, BC: PacMARA. http://pacmara.org/marxan-good-practices-workshop-and-handbook.

———. 2012. *Demonstration Marxan Analysis of the Nunavut Settlement Area: Industrial Use and Ecological Considerations*. Vancouver, BC: PacMARA.

Pattison, D., D. dos Reis, and H. Smillie. 2004. *An Inventory of GIS-Based Decision Support Tools for MPAs*. Washington, DC: National Oceanic and Atmospheric Administration National Marine Protected Areas Center.

Possingham, H. P., I. R. Ball, and S. Andelman. 2000. "Mathematical Methods for Identifying Representative Reserve Networks." In *Quantitative Methods for Conservation Biology*, edited by S. Ferson and M. Burgman, 291–305. New York: Springer-Verlag.

Possingham, H. P., J. L. Smith, K. Royle, D. Dorfman, and T. G. Martin. 2010. "Introduction." In *Marxan Good Practices Handbook*, version 2, edited by J. A. Ardron, H. P. Possingham, and C. J. Klein, 1–11. Victoria, BC: PacMARA. http://pacmara.org/wp-content/uploads/2010/07/Marxan-Good-Practices-Handbook-v2-2010.pdf.

Pressey, R. L., and M. C. Bottrill. 2009. "Approaches to Landscape- and Seascape-Scale Conservation Planning: Convergence, Contrasts, and Challenges." *Oryx* 43:464–75.

Pressey, R. L., C. J. Humphries, C. R. Margules, R. I. Vane-Wright, and P. H. Williams. 1993. "Beyond Opportunism: Key Principles for Systematic Reserve Selection." *Trends in Ecology & Evolution* 8 (4): 124–28.

Rozum, J. S., and S. D. Carr. 2013. *Tools for Coastal Climate Adaptation Planning: A Guide for Selecting Tools to Assist with Ecosystem-Based Climate Planning.* Arlington, VA: NatureServe. https://connect.natureserve.org/sites/default /files/documents/EBM-ClimateToolsGuide-FINAL.pdf.

Sarkar, S., J. Justus, T. Fuller, C. Kelley, J. Garson, and M. Mayfield. 2005. "Effectiveness of Environmental Surrogates for the Selection of Conservation Area Networks." *Conservation Biology* 19 (3): 815–25.

Sarkar, S.,R. L. Pressey, D. P. Faith, C. R. Margules, T. Fuller, D. M. Stoms, A. Moffett, K. A. Wilson, K. J. Williams, P. H. Williams, and S. Andelman. 2006. "Biodiversity Conservation Planning Tools: Present Status and Challenges for the Future." *Annual Review of Environment and Resources* 31:123–59.

Secretariat of the Convention on Biological Diversity. 2011. *An Introduction to National Biodiversity Strategies and Action Plans.* http://www.cbd.int/doc/training/nbsap/b1-train-intro-nbsap-revised-en.pdf. Last accessed December 4, 2014.

Segan, D. B., E. T. Game, M. E. Watts, R. R. Stewart, and H. P. Possingham. 2011. "An Interoperable Decision Support Tool for Conservation Planning." *Environmental Modelling & Software* 26:1434–41.

Simberloff, D. 1998. "Flagships, Umbrellas, and Keystones: Is Single Species Management Passé in the Landscape Era?" *Biological Conservation* 83:247–57.

Sloan, N. 2014. "Gwaii Haanas: From Conflict to Cooperative Management." In *Marine Conservation: Science, Policy, and Management*, edited by G. C. Ray and J. McCormick-Ray, 262–87. New York: John Wiley & Sons.

Stewart, R. R., T. Noyce, and H. P. Possingham. 2003. "Opportunity Cost of Ad Hoc Marine Reserve Design Decisions: An Example from South Australia." *Marine Ecology Progress Series* 253:25–38.

Tallis, H., Z. Ferdaña, and E. Gray. 2008. "Linking Terrestrial and Marine Conservation Planning and Threats Analysis." *Conservation Biology* 22:120–30.

Tufte, E. R. 2001. *The Visual Display of Quantitative Information*, 2nd ed. Cheshire, CT: Graphics Press.

Von Nordheim, H., D. Boedeker, and J. C. Krause, eds. 2006. *Progress in Marine Conservation in Europe: NATURA 2000 sites in German offshore waters.* Berlin: Springer.

Watts, M. E., I. R. Ball, R. S. Stewart, C. J. Klein, K. Wilson, C. Steinback, R. Lourival, L. Kircher, and H. P. Possingham. 2009. "Marxan with Zones: Software for Optimal Conservation-Based Land- and Sea-Use Zoning." *Environmental Modelling & Software* 24 (12): 1513–21.

Watts, M., B. Pressey, and M. Ridges. 2014. The C-Plan Conservation Planning System. http://www.edg.org.au /resources/free-tools/cplan.html. Last accessed June 21, 2014.

Weeks, R., G. R. Russ, A. A. Bucol, and A. C. Alcala. 2010. "Shortcuts for Marine Conservation Planning: The Effectiveness of Socioeconomic Data Surrogates." *Biological Conservation* 143:1236–44.

Wilson, K. A., H. P. Possingham, T. G. Martin, and H. S. Grantham. 2010. "Key Concepts." In *Marxan Good Practices Handbook*, version 2, edited by J. A. Ardron, H. P. Possingham, and C. J. Klein, 18–23. Victoria, BC: PacMARA. http://pacmara.org/wp-content/uploads/2010/07/Marxan-Good-Practices-Handbook-v2-2010.pdf.

CHAPTER 3

Artificial Reefs, Beach Restoration, and Sea Turtle Nesting in Martin County, Florida

Alexandra Carvalho, Kathy Fitzpatrick, Jessica Garland, and Frank Veldhuis

Abstract

The Martin County Engineering Department's Coastal Engineering Division manages several programs, including habitat management and restoration (inshore and offshore artificial reefs, oyster restoration, and natural reefs), beach restoration, Saint Lucie Inlet maintenance, waterway improvements and dredging, anchoring and mooring sites, and a few countywide coastal monitoring projects focusing on the coastal ecosystem, topographic and hydrographic data collection, and rectified aerial photography.

To consolidate and manage the large volumes of information created by the permitting process for each coastal project, and to facilitate the access to historical and current project data, the Martin County Coastal Engineering Division opted, in 2003, to consolidate its coastal data into a GIS. Ten years later, the system now provides access to over 20 years' worth of coastal project data that is in a GIS or stored elsewhere but linked to a GIS and several tools to retrieve and share data. Users refer to the consolidated information and tools as the county Coastal GIS Program.

This chapter briefly describes the history of the Coastal GIS Program, its goals and objectives, the data consolidation and organization strategy, the data model and geodatabase schema, types of GIS applications developed, and the ways system implementation is changing the Engineering Department's project management workflows and helping to plan future projects.

This chapter also describes two of the most active coastal programs: the artificial reefs program and beach management program. The Hutchinson Island Shore Protection Project, used as an example of the county's beach program, illustrates the benefits to the beach management program from GIS work completed to date. A brief overview of program areas and the projects they encompass illustrates the importance of these programs to coastal communities, the types of spatial data they produce, and how the GIS program organizes these datasets. The chapter concludes with a description of the GIS program tools developed to help with project management and the data-sharing needs of the Engineering Department as well as concepts for future tool development.

Dawn J. Wright, ed.; 2015; *Ocean Solutions, Earth Solutions*; http://dx.doi.org/10.17128/9781589483651

Martin County, Florida

Located on the east central coast of Florida, Martin County covers 556 sq mi and supports a population of approximately 150,000 residents. With only four municipalities, this low-density county (approximately 270 residents per sq mi) sits in sharp contrast to the highly developed counties to the north and south. The Atlantic Ocean and Lake Okeechobee bound the east- and west-county shorelines. The Okeechobee Waterway, one of two federal waterways within county boundaries, connects the two water bodies. A second waterway, the Florida Intracoastal Waterway, runs the length of the state along the Indian River Lagoon in Martin County.

Water—its quality and access to it—define the unique lifestyle and economy of Martin County. The Saint Lucie Inlet, sitting at the confluence of the Florida Intracoastal Waterway, Okeechobee Waterway, and Atlantic Ocean, provides important services to boats navigating to destinations toward the north and south of the state between the Atlantic Ocean and Gulf of Mexico. It also serves as a popular departure site for trips to the Bahamas.

Lake Okeechobee sits at the northern extent of North America's largest wetland system, the Everglades, a location included in federally mandated restoration activities. The county mainland includes extensive natural uplands, wetlands, and sections of two barrier islands separated by Saint Lucie Inlet. Federally protected estuaries separate the county mainland and barrier islands. These estuaries, linked to the Atlantic Ocean through Saint Lucie Inlet, are important habitat for commercially and recreationally important fisheries, as well as several threatened or endangered species such as the West Indian manatee and smalltooth sawfish. The beach/nearshore area also supports threatened and endangered species such as loggerhead, green, and leatherback sea turtles; wading birds (e.g., great blue heron and wood stork); and migratory wintering birds (e.g., piping plover). Martin County also sits at the northern extent of the Florida (Coral) Reef Tract.

Martin County considers environmental preservation and restoration its top priorities. That focus has allowed the county to purchase and hold in preservation large tracts of land. Restoration and habitat creation projects, typically focused on the freshwater/estuarine/marine habitat continuum, receive strong support from county taxpayers. For this reason, increased public awareness and information sharing is critical to keeping the public engaged in these interests.

Coastal GIS Program Overview

The county Coastal GIS Program started in 2003 with a vision of a fully integrated, GIS-based coastal information system designed to provide coastal managers, planners, and the public easy access to coastal information (Carvalho et al. 2003).

Goals and Objectives

Since its inception, the goals of the program have been to (1) facilitate coastal engineering project planning and management, (2) consolidate and manage permit and grant information held by multiple contractors and agencies, (3) track monitoring and compliance requirements for each permit and grant, and (4) reduce the requests for information and staff time spent gathering information.

Accomplishing these goals is a continuous process that involves (1) consolidating and organizing the county's coastal data, (2) developing the tools to provide the county and public with easy access to coastal project information, and (3) streamlining contractors' coastal data submissions.

Program History and Current Status

During 2003 and 2005, the county made several efforts to develop a geodatabase schema and data access. Decreased funding reduced the program to a series of coastal program data consolidation efforts in 2007, 2010, 2012, and 2013. In 2012, the county contracted CMar Consulting to carry the program forward.

Figure 3.1 shows an overview of components of the current GIS program. At the beginning of 2013, the county had over 20 years' worth of historical and current project information consolidated into file geodatabases (vector and raster). This first program milestone marked the transition and expansion of the program focus from data organization in GIS (file geodatabases) to tool development providing non-GIS-trained managers and planners easy access to their coastal data.

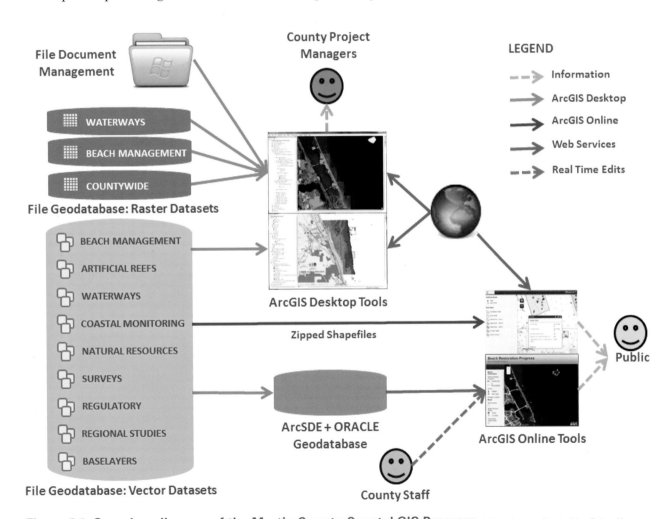

Figure 3.1. **Overview diagram of the Martin County Coastal GIS Program.** Lidar data collected by Brian K. Walker, Nova Southeastern University for the Florida Fish and Wildlife Conservation Commission Fish and Wildlife Research Institute.

In less than a year, county project managers had several customized ArcGIS for Desktop maps available to access historical and current program data and help them plan and manage different projects within each program. Access to the data through the desktop map tools allowed the Coastal Engineering Division staff to start using GIS to manage their own projects. Currently, the ArcGIS for Desktop maps function as both planning and management tools and as data access tools providing easy access to information stored in the geodatabase and the county's file document management system.

In March 2013, the county published its first public ArcGIS Online application to communicate with the public on the daily status and progress of one of its coastal initiatives: the 2013 Hutchinson Island Shore Protection Project. The application went live at the end of March just before the project started. ArcGIS Online tools provide an information pathway to accomplish the county's program goal of providing the public with project information. As this component of the program matures, the county hopes to make the connection between the engineering aspect (e.g., nourishing a beach) and the natural resources (e.g., increasing turtle nesting habitat). An example of this connection is the Story Map app featured later in this chapter, based on the Esri Story Map app, showing that sea turtles return to nest on a nourished beach. The county thinks this will help engage local communities and taxpayers.

Currently, updates to the Coastal GIS Program and desktop mapping tools from ongoing projects occur two to three times a year as datasets become available. Efforts during 2014 continued to focus on data consolidation and development and optimization of data access tools for project planning and management. The addition of hyperlinks, another program enhancement for 2014, aimed to connect the external file document management system for permits and grants and other documentation to the geodatabase via the desktop planning tools.

ArcGIS Software Products

The program team looked at ArcGIS Online service to determine the adequacy of that platform to accommodate the county's data access, visualization, management, and analysis needs. Some of the analysis tools the county uses are part of the ArcGIS 3D Analyst extension, which are not available in ArcGIS Online. Therefore, the project team opted to use ArcGIS for Desktop software to develop the planning, management, and analysis tools and ArcGIS Online to share data with the public.

The county's Information Technology Services Department (county ITS) uses ArcSDE software with Oracle to manage other county GIS data. The Coastal GIS Program stores data in file geodatabases (see figure 3.1). Datasets that support the Coastal GIS Program's ArcGIS Online applications (e.g., Hutchinson Island sea turtle nesting, beach restoration progress, roads) are stored in ArcSDE and published to the web via ArcGIS for Server service.

The following sections of this chapter discuss the strategies and challenges faced by the Coastal GIS team to carry the program forward. Topics include data consolidation and organization strategies and challenges, followed by an overview of the data access and planning and management tools developed to date. The chapter ends with examples of two county programs to illustrate all the components of the Coastal GIS Program discussed in this chapter.

Coastal Data Consolidation Strategies

The county's diverse habitats, large variety of projects, and, most important, reduced funding availability prompted county program managers to identify priorities regarding data consolidation into the GIS. They had to think strategically to rank in proper order which projects and datasets within each program should receive funding. Coastal managers require quick access to specific information (e.g., location of the edge of hard bottom over the years or a measure of impacts from beach nourishment) with sufficient detail to enable informed decision-making. Less often, they need access to detailed information (e.g., biological information collected in 40-plus quadrat sampling stations) to complete the decision-making process.

Prioritizing Data Consolidation

Currently, natural resource datasets, such as sea turtle nesting and shorebird activity, are the only detailed natural resource information contained in the GIS. The massive amount of datasets associated with natural hardbottom reefs, coral reefs, artificial reefs, seagrass beds, and the fish communities that inhabit them will become the focus for future integration efforts.

GIS data consolidation efforts to date have focused on boundaries and limits of topographic and bathymetric surveys, natural resources, regulations, and engineering design (e.g., dredge and fill footprints, equipment location areas). Also included in this current phase are the locations and boundaries of environmental and physical monitoring surveys (e.g., transects, video surveys, and quadrat locations) and of water and sediment sampling locations.

Data Consolidation Challenges

The county receives coastal datasets in many different formats. The same monitoring requirement (e.g., a seagrass survey) in different projects may produce a boundary, a limit or delineation, transects, or detailed biological data quantified through quadrat sampling. In some instances, a single project generates data in all these formats.

Consolidating data for a specific project requires engagement—not only of the county project managers, but also of contractors (and their subcontractors) who produced the data and the permitting agencies that review them. Implementing data submission standards and document tracking protocols up front to allow seamless data transfer will reduce the extensive time currently required for these "after the fact" efforts. As documented later in this chapter, Martin County has already put some standards and protocols in place.

Topographic and hydrographic surveys are the most common datasets produced by coastal projects. The geodatabase currently stores more than 160 surveys, and this number continues to grow every year. As the number of points associated with these surveys grew, it became unfeasible to store all the points in the system. After looking at several alternatives (triangulated irregular networks, or TINs; mosaics; and light detection and ranging, or lidar, tools), the project team decided to convert all x,y,z survey text and lidar point data to 5-by-5-ft rasters. Rasters are created by first converting survey points to a TIN, and then rasterizing the TIN into 5-by-5-ft cells. This raster resolution, much higher than the resolution of most surveys, allows better comparison with lidar surveys (also converted to 5-by-5-foot rasters) conducted in the area.

This survey data format example illustrates the importance of choosing carefully the best option to store a dataset. For coastal planning and management purposes, more important than visualizing all the survey points is the capability to allow users to

- visualize the elevation or depth pattern revealed by a survey (e.g., dune, deeper water);
- delineate an area and quantify area or volume changes relative to an older survey or a reference plan (e.g., authorized channel depth); and
- detect erosion and accretion trends in the study area.

With survey data in raster format, users can quickly create contours, extract cross sections, compare surveys, and quantify volume changes, using commercially off-the-shelf (COTS) tools provided (i.e., 3D Analyst). Furthermore, the x,y,z points for all surveys remain stored outside the geodatabase. Users who want to look at the underlying dataset can follow the hyperlink provided in the survey boundary attribute table to the x,y,z file location. It is a simple step to add that data to the map if desired. Lidar surveys are stored in the original LAS format (also known as *LASer*, as specified by the American Society for Photogrammetry and Remote Sensing) and accessed through data import/export tools available through ArcGIS software.

Natural resource survey data is the second-most common type of dataset and most demanding in terms of physical hard-drive storage. The county collects natural resource data annually. Datasets, important in the permitting process, include sea turtle nesting activity, shorebird activity, sea grasses, hardbottom edges, as well as hardbottom extent, location, and communities encountered along monitoring transects.

Typically, scientists analyze the detailed biological community data collected during these surveys (i.e., quadrat sampling results), summarize the analytic results in technical reports, and submit the reports to the county and permitting agencies. Because the county conducts no further analysis using this data, the project team assigned its incorporation into the GIS a lower priority.

Recent changes in the Florida Department of Environmental Protection (FDEP) natural resource survey data submission requirements may help justify the allocation of funds to add these types of datasets to the GIS. To allow data comparison between all coastal counties in Florida, FDEP is asking individual counties to submit natural resource data in a standardized format for incorporation into the GIS.

Coastal Data Organization Strategies

The data mining and organization required to move the Coastal Engineering Division to a comprehensive geodatabase has required a substantial commitment of time and money. The expected long-term benefits that justify this investment include improved project planning and management and a reduction in staff time spent gathering information for public inquiries. Examples of frequent internal and public requests for information (RFI) that require quick and easy access include:

- data availability and storage location for a project area to satisfy an RFI of "all available data";
- surveyor names, costs, and dates in the project areas;
- natural resources in a project area;

- permit or grant monitoring requirements and timelines;
- names of infrastructure owners or maintenance responsibilities;
- quick methods to find and extract data; and
- best ways to display and organize data in desktop maps (e.g., by permit, date, or project).

Data organization within the Martin County Coastal GIS Program consists of three components: (1) location of all project-related files in the file document management system; (2) organization of data in easily recognizable feature datasets (i.e., HI_Dredge_Template), feature classes (i.e., Hardbottom_Survey_Quadrats), and tables (i.e., 2012_Seagrass_Quadrats) in the geodatabase; and (3) groupings of project data and base layers in data access tools. The next sections of this chapter discuss the last two components in detail.

The county has had an established file document management organization strategy for permit and grant project documentation for several years. Efforts are under way to organize other data eligible for GIS and other ancillary documentation, such as surveys, reports, photos, and videos.

Coastal GIS Data Model and Schemas

Originally based on the Arc Marine data model (Wright et al. 2007), the county coastal data model has evolved to address the needs of the county's coastal programs. The current data model and geodatabase schemas focus on portability and easy access to project information by users untrained in GIS. Because of its role as a public agency, the county required the geodatabase organizers to build in sufficient transparency to simplify responses to public RFI (e.g., all data related to the Hutchinson Island Shore Protection Project).

To accommodate the "All data related with" types of requests, the project team organized specific feature datasets (vector data only) within the main geodatabase for the beach and waterways programs into project-specific feature datasets with identical schemas (see figure 3.1). Four core feature datasets—survey boundaries and locations, regulatory datasets, base layers, and regional studies—remain in the same dataset because they serve multiple projects.

The chosen data model separates rasters for beach, inlet, and waterway projects into project-specific geodatabases. Isolating the rasters as necessary because of the large volume of topographic and hydrography survey data produced by the beach management and Saint Lucie Inlet maintenance programs. Furthermore, county ITS department policy requires storage of rasters in file geodatabases. Raster mosaics display faster, are portable, and do not require the overhead of an enterprise database management system (DBMS) such as Oracle.

Project Planning and Management Tools

The project planning and management tool or "desktop map template" solution in ArcGIS for Desktop provides managers with quick and organized access to project data in a GIS. From within this map, they can copy or remove layers as needed to create individualized desktop maps addressing a particular issue.

ArcGIS for Desktop Map Solution

The desktop mapping approach helps ease the pressure on one of the most valuable county ITS resources: storage space. To date, the entire Coastal GIS Program files occupy 4 GB of disk space, and the majority of the desktop files created based on templates are very small in size.

Generally, project data in the desktop map tools is organized by program (e.g., one map for artificial reefs, one map for waterways management) or, in some cases, by geographic area. The latter is the case for programs with recurrent projects within a specific geographic area (e.g., one map for all the Hutchinson Island Shore Protection Projects). In areas where the same program has more than one recurrent project but that are adjacent, one ArcGIS for Desktop map provides access to both projects. This organization enables easy comparison and analysis between past and current projects; in addition, it provides access to all the available data for the project area.

ArcGIS for Desktop Map Organization

All desktop maps include as base layers regulatory and jurisdictional limits, water access, countywide natural resources, transportation, and countywide parcel data (county ITS web service) base layers. Each map also includes aerial photography web services (i.e., countywide or the rectified coastal aerial photography) provided by county ITS and Esri Ocean and World Imagery basemaps. Some programs have unique base layers (e.g., sand sources for beach nourishment) or basemaps such as a National Oceanic and Atmospheric Administration (NOAA) Raster Nautical Chart (RNC) map service for the artificial reefs, oyster restoration, and waterways maintenance programs.

All desktop maps also include project-specific data such as engineering drawings, topographic and bathymetric surveys, and natural resource monitoring surveys. For the most part, the first two datasets are organized according to the following criteria: (1) project year, (2) project phase, and (3) date. The natural resource monitoring data is organized differently. In this case, datasets from all project phases are grouped together to allow for a temporal interpretation of the data.

Including all these datasets and base layers in the maps rather than adding layers on an as-needed basis allows non-GIS-trained users to find project data quickly without having to sieve through several group layers and folders of project data.

ArcGIS for Desktop Map Optimization

Some recurrent projects have over 15 years' worth of data. Adding all this data to the same desktop map, although advantageous for planning and management purposes, may significantly slow the map's performance.

Establishing visibility ranges for layers or representing data using less complex dataset types (polygon versus raster) when zoomed out has improved the performance of desktop maps. The vast majority of the more detailed datasets become visible only when the user zooms in beyond 1:12,000. Topographic and bathymetric surveys, one of the most common datasets, display the survey boundary when zoomed out beyond 1:24,000 and the survey raster when zoomed in beyond that value.

One raster display property setting named "show map tips" provides one of the most useful data display features for desktop maps. Users can view the raster attribute (e.g., depth, relief) by dragging the mouse over the raster cell. The alternative to rasters, contours with labels, required more resources and provided less accurate data.

Analysis Tools

The Coastal GIS Program project team is developing a "Martin County Toolset" toolbar add-in to facilitate access to the most commonly used analysis and data management tools. Some of the tools are available in toolbars from extensions such as ArcGIS 3D Analyst software, but are not easily found by non-GIS-trained staff. The project team will make the toolbar add-in available for all desktop maps.

The toolbar includes analysis tools such as cross section extraction, volume calculation, and contour extraction tools that help users accomplish some of the topographic and bathymetric data analysis described earlier. Other tools often used include add x,y data, measurement, buffer, graphs, statistics, hyperlink, and hypertext markup language (HTML) pop-up windows. Funding will determine the schedule for adding planning and analysis tools for each program.

Metadata and Data Submission Standards

Metadata records, one of the most important components of the program, is the most challenging to collect. Presently, the majority of the data consolidated into the Coastal GIS arrives at the county in AutoCAD, shapefiles, spreadsheets, text format, or PDFs, often without metadata. Over the years, the county has attempted to implement data submission standards that include metadata. Such standards are difficult to enforce when dealing with contractors unfamiliar with GIS. The county reviewed and simplified its Coastal Data Submission Standards in 2014 and is once again trying to implement these for all contract renewals. The 2014 effort included two important differences from past efforts: the county's willingness to share part of the cost of submitting the data in the preferred format with its contractors and the availability of guidelines and examples from the Coastal GIS Program that the county can provide to its contractors.

County staffers wish to build on relationships formed with contractors during the data-mining process in order to encourage increased GIS integration and, ultimately, to require GIS-formatted submittals. Successful implementation of this policy will allow a more efficient incorporation of data and metadata into GIS. During this transition, open discussion with contractors will help identify compromise solutions to facilitate data delivery and incorporation into the GIS and limit cost increases. Other possible options to reduce contract workload and resulting fee increases include providing the contractor with GIS data files or geodatabase schemas and creating metadata document templates to encourage better data and metadata record submission.

Case Study Examples

The remainder of this chapter presents two case studies that illustrate the application of all the concepts discussed to this point. The first case study focuses on the artificial reef program, the second on the beach management program. The second study draws on lessons learned on the Hutchinson Island Shore Protection Project, one of the active beach restoration projects in Martin County.

The description of each case study focuses on the project's importance to coastal communities, followed by an overview of the types of spatial data produced, and dataset organization and description of the planning, management, and public access tools developed to date. The section concludes with a Story Map app, based on the Esri Story Map app, developed to assist the county in public outreach efforts.

Case Study 1: The Martin County Artificial Reef Program

The Martin County Artificial Reef Program started in the 1970s when a group of retirees and sport fishing enthusiasts, led by Bill Donaldson, began a movement to create self-sustaining marine habitats (Martin County Department of Engineering 2006). Presently, the program includes four active construction sites with a total area of 9,700 acres over 15 sq mi: the Donaldson, Ernst, and Sirotkin reef sites (permitted in 1987), and the South County Reef site (permitted in 2007), which lies adjacent to the Saint Lucie Hump Marine Protected Area (MPA) created in 2009. Depths vary between 50 and 60 ft in the Donaldson Reef site, between 60 and 70 ft in the Ernst Reef site, between 70 and 200 ft in the Sirotkin Reef site, and between 55 and 120 ft in the South County Reef site.

Approximately one-third of the Donaldson Reef site is located in state waters (within 3 mi of shore), and two-thirds lies in federal waters. All other offshore reef sites are located in federal waters. Because of the costs of obtaining and maintaining state artificial-reef permits, the county constructs new reefs exclusively in federal waters.

To date, over one hundred artificial-reef deployments—each constructed at different depths—provide popular fishing and diving spots to locals and tourists. The county also constructs artificial reefs to benefit fisheries by siting a percentage of the reefs in remote locations to minimize their "attractiveness" for human recreation. This selective placement allows the artificial reefs to function as fishery enhancement. In addition to their habitat creation and recreational value, reefs add socioeconomic value to business communities nearby (Johns et al. 2004).

Artificial Reef Program Data

Spatial datasets produced by the artificial reef program include permitted artificial reef site boundaries and the coordinates of all artificial reefs constructed within each site since the 1970s. Other data related to this program comprises spreadsheets with biological and physical data, digital photographs, videos, reports, and other documents (e.g., permit and grant documents). Table attribute data for each reef includes regulatory and cost information, construction and monitoring dates, construction materials and characteristics, water depth, reef relief, and biological data collected before, after, and during construction events.

The county monitors each artificial-reef deployment for marine activity and structural integrity annually for two years after construction, and then adds the deployment to an ongoing three-year monitoring survey rotation. These monitoring efforts produce significant amounts of data (reports with biological and engineering data), photographs, and videos. At this time, lack of funding prevents incorporation of all data. However, the county will add the data from each monitoring event to the geodatabase as funding becomes available.

Because the program collects little bathymetric information, artificial reef program data is stored in the main geodatabase, presenting the county program with fewer data maintenance requirements.

Adding data from new sites requires a simple update of the reef location and attribute table. Once the artificial reef monitoring data is incorporated into GIS, hyperlinks and customized HTML pop-up links to that data will be added to the artificial reef program data access tool.

Artificial Reef Program ArcGIS for Desktop Applications

The artificial reef desktop map application was one of the first data access applications developed for the county. During 2013, the county used the desktop map to plan the construction of new sites. Because artificial reefs at different depths provide different habitats, county staff used the map to identify areas with ideal depth/relief/bottom-type conditions for the planned reef. The ArcGIS for Desktop map application (figure 3.2) provides easy access to the data stored in the geodatabase and within the county file document management system for planning and project management. The map document's table of contents includes the reef site boundaries and all artificial reef projects, background data to support planning and program management, and basemaps. A grant and permit group hyperlinked to the corresponding documents is currently under development. A scaled-down example of the desktop map application in MXD (map document) format is included with the digital content provided with this chapter on the Esri Press "Book Resources" webpage at esripress.esri.com /bookresources. See "Supplemental Resources" at the end of this chapter for more information.

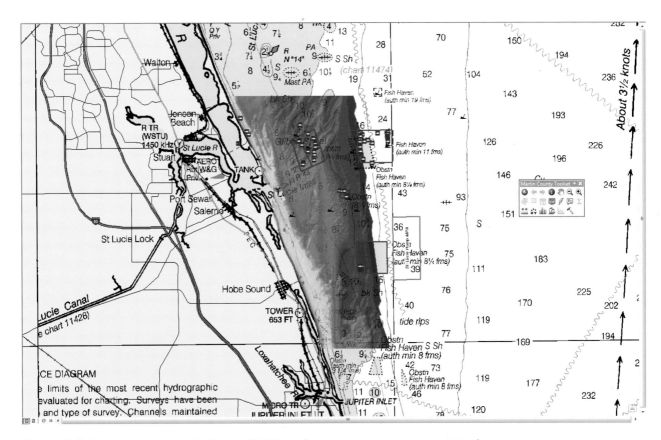

Figure 3.2. Screen capture of the artificial reef program management tool. Data courtesy of Martin County, and Bureau of Ocean Energy Management Mapping and Boundary Branch; lidar data collected by Brian K. Walker, Nova Southeastern University for the Florida Fish and Wildlife Conservation Commission Fish and Wildlife Research Institute; web service courtesy of National Oceanic and Atmospheric Administration Department of Commerce National Ocean Service Special Projects and National Marine Fisheries Service.

Background information used for planning and project management includes:

- water access locations such as beach parks and boat ramps that provide access to the nearshore and offshore reefs;
- jurisdictional limits (federal and state waters, federal marine protected area boundaries, cities);
- nearshore and offshore high-resolution bathymetry lidar, relief, and benthic habitat characterization study (Walker 2012) to help guide the placement of new reefs; and
- offshore sand source studies that show boundaries of areas with good-quality sand for beach nourishment. These areas, if not already designated artificial reef sites, are a competing use for future artificial-reef site establishment.

The desktop map tool uses the NOAA RNC raster chart web service as a basemap. Most maps, plots, or images provided to government agencies or the marine industry that are related to artificial reefs require nautical charts as background. Alternatively, users can also include the Esri Ocean Basemap in the desktop map.

Artificial Reef Program Web GIS Applications

In contrast with the detailed desktop map used for planning and management, the ArcGIS Online web map application (figure 3.3) provides simple and basic location information about the program's artificial reefs to the boating, fishing, and diving community. Developed during 2014 and embedded on the program website www.martinreefs.com, it uses the "One Pane" ArcGIS Online template with the NOAA RNC raster chart web service as a basemap.

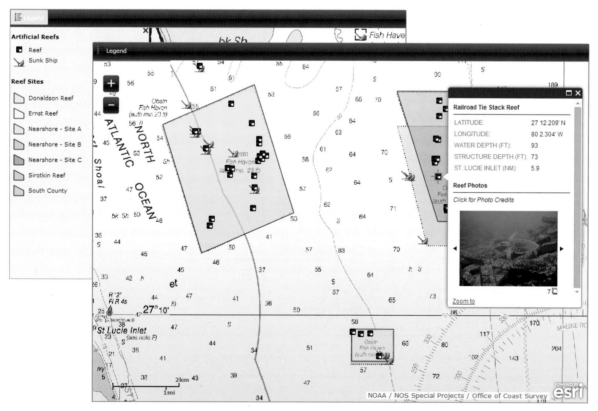

Figure 3.3. **Screen capture of the Artificial Reef Locations web map. See also: http://www.martinreefs.com/pages/locations.html.** Data courtesy of Martin County; photo courtesy of Martin County; National Oceanic and Atmospheric Administration (Department of Commerce National Ocean Service Special Projects.

Information available to the public in this application includes reef name and geographic coordinates, reef depth, water depth, distance to Saint Lucie Inlet, photos, and a link to the monitoring report if available. Future enhancements will include adding more pictures (*see* figure 3.3) and developing more complex web applications to promote the sites. Examples are applications to engage the fishing community and collect information related to use of the reefs, such as preferred sites and species encountered during the visits. The county will rank these enhancements, match them to various grant opportunities, and implement them as funding allows.

Case Study 2: The Martin County Beach Management Program

Shoreline recession and dune loss along coastal states such as Florida threatens infrastructure and beach habitat, public recreation areas, and the overall economy of coastal communities (Houston 2013). As with the artificial reefs, beach restoration provides a socioeconomic added value to the county and business communities near and around the beaches. Beaches expand the tax base through enhanced property values; provide revenue from locals and tourist spending at local businesses; and provide employment and income to local residents (Murley et. al 2005).

The storms and wave action that continually erode the Martin County shoreline create increased vulnerability to storm surge and other storm event damage. The FDEP Bureau of Beaches and Coastal Systems (BBCS) 2012 update of critical erosion areas in Florida states: "18 miles of the Martin County shoreline, out of a total of 22 miles, are critically eroded." According to BBCS (2012), the term *critically eroded* describes areas where erosion and recession of the beach or dune system is such that upland development, recreational interests, wildlife habitat, or important cultural resources are threatened or lost.

In recent years, severe storms (e.g., Hurricanes Frances and Jeanne in 2004, Wilma in 2005, and Sandy in 2012) have altered several sections of the coast; damaged buildings, roads, and other infrastructure; and displaced thousands of people. Proactive coastal management, which requires maintenance of the beach height, length, and width through beach restoration projects, can help minimize, if not avoid, damage from storms. In a typical beach restoration project, a dredge collects sand from an offshore site and pumps a mixture of sand and water onto the beach. Once the water drains away, bulldozers smooth and adjust the new sand until the beach matches the engineered design profile. An option for smaller areas includes trucking sand from upland sand mines to the beach.

The county must carry on the efforts to maintain the coastline (i.e., dredging, transfer, and disposal of sand from inlets and offshore sites into eroded areas) in a manner that improves the shoreline while minimizing impacts to the coastal ecosystem (dunes, intertidal, nearshore, and offshore environments). Many beach restoration projects include construction and vegetation of shoreline dunes to maintain their ecological value and decrease the risk of erosion and infrastructure damage.

Beach Management Program Data

Environmental and physical monitoring efforts required to support beach restoration projects, combined with the engineering and regulatory data associated with each project, make this program one of the most data intensive in the county. Because of environmental permitting requirements, one beach nourishment project may produce over one hundred datasets during the project's life-span (seven to 11 years).

An environmental permit application for a beach nourishment project requires collection of baseline environmental and physical data. Engineers and scientists use this data to prepare construction plans and specifications and establish the physical and environmental resource monitoring baselines. When compared with postconstruction data, the baselines provide the means to document project impacts. Similar data requirements apply during the construction phase to ensure the project meets specifications (e.g., sand size, beach slope). Once completed, the project is monitored during three or more years to document the project's performance. Because these nourishment projects occur on a seven- to 11-year cycle, this process often starts again soon after the three-year post-project-monitoring effort concludes.

Datasets collected before and after construction of a typical beach nourishment project include:
- regulatory data (boundaries, leases, limits);
- engineering data (dredge and fill design templates, infrastructure surveys, monitoring stations and locations);
- topographic and bathymetric surveys of both filled and dredged areas;
- natural resource surveys (shorebirds, sea turtles, nearshore and offshore hard bottom, sea grasses, and invertebrates); and
- sediment and water-quality monitoring locations.

Datasets collected during the construction phase for a typical beach nourishment project include:
- temporary changes to infrastructure and public facilities to support the project;
- topographic and bathymetric surveys of both filled and dredged areas;
- physical and chemical characteristics of sediment and water; and
- natural resource surveys (shorebirds, sea turtles).

Hutchinson Island Shore Protection Project

The Hutchinson Island Shore Protection Project showcases the Martin County beach management program. Congress authorized the project on August 3, 1995, as a 3.75 mi beach restoration project. Project construction occurred between December 1995 and April 1996 with the county funding a quarter-mile extension, resulting in a four-mile-long project. The US Army Corps of Engineers (USACE) constructed smaller segments of the project in 2001 and 2002. In 2005 and 2013, USACE and the county constructed full projects again in response to the erosion caused by Hurricanes Frances and Jeanne in summer 2004 and Hurricane Sandy in fall 2012. The Hutchinson Island Shore Protection Project is the county's oldest, largest, and, given that construction has occurred five times, most data-intensive beach nourishment project.

The 2013 Hutchinson Island Shore Protection Project placed approximately 613,000 cu yd of sand, using a hopper dredge to mine sand from an offshore borrow site (Saint Lucie Shoal) lying approximately seven miles offshore. Once full, the hoppers navigated to one of two pipelines that extended approximately 1,000 ft offshore from the project, connected to the pipeline, and pumped sand from its hold, through the pipeline, onto the beach. Sections of pipe were added to move the sand farther up and down the beach. Construction began March 25, 2013, and ended April 22, 2013. A public outreach web map application (i.e., Story Map app) described later in this chapter summarizes this process.

Beach Management Program ArcGIS for Desktop Applications

As mentioned earlier, the five projects that comprise the Hutchinson Island Shore Protection Project generated large amounts of data. The sheer volume of information produced by each beach restoration project dictated data organization into individual feature datasets in the main geodatabase, one geodatabase with the raster datasets, and one ArcGIS for Desktop map application for all projects.

After several iterations of data organization, the table of contents for the project area's desktop map now includes four groups: project engineering data, physical monitoring data, natural resources, and base layers and basemaps (figure 3.4). A grant and permit group hyperlinked to the corresponding documents is currently under development.

Figure 3.4. **Screen capture of the Hutchinson Island Shore Protection Project management desktop mapping tool.** Data courtesy of US Fish and Wildlife Service; Florida Department of Environmental Protection; Martin County; Ecological Associates Inc.; Taylor Engineering Inc.; CSA Ocean Sciences Inc.; TNC Oceanographic LLC.

Figure 3.4 shows the Hutchinson Island Shore Protection Project desktop mapping tool. Visible groups include bathymetric and topographic surveys; permitting elements such as borrow areas, beach fill areas, and pipeline corridors; natural resource survey locations; and hardbottom monitoring survey results (e.g., substrate type, edge of hardbottom location).

Natural resource layers are organized in three groups: (1) Natural Resources, which contains information collected countywide (e.g., sea turtle nesting information); (2) Hardbottom Surveys, which includes the hardbottom monitoring data for the five nourishment projects (1995–2013); and (3) which, not shown in figure 3.4 but included within the 2013 Beach Nourishment/Construction group, contains information on relocated sea turtle nests during construction and beach scarp formation monitored after project construction but unavailable for past projects. In addition to project-specific data, a typical beach restoration map includes:

- regulatory reference data (e.g., reference monuments, regulatory limits, Coastal Barrier Resources Act units);
- beach access data (e.g., public parks and beach access points);
- shoreline elevation and bathymetric surveys collected annually as part of a countywide coastal monitoring program;
- sand source studies from others;
- land-based sand sources for smaller projects (e.g., sand mine location, truck routes);
- transportation, county parcel information, and aerial photography (the latter updated annually by the county and provided via county ITS as a web service); and
- Esri Ocean and World Imagery basemaps.

The 2013 project dataset is included with the digital content accompanying this chapter on the Esri Press "Book Resources" webpage. See "Supplemental Resources" at the end of this chapter for more information. Datasets are organized in MXD format. Base layer data or studies provided to the county by other organizations are not included. Survey data is represented only by the survey boundaries.

Beach Nourishment Program Web GIS Applications

Before construction of the 2013 project, the project team built one ArcGIS Online application to keep the public informed of the project's progress. Information included the location of equipment offshore, nearshore, and on the beach; aids to navigation characteristics; infrastructure closures (e.g., beach parks often used to store equipment and support the beach projects); and the daily progress of beach sand placement to prevent the public from entering the construction zone.

Because weather and ocean conditions change quickly and affect the timeline for this type of project, the county must stay alert to inform the public of schedule changes. The real-time update of the project status through ArcGIS Online feature editing provided another way to introduce GIS into the county's project planning and management workflows and keep the public informed.

The county developed this experimental application to determine the feasibility of using ArcGIS Online for future projects. E-mails demonstrated that the public used the application. A non-GIS-trained staff successfully updated the application daily. The county will likely continue to use this technology for future projects based on the positive reviews from users and other stakeholders.

2013 Hutchinson Island Shore Protection Project Story Map

To summarize the 2013 project and determine the usefulness and feasibility of these types of applications for future use in the Coastal GIS Program, the project team developed an Esri Story Map app. The next sections provide background information for the types of information presented in the side panels and web maps and navigation instructions.

The Story Map app uses the Esri Story Map Side Accordion template with a splash screen (figure 3.5) to introduce the project. Data sources include web services for the time-enabled data, a combination of feature layers and data stored in ArcGIS Online, and the Esri World Imagery basemap as background.

Hutchinson Island 4 Mile Beach Nourishment Project
Martin County, Florida

Congress authorized the project on August 3, 1995 as a 3.75-mile beach nourishment project. Project was first constructed between December 1995 and April 1996. At the time, the County funded an additional quarter-mile of beach nourishment extending the total length of the project to four miles. The USACE constructed smaller segments of this project in 2001 and in 2002. In 2005 and 2013 the USACE and County constructed full projects again in response to the erosion caused by hurricanes Frances and Jeanne during summer 2004 and Hurricane Sandy in fall 2012.

Source: Martin County 2013

Figure 3.5. The 2013 Hutchinson Island Shore Protection Project Story Map opens with an overview of the project history and 2013 beach nourishment. Photo courtesy of Martin County; splash screen from Story Map app courtesy of CMar Consulting LLC.

The Project Elements web map within the Story Map app (figure 3.6) introduces the project by showing the location of the permitted beach-fill footprint, the permitted Saint Lucie Shoal borrow area (the sand source), and permitted pipeline corridors at both ends of the project.

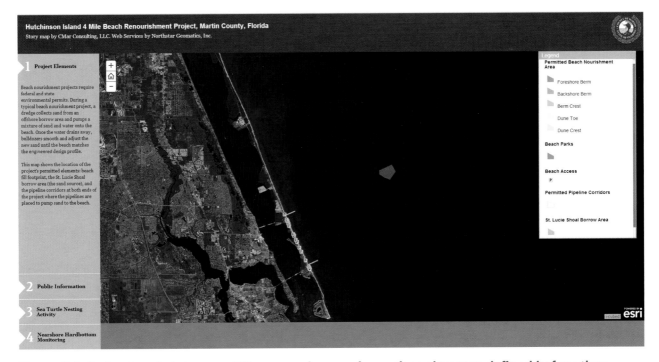

Figure 3.6. Federal and state permitting agencies require project elements defined before they consider awarding environmental permits for beach nourishment projects. Story Map app courtesy of CMar Consulting LLC; time-enabled web services by NorthStar Geomatics Inc.; data courtesy of Martin County and Taylor Engineering Inc.

The Public Information web map within the Story Map app (figure 3.7) shows the datasets displayed and collected through the ArcGIS Online application developed to inform the public during construction (discussed earlier). Time-enabled layers show the timeline of beach park and beach access closures (red prohibited sign) and sand placement on the beach (yellow line) during beach construction (March 25 through April 22, 2013).

Figure 3.7. **Time-enabled reconstruction of the project's progress and infrastructure closures during beach construction.** Story Map app courtesy of CMar Consulting LLC; time-enabled web services by NorthStar Geomatics Inc.; data courtesy of Martin County, Taylor Engineering Inc., and CMar Consulting LLC.

Five species of sea turtles swim in Florida's waters; three of them—the leatherback, loggerhead, and green turtle—nest regularly on Martin County beaches. Per the Endangered Species Act of 1973, all sea turtles are designated either threatened or endangered. Over 2,000 sea turtles nest each year on Martin County beaches. The Martin County Coastal Engineering Division and its contractor, Ecological Associates, maintain a comprehensive sea turtle monitoring and conservation program. The program's intent is to ensure that the design and construction of beach nourishment projects maintains the local beaches as vital sea turtle nesting habitat. Monitoring of daily sea turtle nesting activity occurs annually during sea turtle nesting season (May 1 to October 31) to identify any project-induced changes.

The Sea Turtle Nesting Activity web map within the Story Map app (figure 3.8) shows the 2013 nesting season activity before, during, and after the Hutchinson Island Shore Protection Project (March 14 to October 14). The time slider allows the user to visualize nest locations of the first leatherback turtles (in brown), the progress of the beach nourishment project described in the Public Information web map, and the locations of all leatherback, loggerhead (in orange), and green turtle (in green) nests laid on the beach during the 2013 turtle nesting season.

Figure 3.8. **Time-enabled reconstruction of the 2013 sea turtle nesting along the beach shows the typical seasonal sequence and the peaks of the three species: leatherback, loggerhead, and green sea turtles.** Story Map app courtesy of CMar Consulting LLC; time-enabled web services by NorthStar Geomatics Inc.; data courtesy of Martin County, Ecological Associates Inc., CMar Consulting LLC, and Taylor Engineering Inc.

Users can observe beach scarp formation from the end of May through the summer. Steep scarps can prevent sea turtles from crawling up the beach to lay their eggs. Color-coded, shore-parallel lines symbolize beach scarps by height (highest in red, lowest in light pink) in the web map.

The last web map within the Story Map app, Nearshore Hardbottom Monitoring (figure 3.9), shows the eastern hardbottom monitoring limits established during the 1995 permitting of the first project. Figure details include the location of the nearshore permanent monitoring transects, video segments, and quadrats established in 2010 by CSA Ocean Sciences, a subsidiary of Continental Shelf Associates, to help monitor project-induced beach changes.

Figure 3.9. **Projects are monitored to eliminate or minimize environmental impacts. Comparing the post-project-monitoring data to the baseline data identifies nearshore hardbottom impacts.** Story Map app courtesy of CMar Consulting LLC; photo courtesy of CSA Ocean Sciences Inc.; data courtesy of Martin County, Taylor Engineering Inc., CSA Ocean Sciences Inc., and CMar Consulting LLC.

In addition to the monitoring locations established in 2010, the fourth web map shows the location of the nearshore substrate and hardbottom edge in 2010 and 2013 and the location of the sand patches along the shore-perpendicular transects in 2010 (thicker, darker lines) and 2013 (thinner, lighter lines). A few example photos were added to the map to give users an idea of the type of species and environments found in the area. Zooming further into the map allows viewers to view some of these map elements.

Final Considerations

The data model, geodatabase schemas, maps, and tools described in this chapter show only a few components of the Martin County Coastal GIS Program. The county is close to having a full coastal program data model portable to other coastal counties with similar projects and programs.

Data Model and Geodatabase Schema

The data model and geodatabase schemas discussed in this chapter reflect the data monitoring requirements in the State of Florida; however, any region with similar projects will benefit from the work completed for Martin County to date.

The data model and geodatabase schemas are easy to adapt to other geographic areas, but the Martin County project team recommends that any applications of the data model and geodatabase schema be preceded by a thorough review of the spatial data available for the new area and of an assessment of the GIS skill level of the new users of the data access tools.

GIS specialists on the Martin County project team face ongoing challenges, specifically coordinating the needs of county project managers within the parameters set by county ITS and finding the most efficient way to organize the data and develop the necessary tools. The data model's current organization, the file versus ArcSDE geodatabase option, and the desktop map solutions discussed earlier provide good examples of strategies devised by the development team to allow the Coastal GIS Program to move forward.

As the program matures, the goal is to move all vector data to an enterprise Spatial Database Engine (SDE) geodatabase and provide quarterly updates to the public version managed by county ITS. In the interim, county ITS has agreed to keep data in file geodatabases until the system is fully developed and tested. Based on the progress made in the past two years, the projected schedule for full ArcSDE and Oracle implementation is fall 2015.

Desktop Tools

In spite of the large amounts of information the county accesses, the desktop maps are adequate to provide a good project planning and management tool. Still, the Coastal GIS Program team intends to streamline these desktop maps further to improve project management workflow.

The program team has considered separating the largest desktop maps (i.e., from beach or inlet projects) into less data-intensive individual project maps (i.e., by year or permit). Unfortunately, often as each project rolls into the next, the same data serves as the last monitoring event from one project and the baseline for the next one. In addition, because of the cost of collecting data, it is also common to use the same dataset to address the permit monitoring requirements of multiple projects.

Accessing simple information through ArcGIS Online web maps would help answer some of the most frequent internal or external RIFs (i.e., countywide natural resource and bathymetric survey location), but county staff would not be able to perform analysis on the data.

As county staff continues to use these desktop maps, and transitions to a planning and management workflow that relies more on GIS, options to streamline the desktop maps for project management will become clearer for both county managers and GIS specialists as they monitor how the county uses the desktop tools. One strategy to monitor that use includes a short monthly report on how users employ each desktop tool. Collected information will include (1) program and project details; (2) desktop template used; (3) toolbar buttons used; (4) issue addressed; and (5) a brief description of the work done and data used.

The Esri Story Map

The Esri Story Map approach appears to offer significant benefits to the county GIS program. In this case, the Hutchinson Island Story Map not only tells the story of the project, but also integrates the time variable to show various phases of the project (i.e., beach permitting, construction, and monitoring [turtle nesting, scarps, and hardbottom]). This visualization provides an integrated view of the project, and an opportunity to educate stakeholders and the public on the effects of beach nourishment on turtle nesting and hardbottom reefs.

Typically, permitting agencies receive project information in a lengthy report seldom viewed by the public. The county plans to continue developing new web applications (i.e., story maps) to provide project information to the public and other stakeholders. The capability to tell a project's story and integrate the several project elements and visualize them is very powerful. The Esri Story Map approach has potential to reach government, stakeholders, and the public in a simple and succinct way.

Acknowledgments

Two of the original project visionaries—Dr. Alexandra Carvalho with CMar Consulting and Kathy Fitzpatrick, Martin County's coastal engineer—have been responsible for the majority of the work done since the project's inception, and remain the main forces behind the program. With the team since the beginning and representing Martin County's ITS Department is Frank Veldhuis with NorthStar Geomatics. Veldhuis ensures that the Coastal GIS Program complies with the county's GIS standards and requirements. In his other role as enterprise geodatabase manager, he will eventually assure the system's implementation on the county ITS enterprise platform. Veldhuis also publishes all web services needed for the web and desktop applications.

Baret Barry, former county environmental manager, has been associated with this effort for several years owing to her involvement with the Martin County Artificial Reef Program. Existing data for the reef program became the prototype for this larger program; thanks to Barry's data checking, updates, and corrections, the effort progressed from an idea to the current version. The most recent addition to the team, Jessica Garland, associate county project manager, is the most active user of the system. Quickly becoming a GIS power user, she has been an essential member of the team, providing feedback on every product developed.

The Coastal GIS Program relies on extensive collaboration with all other Martin County contractors, data providers, county collaborators, partners, grantors, and so forth, that warrant acknowledgment. They include agencies such as the Florida Fish and Wildlife Conservation Commission, FDEP, USACE, National Fish and Wildlife Foundation, NOAA, and county contractors such as Taylor Engineering, Atkins North America, Ecological Associates, CSA Ocean Sciences, Morgan & Eklund, Gahagan & Bryant, and TNC Oceanographic.

The comments and suggestions of two anonymous reviewers greatly improved this chapter.

References

BBCS (Bureau of Beaches and Coastal Systems). 2012. *Critically Eroded Beaches in Florida: June 2012 Update.* Tallahassee, FL: Florida Department of Environmental Protection.

Carvalho, A., K. Fitzpatrick, and H. Kostura. 2003. "Streamlining Coastal Monitoring Programs with GIS in Martin County, Florida." In *Proceedings of the Esri International User Conference* 23, Paper 0603. http://gis.esri.com /library/userconf/.

Houston, J. R. 2013. "The Economic Value of Beaches: A 2013 update." *Shore and Beach* 81 (1): 3–11.

Johns, G. M., J. W. Milon, and D. Sayers. 2004. *Socioeconomic Study of Reefs in Martin County, Florida: Final Report.* Hollywood, FL: Hazen and Sawyer Environmental Engineers & Scientists. http://www.dep.state.fl.us /coastal/programs/coral/pub/.

Martin County Department of Engineering. 2006. "Welcome to the Martin County Artificial Reef." http:// www.martinreefs.com. Last accessed June 15, 2014.

Murley, J., L. Alpert, W. Stronge, and R. Dow. 2005. "Tourism in Paradise: The Economic Impact of Florida Beaches." In *Proceedings of the 14th Biennial Coastal Zone Conference*, New Orleans, LA, July 17–21.

Walker, B. K. 2012. *Draft Final Report: Characterizing and Determining the Extent of Coral Reefs and Associated Resources in Southeast Florida through the Acquisition of High-Resolution Bathymetry and Benthic Habitat Mapping.* Florida Fish and Wildlife Conservation Commission, Agreement No. 08014:41.

Wright, D. J., M. J. Blongewicz, P. N. Halpin, and J. Breman. 2007. *Arc Marine: GIS for a Blue Planet.* Redlands, CA: Esri Press.

Supplemental Resources

Dawn J. Wright, ed.; 2015; *Ocean Solutions, Earth Solutions*; http://dx.doi.org/10.17128/9781589483651_d

For the digital content for this chapter, explained below in this section, go to the Esri Press "Book Resources" webpage at esripress.esri.com/bookresources. Then, in the list of Esri Press books, click *Ocean Solutions, Earth Solutions*. On the *Ocean Solutions, Earth Solutions* resource page, click the chapter 3 link to access that webpage and the links to the digital content.

Martin County, Florida, Artificial Reef Planning Tool Example

This project shows an example of an artificial reef program planning tool that consolidates its data for future reference and planning of new projects. This example can be a starting point for other areas with similar programs. For convenience, an XML schema of the geodatabase as well as ArcGIS Diagrammer tool is included together with the MXD and geodatabase containing the data for the location of all the offshore reefs in Martin County, Florida (courtesy of Martin County Engineering Department).

2013 Martin County, Florida, Beach Restoration Planning Tool Example

This project shows an example of typical spatial datasets collected for beach restoration projects in Florida and a planning tool that consolidates its data for future reference and planning of new projects. It includes three phases of the project: permitting, construction, and monitoring of natural resources. This example can be a starting point for beach restoration in other areas. For convenience, an XML schema of the geodatabase as well as ArcGIS Diagrammer tool is included together with the MXD and geodatabase containing the data for the 2013 Hutchinson Island Shore Protection Project (courtesy of Martin County Engineering Department).

Hutchinson Island Story Map

Go to http://www.esri.com/hutchinsonstorymap.

This link is also available on the book resource page.

CHAPTER 4

Tools for Implementing the Coastal and Marine Ecological Classification Standard

Lori Scott, Kathleen L. Goodin, and Mark K. Finkbeiner

Abstract

NatureServe and the National Oceanic and Atmospheric Administration have led the development of the Coastal and Marine Ecological Classification Standard, which provides a comprehensive national framework for organizing information about coasts and oceans and their living systems. This Federal Geographic Data Committee–approved standard includes the physical, biological, and chemical data used collectively to define coastal and marine ecosystems. Designed for use within all waters ranging from the head of tide to the deep ocean, the Coastal and Marine Ecological Classification Standard offers a common framework and language to unify inventory and mapping approaches for coastal and marine ecosystem data. It is compatible with existing upland and wetland classification standards and can be used with most, if not all, data collection technologies. Because of these characteristics, this standard allows scientists to use and compare data from various sources and time frames. As a standard lexicon, this classification standard makes aggregating the data more efficient and will ultimately support more compatible coastal and marine assessments and resource management efforts in nearshore and offshore environments. Users now need Coastal and Marine Ecological Classification Standard–compliant data collection, data entry, and data management tools that support consistent application of the standard. This chapter describes how the coastal and marine user community can apply the NatureServe observation toolkit and the National Oceanic and Atmospheric Administration's crosswalk tool for the Coastal and Marine Ecological Classification Standard to support more effective and standardized field data collection, upload, validation, aggregation, integration with existing data, and publication. The online toolkit allows users to enter, manage, and share Coastal and Marine Ecological Classification Standard data efficiently and flexibly, and a suite of data templates specify the fields essential for Coastal and Marine Ecological Classification Standard–compliant observations. Users can customize the templates as needed to

Dawn J. Wright, ed.; 2015; *Ocean Solutions, Earth Solutions*; http://dx.doi.org/10.17128/9781589483651

accommodate additional survey-specific data. The design is compatible with Federal Geographic Data Committee and International Organization for Standardization metadata standards.

Introduction

Observational data of species and ecosystems is fundamental for scientific inventory and monitoring, conservation planning, habitat management, invasive-species assessments, predictive distribution modeling, and more (National Oceanographic Partnership Program 2010; Duffy et al. 2013). The structure of these datasets can vary greatly depending on the source and intended use, making it difficult to gain a comprehensive picture of biodiversity within an area. Integrating disparate biodiversity data to maximize its utility for supporting decision-making continues to challenge resource managers and conservation practitioners.

The first step in integrating observation data is a consistent classification of the units to be aggregated. Until recently, there was no consistent US national classification for observing, mapping, and integrating marine ecosystem observation data. The Coastal and Marine Ecological Classification Standard (CMECS) now provides the classification framework and definitions needed to consistently identify and name ecosystem units (FGDC 2012).

The second step is documenting observations in the field and applying the correct standard name to the observed units. Identification of the "what is it?" for ecosystem observations is not trivial: it is hard work to document the attributes (or *classifiers*) that allow you to assign the right name to a given ecosystem.

The third step in integrating observation data is to identify existing data and data products such as maps that are labeled with nonstandard names. Or you can translate or "crosswalk" (i.e., convert) these data products into the standard so that they can be integrated with other CMECS-compliant observation data.

The fourth step is to compile these datasets into a common database, supplement them with supporting data as needed, and make the data and data products widely available to support decision-making. NatureServe and the National Oceanic and Atmospheric Administration (NOAA) are partnering to streamline the data integration process by developing the standards, methods, and tools needed to integrate and share coastal and marine ecosystem observation data on national and international scales. In this chapter, we describe the tools of CMECS and how they can be used to collect, validate, aggregate, integrate, share, and publish marine ecosystem observation data.

Background

NatureServe represents an international network of biological inventories, known as natural heritage programs or conservation data centers, operating in all 50 US states, Canada, Latin America, and the Caribbean. Together, it not only collects and manages detailed local information on plants, animals, and ecosystems, but also develops information products, data management tools, and conservation services to help meet local, national, and global conservation needs.

NatureServe's Role as a Standardized Biodiversity Observation Network

To advance its mission of providing the scientific basis for effective conservation action, NatureServe develops standardized methods and associated tools designed to improve the ability of field biologists to record, manage, and share geospatially explicit observation data. Benefits of NatureServe's observation data toolkit include (1) improving the accuracy, precision, and documentation of geospatial data for inventories of species and ecological communities and decreasing the time from data capture to data sharing; (2) increasing the efficiency and productivity of field researchers by introducing elements of location-aware computing into the field data collection process; and (3) enhancing information sharing and interoperability among observation data collections, observation-oriented data networks, and communities of practice.

NOAA Office for Coastal Management's Role in CMECS Tool Development

NOAA has broad national responsibilities related to fisheries management, habitat conservation, navigation, geodesy, and meteorology. Many efforts are under way within NOAA in which CMECS is relevant and NOAA is the steward agency for the standard. Within NOAA, the mission of the Office for Coastal Management is to link science and best practices to coastal management. The center routinely partners with agencies at the federal, state, and local level, as well as academic institutions, nonprofit organizations, and the private sector to help ensure that managers have access to the best tools and data needed for their work and researchers have an understanding of how their results can inform management decisions.

Geospatial tools that support data synthesis, analysis, and product development are a key element of the center's mission. The Office for Coastal Management has developed a GIS tool to automate the translation and conversion of existing spatial ecosystem data into CMECS. The center will ensure that this tool can be integrated with other geospatial tools, applied in commonly used software environments, and crafted so that it can be incorporated into existing workflow processes, including the NatureServe observation toolkit.

Overview of CMECS

CMECS provides a comprehensive framework for organizing information about coasts and oceans and their living systems. This information includes the physical, biological, and chemical data collectively used to define coastal and marine ecosystems. The standard was designed to meet the needs of many users, including resource managers and planners, conservation practitioners, engineers, mappers, and researchers from government, industry, and academia that need to describe, map, or assess coastal and marine ecosystems. CMECS is the result of an ongoing collaboration among several organizations, including NatureServe, NOAA, the US Environmental Protection Agency, US Geological Survey (USGS), and University of Rhode Island with input from dozens of professional scientists and marine managers (see appendix J of FGDC [2012] for a complete listing). The Federal Geographic Data Committee (FGDC) has approved CMECS as a US federal standard (2012).

The CMECS framework is organized into two settings, biogeographic and aquatic, and four components: water column, geoform, substrate, and biotic (figure 4.1). Each describes a separate aspect of the environment and biota. Settings and components can be used in combination or independently to describe or map ecosystem features. The hierarchical arrangement of units within the settings and components allows users to apply CMECS to the scale and specificity that best suits their interests. Modifiers allow users to customize the classification to meet specific needs. Biotopes can be described when there is a need for more detailed information on the biota and their environment. An online catalog of the units with complete definitions is also available at NatureServe (2013).

Figure 4.1. The Coastal Marine Ecological Classification Standard organizing framework showing the structure of the settings and components.

Designed for use within all waters ranging from the head of tide (as defined in FGDC [2001]) to the deep ocean, CMECS offers a common framework and language to unify inventory and mapping approaches for coastal and marine ecosystem data. It is compatible with existing upland and wetland classification standards such as the National Vegetation Classification Standard (FGDC 2008) and Classification of Wetlands and Deepwater Habitats of the United States (FGDC 2013) and can be used with most, if not all, data collection technologies. Because of these characteristics, the standard allows scientists to use and compare data from various sources and time frames.

As a standard lexicon, CMECS makes aggregating the data more efficient and will ultimately support more compatible coastal and marine assessments and resource management efforts in nearshore and offshore environments. Users now need CMECS–compliant data collection, data entry, and data management tools that support consistent application of the standard as well as data sharing.

Modeling Observation Data

Well-established data standards for biological observations and an active biodiversity informatics community contribute to the ongoing development and adoption of these standards. The Biodiversity Information Standards Taxonomic Databases Working Group (TDWG) is one such initiative that promotes standards for the exchange of biodiversity data (TDWG 2014). TDWG recognizes Darwin Core (TDWG 2009b) and Access to Biological Collections Data (ABCD) (TDWG 2009a) as the two most widely deployed formats for biodiversity occurrence data.

Historically, these standards have been developed and applied retrospectively to existing observation data, usually found in museum specimen collections. Doing so has enabled researchers and other users to query, using well-defined access protocols, hundreds of millions of occurrence records mapped to the Darwin Core schema and published by institutions worldwide through the Global Biodiversity Information Facility (GBIF) (2014). A limitation of these schemas is their inability to integrate observations representing different levels of biological diversity, from genes to species to ecosystems. The Darwin Core schema is typically used to describe flat data structures, which works well for an observer describing an observation of a species present at a given place and time. However, the specimen-based schemas do not adequately address more complicated biological assemblages (e.g., vegetation, biotic communities, and so forth) and ecosystem observations that require documentation of detailed biotic data (e.g., species composition, structure, individual species traits) as well as detailed abiotic data (e.g., substrate, geologic formations, hydrologic attributes [Wiser et al. 2011]). These types of observations may also entail more complicated sampling methodologies that include multiple nested sampling units (such as points along transects or plots and subplots).

The observation toolkit's underlying schema accommodates a wide diversity of data models, including those for biological assemblages that require hierarchical or nested observation data structures. Figure 4.2 shows how the toolkit accommodates data models ranging from simple species observations to more complex models that incorporate all the CMECS components. The core data model provides a means of recording an observation that, at a minimum, documents (1) what biological entity occurs, (2) at what location, (3) at what time, and (4) as observed by whom. To document the biotic component of CMECS, the model captures supporting data for the information needed to appropriately assign a standard name to the biotic assemblage, and then provides for additional locational, environmental, and taxonomic information that allows further characterization and classification.

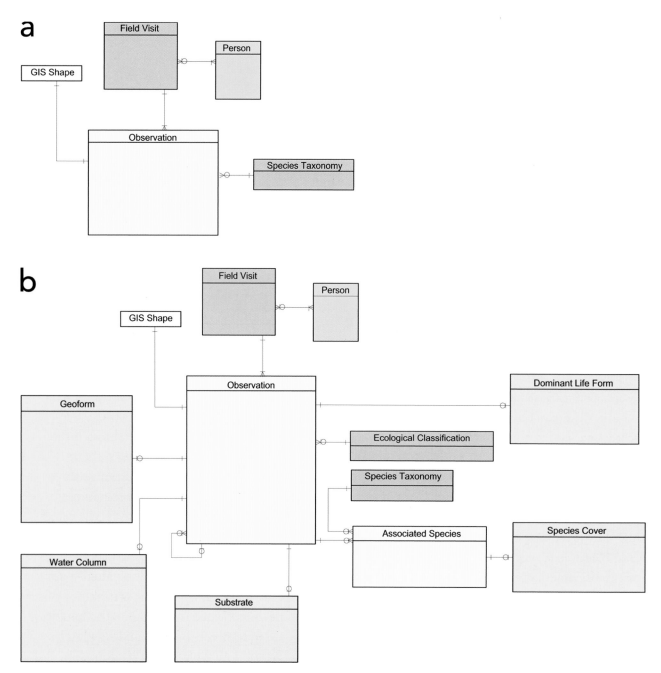

Figure 4.2. The (a) core data model for species observation compared with (b) enhanced data model for Coastal Marine Ecological Classification Standard component observations.

The established standards also have a limited ability to integrate data from disparate data sources without losing potentially valuable, unique fields of information that do not fit the standard's core.

By definition, data standards involve trade-offs that must balance the value of unique information housed in individual data collections versus the value that comes from being able to combine these pieces of information into a common database with a shared schema. Over time, communities of practice have adopted extensions to the Darwin Core to accommodate these differences (for example, the Marine Biogeography Extension developed by Ocean Biogeographic Information System-USA [OBIS-USA] [USGS 2014]). The observation toolkit offers a unique solution that supports retrospective crosswalking of legacy data to a standard classification system and data model without leaving behind nonstandard but project-relevant information. It also offers tools for communities of practice to create and share standardized templates that help harmonize new data collection efforts.

Another impediment to large-scale synthesis of biodiversity observation data is the inability to retrospectively reconcile the taxonomy for species names documented in biological assemblage observations. Taxonomic standards can vary over time and space, and different individuals can use the same name to represent different taxonomic entities (Wiser et al. 2011). When only names are included in the observation dataset instead of taxonomic concepts (i.e., references for the species circumscription), the meaning of the species listed can be ambiguous. This uncertainty in species taxonomy can impede efforts to integrate data. The observation toolkit relates individual observation records to concept-based taxonomic references using dynamic web services, ensuring that information is not lost when taxonomic changes occur.

Observation Toolkit Components

NatureServe's observation toolkit consists of several components that allow data from across levels of biological organization to be integrated into a single biodiversity observation database (figure 4.3). The components are summarized here and described in detail below.

- *Template Library:* an online collaboration tool for creating and sharing data collection survey forms that meet community standards while supporting project-specific customizations.
- *Kestrel:* an online, spatially enabled observation database that supports import or manual data entry into Template Library–authored survey formats and facilitates data sharing among users.
- *Mobile tools for field data capture:* the toolkit is compatible with Windows Mobile GPS-enabled devices and integrates with ArcPad software for field access to reference maps.
- *ArcGIS for Desktop Editor:* utilities for offline data entry, validating, and exporting results.

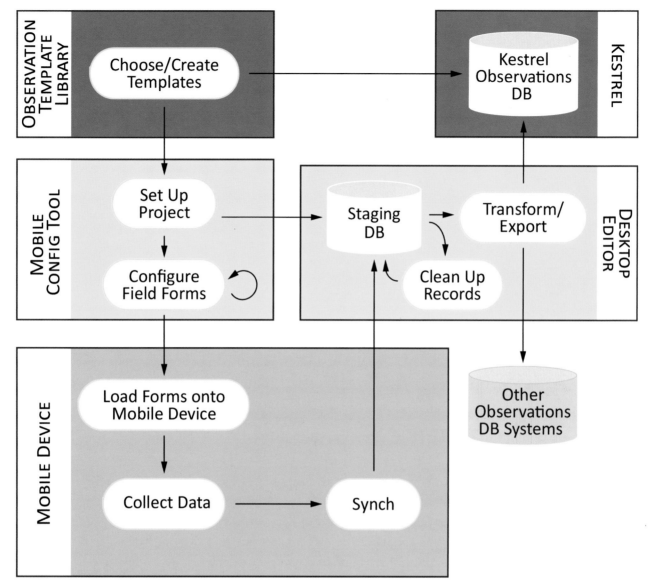

Figure 4.3. The observation toolkit includes technology components that support more efficient field collection, management, and sharing of observation data.

Observation Data Templates: Support Standardized Data Collection and Validation

Observational data formats can vary widely to accommodate unique data collection protocols and study designs. However, a small subset of core attributes minimally defines any observation: what was seen, when, where, and by whom. The observation toolkit embraces diversity among observation data formats while promoting greater consistency and standardization among data collections. It does this by providing tools to allow users to discover, adapt, reuse, and share observation data templates. A template is a collection of individual data fields or attributes and associated rules that define the minimal set of information that must be recorded for an observation according to a data collection protocol, as well as any necessary validation rules. For example, a template can define which attributes are optional and which are required, and a given attribute definition can specify its allowable field values or ranges. It also provides multilingual labels and help text for each attribute.

The template structure, attributes, and validation rules are supported by a standardized schema, allowing the template to be expressed, downloaded, and shared in an extensible markup language (XML) package.

New CMECS-compatible templates will assist users in entering data that is compliant with CMECS by including field values consistent with the standard's unit definitions, cascading drop-down lists that accommodate the standard's hierarchical data fields, and data validation rules that enforce valid combinations of standard units. Users can develop and share customized templates to accommodate different sampling gear and sampling methods and additional project-specific data.

Templates can be deployed via offline and/or mobile devices for documentation of field-collected samples in the lab and via online tools, including Kestrel. Templates can be designed to accommodate FGDC and International Organization for Standardization (ISO) metadata standards (ISO 2014) as well as the Darwin Core species observation data standard, among others.

Template Library: Supports Authoring and Sharing Reusable Templates

Templates are created and made available via the Template Library. The Template Library is a companion online resource for toolkit users to author, share, and reuse customizable survey templates for use with Kestrel and mobile field data collection tools. The Template Library supports a wide variety of observation data structures and data collection protocols, including rare-species sightings; species lists and environmental data collected in quadrats, along transects, or using other sampling schemes; time-series monitoring data; and ecological-integrity assessment data. Users can select and modify an existing template, or create their own—which they can save to use again later and share with others. The Template Library promotes greater standardization among data collectors by encouraging reuse of individual attributes and templates. Learn about the Observation Template Library at NatureServe (2014c).

Kestrel Online Observation Database: Supports Data Management, Integration, Mapping, and Sharing

Kestrel (NatureServe 2014a) is an online tool for biologists, researchers, and natural resource managers to map and manage observational data. Kestrel promotes greater standardization among observational datasets while allowing users to meet specific project needs.

Kestrel provides a secure, hosted environment for observational data management and sharing. It supports diverse observational data formats, all sharing a set of core attributes that minimally define an observation: what was seen, when, where, and by whom. Users can define the structure of the observation data they wish to collect and manage in Kestrel by modifying one of the existing templates published in the library or constructing a new template from scratch.

Through Kestrel's upload tool, users can quickly import and share data. To protect sensitive data, Kestrel allows survey data owners to restrict access by user, organization, or project at either the survey level or individual record level. For example, a state CMECS partner could create a survey to support local project data collection and share that survey and its associated observations within their restricted user group, whereas other partners could choose to share their surveys and observations with the entire CMECS community.

Kestrel uses an ArcGIS for Server-based mapping platform for managing observation locational data. Its multilingual interface supports English-, French-, and Spanish-speaking users. Kestrel allows for integration with taxonomic sources via web services to ensure consistent identification and naming of biodiversity elements across data collections and geographic jurisdictions. NatureServe is now implementing a CMECS taxonomy web service to integrate with Kestrel and support recording of observations against any standard CMECS classification unit.

Kestrel has tools for importing and exporting observation data collections in spatial and tabular formats. Its mapping interface allows users to visualize observation locations in the spatial context of other mapped reference layers such as protected areas, shorelines, and species and habitat ranges. The platform also allows users to output reports of observations by location, species or classification unit, date range, or observer and supports publishing to government data portals via web and map services.

Mobile Tools: Support Data Capture in the Field

Handheld data collection has been demonstrated to be faster and more accurate than paper-based methods (Spain et al. 2001). Increasingly, environmental data collected by instruments (e.g., satellites, buoy markers, remote monitoring stations) is being integrated into data collection methodology. Global Positioning System (GPS) units have vastly improved the ability of field workers to accurately georeference their observations. Similarly, portable computing devices—ranging from laptops and tablets to smartphones—enable field workers to record data directly in digital form, bypassing the need to transcribe field data. The observation toolkit's mobile configuration software (NatureServe 2014b) is compatible with Windows Mobile GPS-enabled devices, such as the Trimble Juno, and integrates with ArcPad for field access to reference maps. Planned upgrades for the mobile tool component will expand support for more device types, including tablets. Smartphone applications are also possible, although their small screens may not be practical for entering complex data formats such as those associated with CMECS and similar ecological sampling survey designs.

Desktop Editor: Supports Offline Data Management, Validation, and Export

The ArcGIS for Desktop Editor toolkit supports offline data entry and analysis through a desktop editor. This can be useful for coastal and marine observation inventory and monitoring data collection protocols, which often involve the collection of large samples that are labeled and later assessed in the lab. The desktop editor supports data entry and staging; record cleanup; data aggregation, integration, and management; and export to Kestrel or other long-term storage databases. The desktop editor can also be used on a laptop in the field or at sea when Internet access is not available.

Crosswalk Tool: Utilities to Integrate Legacy Data

Many users seeking to implement CMECS want to integrate existing data into their analysis. In most cases, this data has been classified or attributed using systems other than CMECS and requires translation to the CMECS system for consistent query or analysis procedures. This process of translating data between classification systems is known as *crosswalking*. The first step of the crosswalking process is to establish the relationships between the units in the source classification and CMECS to develop a conceptual crosswalk (as described in appendix H in the CMECS documentation in FGDC [2012]).

Once the conceptual crosswalk has been established, users are ready to make the actual spatial data translation. The NOAA Office for Coastal Management has developed a GIS tool to automate this conversion to CMECS as a way to facilitate the crosswalk process. This tool, which operates in an ArcMap environment as an ArcToolbox (.tbx) file, relies on Look-Up Tables (LUTs) that articulate the source-to-CMECS unit relationships. A set of LUTs has been developed for some of the common classifications such as the Florida System for Classification of Habitats in Estuarine and Marine Environments (SCHEME) (Madley et al. 2002) and the Classification of Wetlands and Deepwater Habitats of the United States (FGDC 2013), but these can be modified by the user to adjust relationships or incorporate new classification systems. The tool allows input of Esri shapefile data (both point and polygon) and produces an output Esri File Geodatabase with multiple feature layers (by CMECS component).

Figures 4.4 to 4.7 illustrate how spatial topology of the input data is preserved while producing separate component feature layers with CMECS attributes as feature layers in an Esri geodatabase designed in advance by the user. The maps show source SCHEME data for Humboldt Bay, California, as a single polygon shapefile overlaying digital aerial imagery (figure 4.4) and output CMECS component feature layers for the Biotic component (figure 4.5), Substrate component (figure 4.6), and Geoform component (figure 4.7). Transparent areas are null polygons.

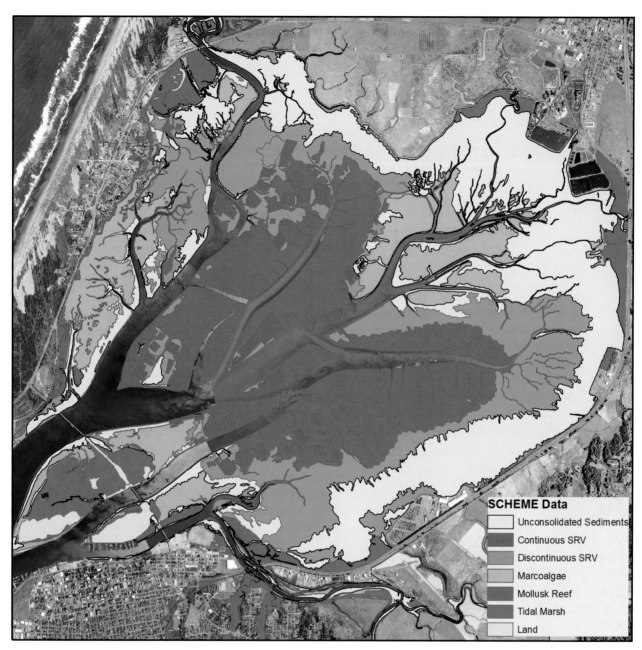

SCHEME Data
- Unconsolidated Sediments
- Continuous SRV
- Discontinuous SRV
- Marcoalgae
- Mollusk Reef
- Tidal Marsh
- Land

Figure 4.4. Florida System for Classification of Habitats in Estuarine and Marine Environments (SCHEME) input dataset in Esri shapefile format for Humboldt Bay used to produce separate Coastal and Marine Ecological Classification Standard component feature layers. Source imagery is digital mapping camera aerial imagery (false-color composite, bands 4, 3, 2) collected June 27, 2009. Resolution is ½ m pixels. Image collected at negative low-tide conditions. Overlying vector dataset is a Humboldt Bay, California, benthic cover dataset, developed by PhotoScience under contract to NOAA NOS Office for Coastal Management, formerly the NOAA NOS Coastal Services Center. Classification is in Florida System for Classification of Habitats in Estuarine and Marine Environments (Madley et al. 2002). The minimum mapping unit is 100 sq m minimum mapping unit. Datasets are in Esri shapefile format. Source for imagery and vector data: National Oceanic and Atmospheric Administration National Ocean Service Office for Coastal Management (2014). Courtesy of NOAA NOS Office for Coastal Management.

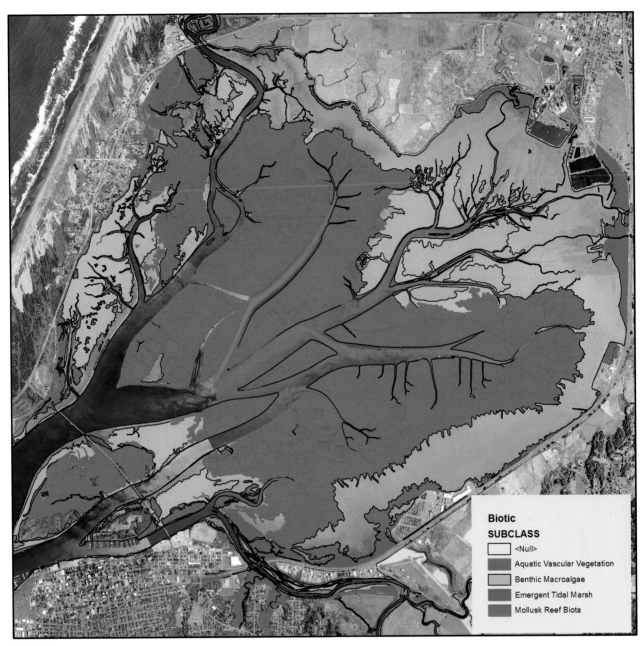

Figure 4.5. Florida System for Classification of Habitats in Estuarine and Marine Environments input dataset in Esri shapefile format for Humboldt Bay used to produce Biotic component feature layer output. Source imagery is digital mapping camera aerial imagery (false-color composite, bands 4, 3, 2) collected June 27, 2009. Resolution is ½ m pixels. Image collected at negative low-tide conditions. Overlying vector dataset is Coastal and Marine Ecological Classification Standard–compliant crosswalk tool output for the Biotic component as a separate feature layer within the California Humboldt Bay geodatabase. Minimum mapping unit is the same as the source Florida System for Classification of Habitats in Estuarine and Marine Environments dataset. Clear (null) polygons are areas where no automated crosswalk was possible between Florida System for Classification of Habitats in Estuarine and Marine Environments and Coastal and Marine Ecological Classification Standard Biotic attributes. Source for imagery and vector data: National Oceanic and Atmospheric Administration National Ocean Service Office for Coastal Management (2014). Courtesy of NOAA NOS Office for Coastal Management.

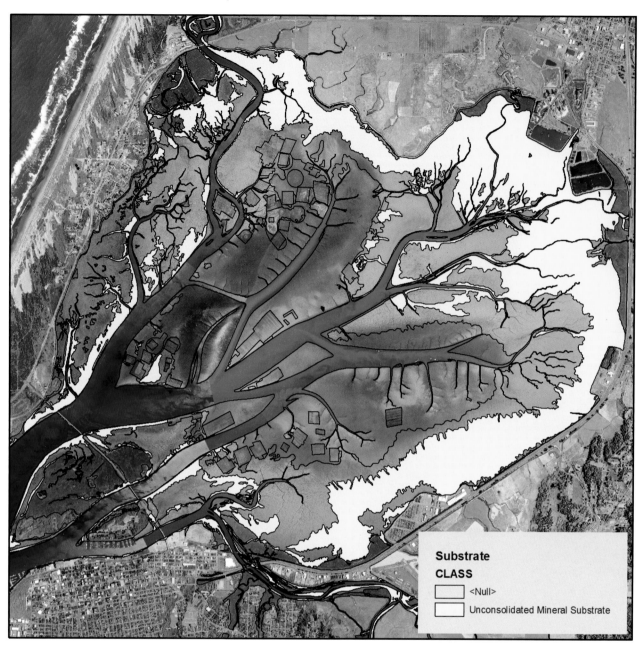

Substrate
CLASS
☐ <Null>
☐ Unconsolidated Mineral Substrate

Figure 4.6. Florida System for Classification of Habitats in Estuarine and Marine Environments input dataset in Esri shapefile format for Humboldt Bay used to produce Substrate component feature layer output. Source imagery is digital mapping camera aerial imagery (false-color composite, bands 4, 3, 2) collected June 27, 2009. Resolution is ½ m pixels. Image collected at negative low-tide conditions. Overlying vector is Coastal and Marine Ecological Classification Standard–compliant crosswalk tool output for the Substrate component as a separate feature layer within the California Humboldt Bay geodatabase. Minimum mapping unit is the same as the source Florida System for Classification of Habitats in Estuarine and Marine Environments dataset. Clear (null) polygons are areas where no automated crosswalk was possible between Florida System for Classification of Habitats in Estuarine and Marine Environments and Coastal and Marine Ecological Classification Standard Substrate attributes. Source for imagery and vector data: National Oceanic and Atmospheric Administration National Ocean Service Office for Coastal Management (2014). Courtesy of NOAA NOS Office for Coastal Management.

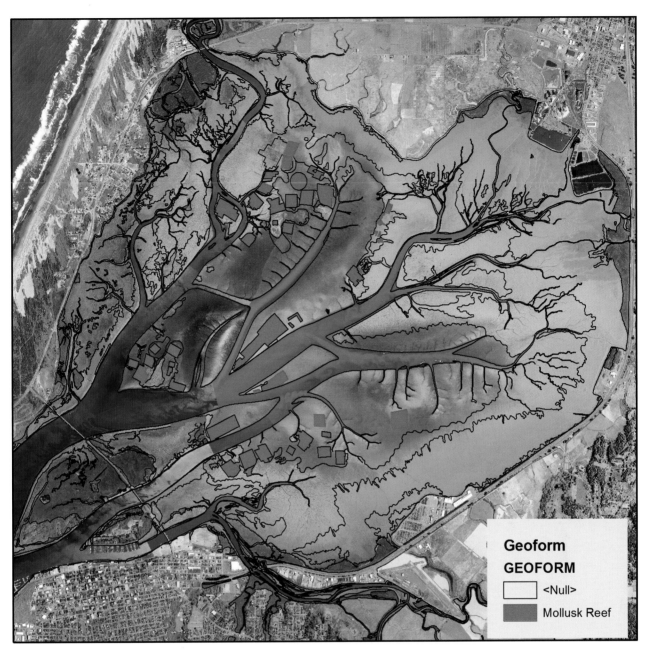

Geoform
GEOFORM
☐ <Null>
⬛ Mollusk Reef

Figure 4.7. Florida System for Classification of Habitats in Estuarine and Marine Environments input dataset in Esri shapefile format for Humboldt Bay used to produce Geoform component feature layer output. Source imagery is digital mapping camera aerial imagery (false-color composite, bands 4, 3, 2) collected June 27, 2009. Resolution is ½ m pixels. Image collected at negative low-tide conditions. Overlying vector is Coastal and Marine Ecological Classification Standard–compliant crosswalk tool output for the Geoform component as a separate feature layer within the California Humboldt Bay geodatabase. Minimum mapping unit is the same as the source Florida System for Classification of Habitats in Estuarine and Marine Environments dataset. Clear (null) polygons are areas where no automated crosswalk was possible between Florida System for Classification of Habitats in Estuarine and Marine Environments and Coastal and Marine Ecological Classification Standard Geoform attributes. Source for imagery and vector data: National Oceanic and Atmospheric Administration National Ocean Service Office for Coastal Management (2014). Courtesy of NOAA NOS Office for Coastal Management.

The crosswalk tool does not attempt to assign CMECS output values for those features that do not have CMECS-equivalent units or for which the confidence in an automated process might be low. Rather, it preserves the original feature with a null value, which the user can then attribute through a more individualized process. By allowing the user to modify LUTs, test runs of the tool can be made to assure that the resulting spatial data meets expectations. Efforts are under way to provide lists of choices for potential equivalent CMECS units as a way to guide the user in conducting the crosswalk. The output attribute table produced by this tool draws on the CMECS code set and integrates with the NatureServe observation toolkit. This tool helps establish consistency in data translation and processing of large amounts of data. Once the crosswalk tool has been used to translate mapped, nonstandard observations to CMECS, the legacy data can be integrated with other CMECS-compliant data in Kestrel.

Case Study

Implementation of the observation toolkit by Parks Canada provides a case study of how the toolkit has been used to collect and manage new observation data and integrate it with existing data to inform decision-making. In this case, Parks Canada needed to manage biodiversity information to meet its mandate to "protect and present significant examples of Canada's natural and cultural heritage" and help them implement and enforce the Species at Risk Act (SARA) (Oliver 2004).

Parks Canada had the objective of standardizing new species observation data collections across a network of parks with many observers and integrating the new data with volumes of legacy species observation data that had been collected over many years using multiple standards in parks throughout Canada. The observation toolkit helped Parks Canada compile a standardized, comprehensive observation dataset on at-risk species in its parks.

Parks Canada adopted the standard species taxonomy employed and managed by NatureServe and its network of provincial conservation data centers, which is a compilation of taxonomic concepts derived from multiple authoritative species groups throughout North America. Using the Template Library, it designed data templates to standardize new observation data collected by the parks. The templates included the core data model fields for Species, Location, Observation Date, and Primary Observer, as well as additional data fields needed for each type of survey (e.g., Critical Habitat). The templates also supported customization to allow for park-specific data needed by the park managers. Templates were also designed to crosswalk, validate, and integrate legacy data within the shared Kestrel data management platform.

Species observations created or uploaded in Kestrel were linked to the standard taxonomy via web services to provide all users access to the standard species names. The Kestrel observation database was made available to parks throughout Canada. This database is regularly updated with new observations and is used to identify critical habitat for at-risk species within the parks to help Parks Canada plan for their management and protection. Although this case focuses on species data in both marine and terrestrial environments, the basic steps and tools for implementation are analogous to those that can be employed for the collection and integration of marine ecosystem observation data.

Conclusion

The need for marine biodiversity observation data to guide resource management and conservation decision-making is receiving increasing recognition, and many national, regional, and local efforts are under way to produce this data. In the United States, NOAA, USGS, and the National Aeronautics and Space Administration are leading a new initiative to create a national Marine Biodiversity Observation Network (MBON), and MBONs are already in place in Europe, Asia-Pacific, and the Arctic (National Oceanographic Partnership Program 2010; Duffy et al. 2013). The National Park Service has recently called for the production of nearshore ecosystem maps to help managers understand the impacts of Hurricane Sandy, and efforts are under way to create a regional habitat map for the Northwest Atlantic. The State of Oregon is also currently remapping its coastal ecosystem. All these efforts call for the need to develop new observations of ecosystems, and most call for integrating the new data with legacy data. All are proposing to use CMECS as the classification standard.

NatureServe's observation toolkit and NOAA's crosswalk tool can support these and future efforts to collect, upload, validate, aggregate, integrate, share, and publish marine ecosystem observation data. The toolkit will provide direct online access to data collection protocols and tools across the growing community of CMECS users and promote easier adoption of the standard. The data templates, online observation database platform, and associated quality control utilities will enforce consistent application of the standard and support the integration of new and legacy data inputs to allow more meaningful information outputs, while allowing flexibility for users to enter data beyond that required for the standard.

Acknowledgments

The comments and suggestions of two anonymous reviewers greatly improved this chapter.

References

Duffy, J. E., L. A. Amaral-Zettler, D. G. Fautin, G. Paulay, T. A. Rynearson, H. M. Sosik, and J. J. Stachowicz. 2013. "Envisioning a Marine Biodiversity Observation Network." *BioScience* 63 (5): 350–61.

FGDC (Federal Geographic Data Committee). 2001. FGDC-STD-001.2-2001. *Metadata Profile for Shoreline Data*. Reston, VA: USGS.

———. 2008. FGDC-STD-005-2008. *National Vegetation Classification Standard*, version 2. Reston, VA: USGS.

———. 2012. FGDC-STE-18-2012. *Coastal and Marine Ecological Classification Standard*. Reston, VA: USGS.

———. 2013. FGDC-STD-004-2013. *Classification of Wetlands and Deepwater Habitats of the United States*. Reston, VA: USGS.

GBIF (Global Biodiversity Information Facility). 2014. GBIF Portal—Home. http://www.gbif.org. Last accessed June 22, 2014.

ISO (International Organization for Standardization). 2014. ISO—International Organization for Standardization. http://www.iso.org/iso/home.html. Last accessed June 22, 2014.

Madley, K. A., B. Sargent, and F. J. Sargent. 2002. *Development of a System for Classification of Habitats in Estuarine and Marine Environments (SCHEME) for Florida.* Report to the US Environmental Protection Agency Gulf of Mexico Program (Grant Assistance Agreement MX-97408100). St. Petersburg, FL: Florida Fish and Wildlife Conservation Commission Fish and Wildlife Research Institute.

NOAA (National Oceanic and Atmospheric Administration) Office for Coastal Management. 2014. Data Registry—Digital Coast. http://www.csc.noaa.gov/digitalcoast/dataregistry/#/. Last accessed June 22, 2014.

National Oceanographic Partnership Program. 2010. *Report to the US Congress on the National Oceanographic Partnership Program: Fiscal Year 2010.* http://www.nopp.org/publications-and-reports. Last accessed June 21, 2014.

NatureServe. 2013. Coastal and Marine Ecological Classification Standard Catalog of Units. http://www.cmecscatalog.org. Last accessed June 22, 2014.

———. 2014a. Kestrel. http://www.natureserve.org/conservation-tools/data-maps-tools/kestrel. Last accessed June 22, 2014.

———. 2014b. ns-mos—NatureServe Mobile Observations System. https://code.google.com/p/ns-mos/. Last accessed June 22, 2014.

———. 2014c. Observation Template Library. http://www.natureserve.org/conservation-tools/data-maps-tools/observation-template-library. Last accessed June 22, 2014.

Oliver, D. 2004. *Parks Canada Biodiversity Business Process Review: Final Report.* A Report by Skylark Information Systems. Gatineau, Quebec: Parks Canada.

Spain, K. A., C. A. Phipps, M. E. Rogers, and B. S. Chapparo. 2001. "Data Collection in the Palm of Your Hand: A Case Study." *International Journal of Human-Computer Interaction* 13: 231–43.

TDWG (Taxonomic Databases Working Group). 2009a. Access to Biological Collections Data, version 2.06 (cover page). http://www.tdwg.org/standards/115. Last accessed June 22, 2014.

———. 2009b. Darwin Core (cover page). http://www.tdwg.org/standards/450. Last accessed June 22, 2014.

———. 2014. Taxonomic Databases Working Group (home page). http://www.tdwg.org. Last accessed June 22, 2014.

USGS (US Geological Survey). 2014. OBIS-USA Marine BioGeography (MBG) Terminology Definitions. http://snapper.colorado.edu/ObisUsa/portal/XsdReader.php. Last accessed June 22, 2014.

Wiser, S. K., N. Spencer, M. De Caceres, M. Kleikamp, B. Boyle, and R. K. Peet. 2011. "Veg-X: An Exchange Standard for Plot-Based Vegetation Data." *Journal of Vegetation Science* 22 (2011): 598–609.

CHAPTER 5

How Does Climate Change Affect Our Oceans?

Peter Kouwenhoven, Yinpeng Li, and Peter Urich

Abstract

Global climate change affects not only terrestrial ecosystems, but also has profound consequences for the oceans. The latest results of the numerical models that were run for the Fifth Assessment Report of the Intergovernmental Panel on Climate Change include many ocean variables. The outputs of these models are difficult to access, assess, use, and interpret. This chapter seeks to transform these outputs into more usable and useful results via global maps highlighting the changes in specific oceanographic variables by the year 2050. It describes the methodology used to extract the relevant information from the models and looks at the consequences for selected ocean variables under projected climate change. These include pH (which is relevant to the survival of coral reefs) and net primary production (which forms the base of the entire oceanic food web and is also relevant to eutrophication issues in coastal areas). In addition, the use of the SimCLIM for ArcGIS/Marine add-in for ArcGIS for Desktop software enabled the creation of global output variables for the ArcGIS platform. The results lend insight into what might happen in the oceans as a result of climate change, and may thus inform priorities in current and future research as well as policy making.

Introduction

The environment is changing on a global scale as the result of changes in the atmospheric concentrations of carbon dioxide (CO_2) and other greenhouse gases caused directly by anthropogenic emissions and changes in land use, and indirectly by processes such as the melting of the permafrost, which releases methane. This increase in atmospheric concentrations causes an increase in air temperatures, commonly referred to as *global warming*, or more generically, *global climate change*. There is no longer any doubt that human activities are causing an accelerated change (IPCC 2013). It is increasingly realized that climate change affects not only terrestrial ecosystems, but also has a profound effect on the oceans. This is even more so as many of the ocean's biogeochemical cycles and ecosystems are becoming increasingly stressed by at least three factors that are driven by climate change: *rising temperatures, ocean acidification,* and *ocean deoxygenation.* Each of these variables can cause substantial change in the physical, chemical, and biological environment, affecting the ocean's biogeochemical cycles and ecosystems in ways that we are only beginning to fathom (Gruber 2011; Doney et al. 2012; Bopp et al. 2013).

Ocean warming is the result of the heat transfer between the warming air and the ocean's surfaces. It affects organisms and biogeochemical cycles directly as well as through an increase of upper-ocean stratification.

Ocean acidification results from the fact that about one-quarter of the anthropogenic CO_2 emissions dissolves in the ocean, making it more acidic (Mora et al. 2013). The changes in the ocean's carbonate chemistry induced by this uptake negatively affect many organisms and processes, including shellfish, phytoplankton, and most profoundly, coral reefs.

Ocean deoxygenation is the decrease of dissolved oxygen levels in the ocean. It is bound to occur in a warming ocean and is exacerbated by stratification, causing stress to macroorganisms that critically depend on sufficient levels of oxygen.

These three stressors—warming, acidification, and deoxygenation—operate globally but with distinct regional differences. The impacts of ocean acidification tend to be strongest in the high latitudes, whereas the already low-oxygen regions around the equator are most vulnerable to climate-change-induced deoxygenation. Specific regions, such as the eastern boundary upwelling systems, will be strongly affected by all three stressors, making them potential hot spots for impacts from climate change. Of additional concern are synergistic effects such as ocean acidification-induced changes in the type and magnitude of the organic matter exported to the ocean's interior, which then might cause substantial changes in the oxygen concentration there. Ocean warming, acidification, and deoxygenation are mostly irreversible on decadal time scales (Andrews et al. 2013). Once these changes have set in, it will take centuries for the oceans to recover. With the emissions of CO_2 as the primary driver of all three stressors, the global strategy should be to reduce these emissions through mitigation, and offset them through sequestration. Given that global deforestation accounts for one-third of the increase in atmospheric CO_2 levels (IPCC 2013), reforestation looks to be a most important strategy.

Natural resource management must also remain flexible to absorb the sudden and nonlinear changes that are likely to characterize the behavior of most ecosystems in the future. Overall, however, reducing greenhouse gas emissions remains the priority, not only because it will reduce the huge costs of adaptation, but also because it will reduce the growing risk of pushing our planet into an unknown and highly dangerous state (Hoegh-Guldberg and Bruno 2010).

The numerical models used for projecting the effects induced by changes in atmospheric greenhouse gases produce outputs that are difficult to access, assess, use, and interpret. This chapter seeks to transform these outputs into more usable and useful results, via global maps highlighting the changes in specific oceanographic variables by the year 2050.

Methodology

The Intergovernmental Panel on Climate Change (IPCC) produces reports on the effects and impacts of climate change. The most recent set of reports is referred to as the Fifth Assessment Report (AR5), Working Group I, Chapter 6 for Carbon and Other Biogeochemical Cycles (Ciais et al. 2013) and Chapter 12 for Long-Term Climate Change: Projections (Collins et al. 2013), and is based on multiple model runs of generalized circulation models or global climate models (GCMs). Detailed results of these models are released through the Climate Models Intercomparison Program (CMIP), which has now reached its fifth cycle, or CMIP5 (World Climate Research Programme 2011; Taylor et al. 2012).

Some GCMs contain ocean biogeochemical models that are mathematical representations of the interactions between chemical and biological components of an ecosystem. Ocean biogeochemical models (Gettelman et al. 2012) (figure 5.1) can be used to study the dynamics between dissolved gases (e.g., oxygen and CO_2) and inorganic nutrients (e.g., phosphate and nitrate), as well as the lower trophic levels of the ecosystem (e.g., bacteria to plankton). The biogeochemical models can be linked with hydrodynamic models and higher trophic-level models to give a mathematical representation of the whole ecosystem (Quere et al. 2005; Fennel 2009).

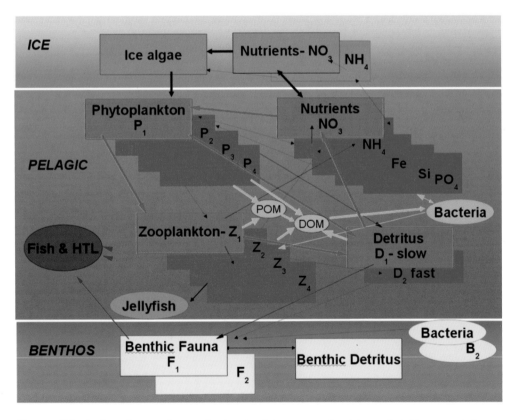

Figure 5.1. **A simple nutrient, phytoplankton, zooplankton, detritus model.**

As GCMs project future climate, assumptions need to be made about future greenhouse gas emissions. AR5 introduced the concept of representative concentration pathways (RCPs). The four greenhouse gas concentration (not emissions) trajectories are RCP2.6, RCP4.5, RCP6.0, and RCP8.5, named after a possible range of radiative forcing values in the year 2100 (2.6, 4.5, 6.0, and 8.5 W/m^2 respectively) (table 5.1).

Table 5.1. **Overview of Representative Concentration Pathways (Moss et al. 2010; van Vuuren et al. 2007; Rogelj et al. 2012)**

Description*		CO_2 Equivalent	SRES Equivalent
RCP8.5	Rising radiative forcing pathway leading to 8.5 W/m^2 in 2100.	1370	A1FI
RCP6.0	Stabilization without overshoot pathway to 6 W/m^2 at 2100.	850	B2
RCP4.5	Stabilization without overshoot pathway to 4.5 W/m^2 2100.	650	B1
RCP2.6	Peak in radiative forcing at ~ 3 W/m^2 before 2100 and decline.	490	None

* Approximate radiative forcing levels were defined as ±5% of the stated level in W/m2 relative to preindustrial levels. Radiative forcing values include the net effect of all anthropogenic greenhouse gases and other forcing agents.

From the GCMs (see table 5.2), selected ocean biogeochemical variables were processed using monthly outputs, determining the global area weighted means of each year from 2006 to 2100 for all the available GCMs' RCP runs. To determine a global average change per variable (GV), the GCM ensemble mean of each RCP was calculated. Finally, the 95-year values were fitted to smooth, curved polynomial lines. The difference between 1995 and a projection year is called ΔGV. Figure 5.2 shows sea surface temperature (SST) as an example of one of the selected ocean biogeochemical variables. In the pattern-scaling approach that is used, the global curves are essential information as local changes are expressed per global change (i.e., local SST changes X°C per 1°C of change in global average SST).

Table 5.2. Variables Available per Global Climate Model

GCM name	DFE	INTPP	NO$_3$	O$_2$	PH	PO$_4$	SI	SST	TALK
ACCESS1-0								X	
ACCESS1-3								X	
CANESM2		X	X		X			X	X
CCSM4								X	
CESM1-BGC	X	X	X	X	X	X	X	X	X
CESM1-CAM5								X	
CESM1-CAM5								X	
CNRM-CM5	X	X	X	X		X	X	X	X
CSIRO-MK3-6-0								X	
EC-EARTH								X	
GFDL-CM3								X	
GFDL-ESM2G	X	X	X	X	X	X	X	X	X
GFDL-ESM2M	X	X	X	X	X	X	X	X	X
GISS-E2-H								X	
GISS-E2-R								X	
HADGEM2-AO								X	
HADGEM2-CC	X	X	X	X	X		X	X	X
HADGEM2-ES	X	X	X	X	X		X	X	X
INMCM4								X	
IPSL-CM5A-LR	X	X	X	X	X	X	X	X	X
IPSL-CM5A-MR	X	X	X	X	X	X	X	X	X
IPSL-CM5B-LR	X	X	X	X	X	X	X	X	X
MIROC5								X	
MPI-ESM-LR	X	X	X	X	X	X	X	X	X
MPI-ESM-MR	X	X	X	X	X	X	X		X
MRI-CGCM3								X	
NORESM1-M								X	
NORESM1-ME	X	X	X	X	X	X	X		X

DFE: dissolved iron concentration at surface (μmol/m^3)

INTPP: net primary productivity of carbon by phytoplankton (gC/m^3/day)

NO$_3$: dissolved nitrate concentration at surface (mmol/m^3)

O$_2$: dissolved oxygen concentration at surface (unit: mol/m^3)

PH: pH at surface (no unit)

PO$_4$: dissolved phosphate concentration at surface (mmol/m^3)

SI: dissolved silicate concentration at surface (mmol/m^3)

SST: sea surface temperature (°C)

TALK: total alkalinity at surface (mol/m^3)

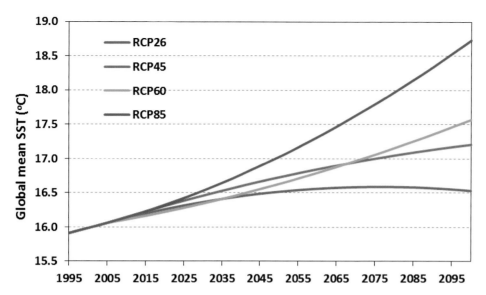

Figure 5.2. Global curves for sea surface temperature calculated from representative concentration pathways 2.6, 4.5, 6.0, and 8.5.

For a given climate variable V, its anomaly ΔV^* for a particular grid cell (i), month (j), and year or period (y) under an RCP is

$$\Delta V_{yij}^* = \Delta GV_y \cdot \Delta V_{ij}', \qquad (5.1)$$

where ΔGV is the annual global mean change of variable V.

The local change pattern value ($\Delta V_{ij}'$) was calculated from the GCM simulation anomaly (ΔV_{yij}), using linear least squares regression—that is, the slope of the fitted linear line, as

$$\Delta V_{ij}' = \frac{\sum_{y-1}^{m} \Delta GV_y \cdot \Delta V_{yij}}{\sum_{y-1}^{m} (\Delta GV_y)^2}, \qquad (5.2)$$

where m is the number of future sample periods used, with a five-year average as a period, and 19 periods from 2006 to 2100.

The global curves and local change patterns were made operational in ArcGIS through an add-in (the SimCLIM for ArcGIS/Marine toolbar) to analyze effects on ocean variables for projected changes in climate. For a selected variable, RCP, and projection year, equation (5.1) is applied to calculate the changes between the baseline period and projection year.

Baselines for the various variables were constructed using different data sources:

- Dissolved nitrate concentration at surface, dissolved oxygen concentration at surface, dissolved phosphate concentration at surface, and dissolved silicate concentration at surface used in the World Ocean Atlas (WOA) (National Oceanographic Data Center 2010).
- The SST baseline was constructed from the Hadley Centre Sea Ice and SST dataset (Met Office Hadley Centre for Climate Change 2009) using the mean of the 1981–2010 data.
- For the remaining variables (dissolved iron concentration at surface, net primary productivity at surface, pH at surface, and total alkalinity at surface), the ensemble mean of 1985–2005 GCM historical runs was used.

All the baseline and change patterns were spatially interpolated from the GCM original resolutions to a finer 0.25° × 0.25° grid using a bilinear interpolation method.

Not all the GCMs have ocean biogeochemical components, and not every model has all the RCP experiments. Moreover, there are sometimes data quality issues such as missing data. Table 5.2 lists the variables that are available per GCM.

The analysis focused on the differences between the baseline situation (2006) and climate-changed situation in 2050. The global atmosphere is currently close to 400 (ppm CO_2 equivalents, and concentrations of CO_2 and non-CO_2 gases are increasing at a rate that is of concern (Prinn 2013). To demonstrate maximum impacts, the RCP8.5 emission pathway was selected in the SimCLIM for ArcGIS/Marine tool, and a scenario was generated. For each scenario, the median value of the full set of GCMs was used (with the actual set varying per variable, depending on availability). All results focused on annual averages (from monthly data), except for SST which focused on the warmest month.

Results

Sea Surface Temperature

SST is relevant for warm-water corals. At certain levels, they can trigger a coral-bleaching event. Figure 5.3 shows the contour lines for SST for the baseline (black) and 2050 (red), for the warmest month, which can be a different month for different locations (e.g., the Northern versus the Southern Hemisphere). As the temperatures around 28°C to 31°C can trigger coral-bleaching episodes, only these contours are shown. A considerable expansion of the area at risk is clearly visible, with core coral zones reaching temperatures of over 31°C, thus endangering all warm-water corals.

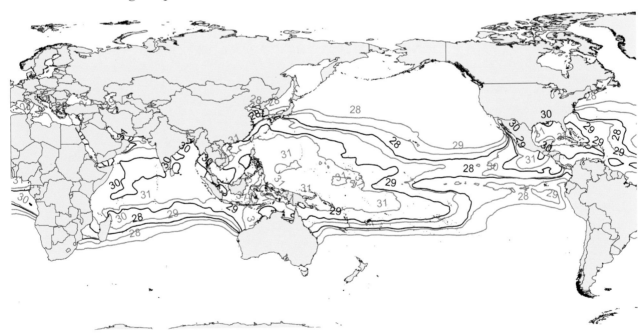

Figure 5.3. Sea surface temperature contours for 2006 (black) and 2050 (red) under representative concentration pathway 8.5. Data from CLIMsystems.

Net Primary Productivity of Carbon by Phytoplankton

The change in net primary production is presented in figure 5.4 as a percentage change (Δ%) from the baseline (units of gC/m³/day, see table 5.2). The bluish areas show a decrease (of up to -10%) in net primary production, while the yellow/orange/red areas indicate an increase (of up to 10%). The results show a decrease in primary production over much of the global ocean, with the exception of parts of the southern ocean and Arctic which have an increasing trend because of the strong increase in temperatures in these regions. Figure 5.5 shows the global average change in primary production.

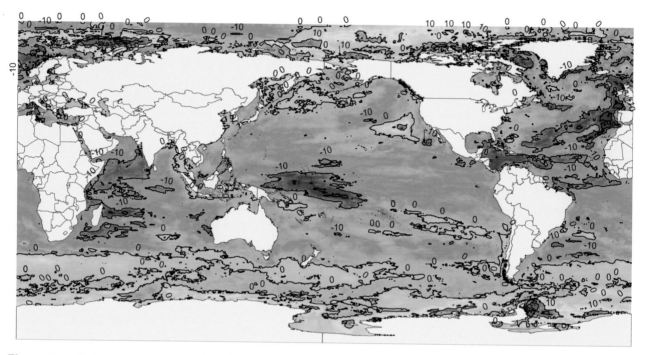

Figure 5.4. Color contour map showing percentage change in net primary production between 2006 and 2050 under representative concentration pathway 8.5 (largest decreases up to 10% shown in blue, largest increases up to 10% shown in red). Data from CLIMsystems.

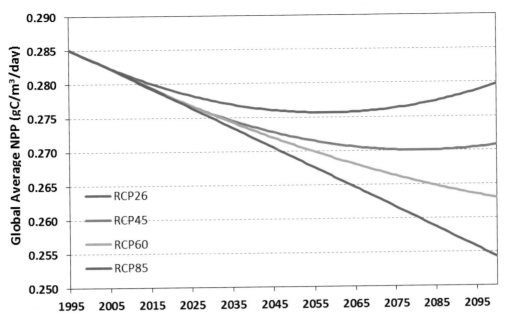

Figure 5.5. Global average change in primary production calculated from representative concentration pathways 2.6, 4.5, 6.0, and 8.5. Data from CLIMsystems.

Dissolved Oxygen Concentration at Surface

Figure 5.6 shows that changes in dissolved oxygen concentration (Δ% from the baseline) are small (less than 3%) and negative for all oceans. The strongest decrease (-5%) is in the red areas, which coincide with important fishing grounds, prompting change in management.

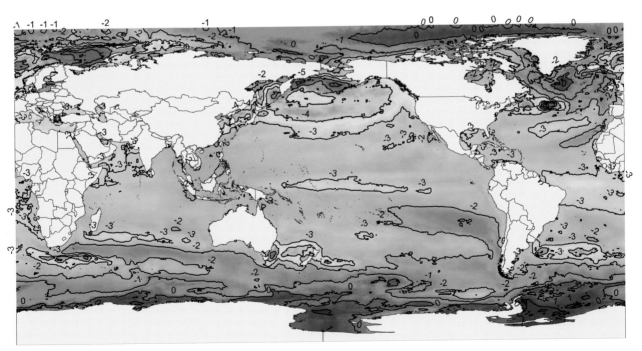

Figure 5.6. Color contour map showing percentage change in dissolved oxygen between 2006 and 2050 under representative concentration pathway 8.5 (largest decreases in red, no change or small increases in blue). Data from CLIMsystems.

Ocean pH and Total Alkalinity

Ocean acidification is a critical effect for coral reefs. Coral reefs are highly productive and biologically diverse ecosystems that are either showing signs of deterioration or undergoing community structure changes because of a host of anthropogenic and natural factors, such as bleaching, resource depletion, changing sedimentation rates and turbidity, eutrophication, cyclone damage, and natural climate variability such as El Niño Southern Oscillation. In addition to these environmental pressures, the ability of coral reefs to calcify, produce calcium carbonate ($CaCO_3$), and provide framework structures as habitat may also be adversely affected by the oceanic uptake of anthropogenic CO_2 (Sabine et al. 2004) and gradual ocean acidification.

Changes in temperature, oxygen content, and other ocean biogeochemical properties directly affect the ecophysiology of water-breathing organisms. Previous studies (Edwards and Richardson 2005; Perry et al. 2005; Dulvy et al. 2008; Cheung et al. 2009; Cheung et al. 2010; Genner et al. 2010) suggest that the most prominent biological responses are changes in distribution, phenology, and productivity. Both theory and empirical observations also support the hypothesis that warming and reduced oxygen will reduce the body size of ocean fishes. The assemblage-averaged maximum body weight is expected to shrink by 14–24% globally from 2000 to 2050 under a high-emission scenario. About half of this shrinkage is because of change in distribution and abundance, the remainder to changes in physiology. The tropical and intermediate latitudinal areas will be heavily impacted, with an average reduction of more than 20%.

Acidification-driven changes in carbonate chemistry can cause diverse physiological responses among organisms. Organisms vary in their response to acidification; negative effects are evident, particularly among calcifying (shell-building) species. The process of shell formation and maintenance in ocean organisms is vulnerable to acidification. Co-occurring environmental stressors can modify or exacerbate the effects of acidification.

Biological processes influence seawater chemistry: for example, photosynthesis and respiration can influence pH on a day and night basis and over longer periods. The processes also contribute to carbon cycling in the ocean.

Biological adaptation to ocean acidification conditions has been demonstrated for some species, but the current rate of acidification is unprecedented over the past three hundred million years (Hoenisch et al. 2012) and similar past events have been accompanied by major extinctions of ocean organisms.

Because pH is a nonlinear variable, the difference between the 2050 situation and baseline is presented in figure 5.7 in absolute units. Because of the fact that about one-quarter of the global carbon emissions dissolves in the oceans, it makes them slightly more acidic, decreasing the pH. There is also an impact from the increase in temperature, which is biggest in the Arctic area. The small shift of around -0.18 is enough to cause major problems for coral reefs.

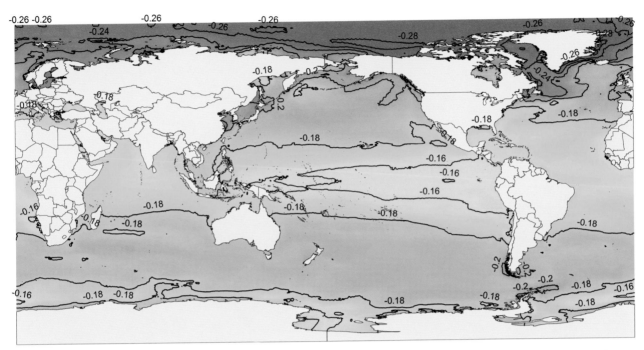

Figure 5.7. Color contour map showing change in pH between 2006 and 2050 under representative concentration pathway 8.5 (largest decreases in pH or more acidic conditions in red, no change or smallest decreases in pH in yellow). Data from CLIMsystems.

Another way to show the effects of ocean acidification is through the total alkalinity at the surface. Figure 5.8 shows that the Δ% in total alkalinity is very small, less than 0.4% (down in the blue areas, up in the purple) but still shows broad regional increases for areas such as the Gulf of Mexico and western part of the Northern Atlantic.

Figure 5.8. Color contour map showing percentage change in total alkalinity between 2006 and 2050 under representative concentration pathway 8.5 (largest increase in total alkalinity in purple, no change or smallest decrease in dark blue). Data from CLIMsystems.

Dissolved Nitrate Concentration at Surface

Nitrification is a central process in the nitrogen cycle that produces both the greenhouse gas (GHG) nitrous oxide and oxidized forms of nitrogen used by phytoplankton and other microorganisms. As anthropogenic CO_2 invades the ocean, pH-driven reductions in ammonia oxidation rates could fundamentally change how nitrogen is cycled and used by organisms in the sea. Figure 5.9 shows the Δ% from the baseline in nitrate concentrations. Because many areas have very low concentrations, small absolute changes translate into large and critical relative changes. With some exceptions (red areas), nitrate concentrations are projected to decrease across most oceans.

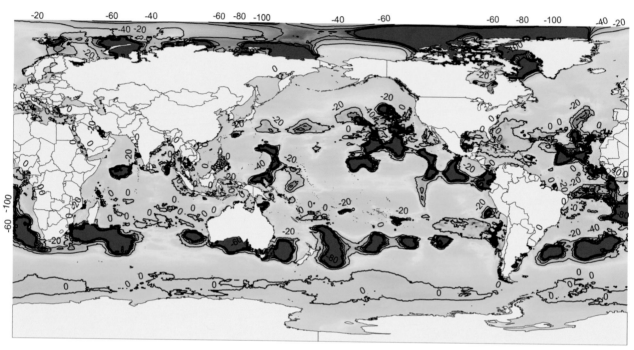

Figure 5.9. Color contour map showing percentage change in dissolved nitrate between 2006 and 2050 under representative concentration pathway 8.5 (largest decrease in dissolved nitrate in dark blue, no change or smallest increases in orange and red). Data from CLIMsystems.

Dissolved Phosphate, Silicate, and Iron at Surface

Figure 5.10 shows that the concentration of dissolved phosphate is decreasing everywhere, marginally (0–10%) in most oceans (red), with dispersed exceptions showing a decrease of up to 60% (blue). Dissolved phosphate and silicate are both important nutrients in the oceans, and thus can have an important impact on primary productivity, as well as the distribution of species and structure of ecosystems. In some parts of the ocean, they are more critical than dissolved nitrogen or dissolved iron as a limiting factor, and thus are becoming more widely recognized as studies of phosphate in the oceans expands (Paytan and McLaughlin 2006).

Iron, along with nitrogen, is one of the better understood elements in the ocean, and it is recognized as a limiting factor over wider areas than phosphate. Figure 5.11 shows dissolved iron concentration at the surface remains about the same (blue) or at an increase of up to 30% in specific hot spots (red). Conversely, figure 5.12 shows an overall decrease in silicate concentrations of 0–20%, but in some areas (red) the decrease is 50%.

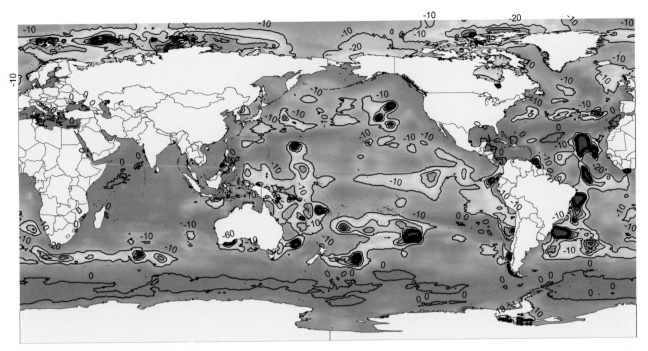

Figure 5.10. Color contour map showing percentage change in dissolved phosphate between 2006 and 2050 under representative concentration pathway 8.5 (decreases in dissolved phosphate of 0–10% in red, largest decreases up to 60% in blue). Data from CLIMsystems.

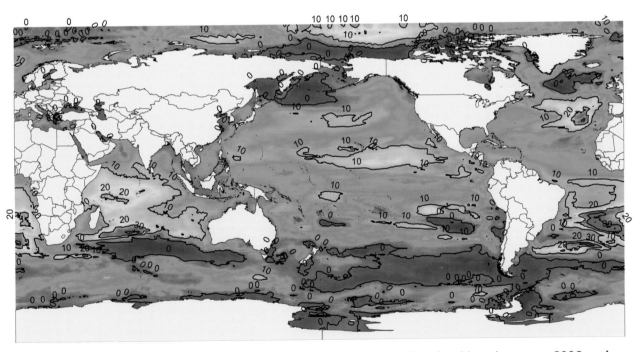

Figure 5.11. Color contour map showing percentage change in dissolved iron between 2006 and 2050 under representative concentration pathway 8.5. Red areas show hot spots of the greatest increases, up to 30%. Data from CLIMsystems.

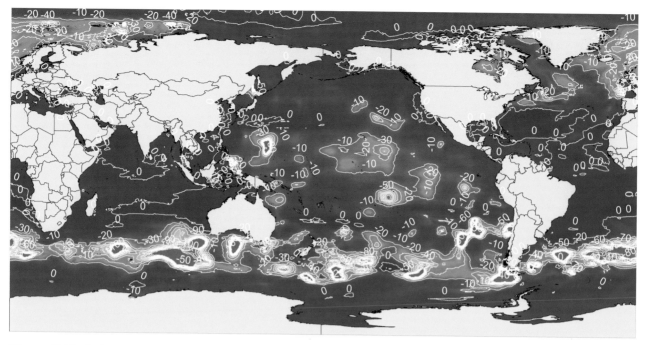

Figure 5.12. **Color contour map showing percentage change in dissolved silicate between 2006 and 2050 under representative concentration pathway 8.5. Red areas show hot spots of the greatest decreases, up to 50%.** Data from CLIMsystems.

Discussion and Conclusion

The rapid ecological shifts that might occur in the world's oceans as a result of climate change present major challenges for managers and policy makers. Understanding and reducing risk exposure will become increasingly important as conditions change and the likelihood of major ecological shifts increases. These changes will decrease the relevance of current models and practices for managing ecological resources and fisheries stocks, leading the management of many ocean resources into "uncharted waters." Nonetheless, management strategies that reduce the impact of local stressors while maintaining ecological resilience will play an increasingly important role as the climate changes. Actions that reduce the flow of nutrients and sediments from coastal catchments, for example, as well as those that reduce activities such as the deforestation of mangroves and overfishing of key ecological species (e.g., herbivores), will become increasingly important as the impacts of climate change mount.

It is desirable to help decision-makers, especially in developing countries, better understand and assess the risks posed by climate change, and better design strategies to adapt their fishing sectors to climate change. The provision of scientifically robust climate change data in accessible formats (such as the results presented in this study) is a beginning. Training and capacity building is also critical, not only in the application of the data and tools but also in the interpretation of results. This stage is often overlooked, as it involves deeper understanding of the integration of data with management and policy and can often involve a myriad of stakeholders with different perspectives. The information derived from accessible data and tools can greater enhance the latter process. It is especially desirable to have the capacity to integrate analysis of, for example, dissolved iron, nitrogen, and phosphorus with SST or fish stock movements. Here, critical zones can be identified where one or more variables, when combined, could show heightened risk or threatened resilience.

Another objective should be to develop global estimates of adaptation costs in, for example, the fisheries sector of countries to inform the international community's efforts; provide access to adequate, predictable, and sustainable support; and provide new and additional resources to help the most vulnerable developing countries meet adaptation costs. Results such as those presented in this study may be used to guide the prioritization of such adaptation options. Adaptation is here understood to mean any action taken to reduce the risk posed by the impact of climate change in a given sector of the economy (for example, fisheries). Adaptation cost is then the cost of taking such action. Yet the impact on ocean changes will extend well beyond fisheries and food security. Increasingly, tourism sectors represent important foreign-exchange earning engines for developing and developed economies. A sustained coastal ecosystem and its intrinsic values, and how they are threatened by loss of species diversity, can also be considered.

Ultimately, the ocean and terrestrial boundary offers countless local economic systems that have historically exploited their juxtaposition. These zones now extend to the farthest extent of a nation's Exclusive Economic Zone and beyond. Climate impacts are likely to affect both systems but in different forms and at variables rates. However, this in no way diminishes the need to see them holistically and visualize them as under threat from a changing climate. The transformation of voluminous climate data to knowledge products through the application of scientific methods and rapidly evolving spatial and analytical tools is more than important. It is critical.

Acknowledgments

The modeling for this work was carried out using SimCLIM 2013 and the SimCLIM for ArcGIS/Marine add-in for ArcGIS for Desktop. Both tools can be tested in a free trial by registering at http://climsystems.com. We thank two anonymous reviewers for their comments and suggestions.

References

Andrews, O. D., N. L. Bindhoff, P. R. Halloran, T. Ilyina, and C. Le Quéré. 2013. "Detecting an External Influence on Recent Changes in Oceanic Oxygen Using an Optimal Fingerprinting Method." *Biosciences* 10:1799–1813.

Bopp, L., L. Resplandy, J. C. Orr, S. C. Doney, J. P. Dunne, M. Gehlen, and M. Vichi. 2013. "Multiple Stressors of Ocean Ecosystems in the 21st Century: Projections with CMIP5 Models." *Biogeosciences Discussions* 10 (2): 6225–45.

Ciais, P., C. Sabine, G. Bala, L. Bopp, V. Brovkin, J. Canadell, A. Chhabra, R. DeFries, J. Galloway, M. Heimann, C. Jones, C. Le Quéré, R. B. Myneni, S. Piao, and P. Thornton. 2013. "Carbon and Other Biogeochemical Cycles." In *Climate Change 2013: The Physical Science Basis. Contribution of Working Group I to the Fifth Assessment Report of the Intergovernmental Panel on Climate Change*, edited by T. F. Stocker, D. Qin, G.-K. Plattner, M. Tignor, S. K. Allen, J. Boschung, A. Nauels, Y. Xia, V. Bex, and P. M. Midgley. Cambridge: Cambridge University Press.

Cheung, W. W. L., V. W. Y. Lam, J. L. Sarmiento, K. Kearny, R. Watson, and D. Pauly. 2009. "Projecting Global Marine Biodiversity Impacts under Climate Change Scenarios." *Fish and Fisheries* 10:235–51.

Cheung, W. W. L., V. W. Y. Lam, J. L. Sarmiento, K. Kearny, R. Watson, D. Zeller, and D. Pauly. 2010. "Large-Scale Redistribution of Maximum Fisheries Catch Potential in the Global Ocean under Climate Change." *Global Change Biology* 16:24–35.

Collins, M., R. Knutti, J. Arblaster, J. L. Dufresne, T. Fichefet, P. Friedlingstein, X. Gao, W. J. Gutowski, T. Johns, G. Krinner, M. Shongwe, C. Tebaldi, W. J. Weaver, and M. Wehner. 2013. "Long-Term Climate Change: Projections, Commitments and Irreversibility." In *Climate Change 2013: The Physical Science Basis. Contribution of Working Group I to the Fifth Assessment Report of the Intergovernmental Panel on Climate Change*, edited by W. J. Stocker, D. Qin, G. K. Plattner, M. Tignor, S. K. Allen, J. Boschung, A. Nauels, Y. Xia, V. Bex, and P. M. Midgley. Cambridge: Cambridge University Press.

Doney, S. C., M. Ruckelshaus, J. E. Duffy, J. P. Barry, F. Chan, C. A. English, H. M. Galindo, J. M. Grebmeier, A. B. Hollowed, N. Knowlton, J. Polovina, N. N. Rabalais, W. J. Sydeman, and L. D. Talley. 2012. "Climate Change Impacts on Marine Ecosystems." *Annual Review of Marine Sciences* 4:11–37.

Doney, S. C., B. Tilbrook, S. Roy, N. Metzl, C. Le Quéré, M. Hood, R. A. Feely, and D. Bakker. 2009. "Surface-Ocean CO_2 Variability and Vulnerability." *Deep Sea Research II*. doi:10.1016/J.dsr2.2008.12.016.

Dulvy, N. K., S. I. Rogers, S. Jennings, V. Stelenmueller, S. R. Dye, and H. R. Skjoldal. 2008. "Climate Change and Deepening of the North Sea Fish Assemblage: A Biotic Indicator of Warming Seas." *Journal of Applied Ecology* 45:1029–39.

Edwards, M., and A. J. Richardson. 2005. "Impact of Climate Change on Marine Pelagic Phenology and Trophic Mismatch." *Nature* 430:881–84.

Fennel, W. 2009. "Parameterizations of Truncated Food Web Models from the Perspective of an End-to-End Model Approach." *Journal of Marine Systems* 76 (1): 171–85.

Genner, M. J., N. C. Halliday, S. D. Simpson, A. J. Southward, S. J. Hawkins, and D. W. Sims. 2010. "Temperature-Driven Phenological Changes within a Marine Larval Fish Assemblage." *Journal of Plankton Research* 32:699–708.

Gettelman, A., J. E. Kay, and J. T. Fasullo. 2012. "Spatial Decomposition of Climate Feedbacks in the Community Earth System Model." *Journal of Climate*. doi: 10.1175/JCLI-D-12-00497.1.

Gruber, N. 2011. "Warming Up, Turning Sour, Losing Breath: Ocean Biogeochemistry under Global Change." *Philosophical Transactions of the Royal Society A: Mathematical, Physical and Engineering Sciences* 369 (1943): 1980–96.

Hoegh-Guldberg, O., and J. F. Bruno. 2010. "The Impact of Climate Change on the World's Marine Ecosystems." *Science* 328 (5985): 1523–28.

Hoenisch, B., A. Ridgwell, D. N. Schmidt, E. Thomas, S. J. Gibbs, A. Sluijs, R. Zeebe, L. Kump, R. C. Martindale, S. E. Greene, W. Kiessling, J. Ries, J. C. Zachos, L. Royer, S. Barker, T. M. Marchitto Jr., R. Moyer, C. Pelejero, P. Ziveri, G. L. Foster, and B. Williams. 2012. "The Geological Record of Ocean Acidification." *Science* 335:1058–63.

Intergovernmental Panel on Climate Change. 2013. "Summary for Policymakers." *In Climate Change 2013: The Physical Science Basis. Contribution of Working Group I to the Fifth Assessment Report of the Intergovernmental Panel on Climate Change*, edited by T. F. Stocker, D. Qin, G.-K. Plattner, M. Tignor, S. K. Allen, J. Boschung, A. Nauels, Y. Xia, V. Bex, and P. M. Midgley. Cambridge: Cambridge University Press.

Le Quéré, C., S. P. Harrison, I. C., Prentice, E. T. Buitenhuis, O. Aumont, L. Bopp, and D. Wolf-Gladrow. 2005. "Ecosystem Dynamics Based on Plankton Functional Types for Global Ocean Biogeochemistry Models." *Global Change Biology* 11 (11): 2016–40.

Met Office Hadley Centre for Climate Change. 2009. Met Office Hadley Centre Observations Datasets. http://www.metoffice.gov.uk/hadobs/hadisst/. Last accessed June 15, 2014.

Mora, C., C. L. Wei, A. Rollo, T. Amaro, A. R. Baco, D. Billett, L. Bopp, Q. Chen, M. Collier, R. Danovaro, A. J. Gooday, B. M. Grupe, P. R. Halloran, J. Ingels, D. O. B. Jones, L. A. Levin, H. Nakano, K. Norling, E. Ramirez-Llodra, M. Rex, H. A. Ruhl, C. R. Smith, A. K. Sweetman, A. R. Thurber, J. F. Tjiputra, P. Usseglio, L. Watling, T. Wu, and M. Yasuhara. 2013. "Biotic and Human Vulnerability to Projected Changes in Ocean Biogeochemistry over the 21st Century." *PLoS Biology* 11 (10): e1001682. doi:10.1371/journal.pbio.1001682.

Moss, M., J. A. Edmonds, K. A. Hibbard, M. R. Manning, S. K. Rose, D. P. van Vuuren, T. R. Carter, S. Emori, M. Kainuma, T. Kram, G. A. Meehl, J. F. B. Mitchell, N. Nakicenovic, K. Riahi, S. J. Smith, R. J. Stouffer, A. M. Thomson, J. P. Weuant, and T. J. Wilbanks. 2010. "The Next Generation of Scenarios for Climate Change Research and Assessment." *Nature.* doi:10.1038/nature08823.

National Oceanographic Data Center of the National Oceanic and Atmospheric Administration. 2010. World Ocean Database 2009. http://www.nodc.noaa.gov/OC5/WOA09/ pubwoa09.html. Last accessed June 15, 2014.

Paytan, A., and K. McLaughlin. 2006. "The Oceanic Phosphorus Cycle." *Chemical Reviews* 2007 (107): 563–76.

Perry, A. L., P. J. Low, J. R. Ellis, and J. D. Reynolds. 2005. "Climate Change and Distribution Shifts in Marine Fishes." *Science* 308:1912–15.

Prinn, R. 2013. 400 ppm CO2? Add Other GHGs, and it's Equivalent to 478 ppm. *http://oceans.mit.edu/featured -stories/5-questions-mits-ron-prinn-400-ppm-threshold.* Last accessed December 12, 2013.

Rogelj, J., M. Meinshausen, and R. Knutti. 2012. "Global Warming under Old and New Scenarios Using IPCC Climate Sensitivity Range Estimates." *Nature Climate Change.* doi:10.1038/NCLIMATE1385.

Sabine, C. L., R. A. Feely, N. Gruber, R. M. Key, K. Lee, J. L. Bullister, R. Wanninkhof, C. S. Wong, D. W. R. Wallace, B. Tilbrook, F. J. Millero, T. H. Peng, A. Kozyr, T. Ono, and A. F. Rios. 2004. "The Oceanic Sink for Anthropogenic CO_2." *Science* 305:367–71.

Taylor, K. E., R. J. Stouffer, and G. A. Meehl. 2012. "An Overview of CMIP5 and the Experiment Design." *Bulletin of the American Meteorology Society* 93:485–98.

Van Vuuren, D., M. den Elzen, P. Lucas, B. Eickhout, B. Strengers, B. van Ruijven, S. Wonink, and R. van Houdt. 2007. "Stabilizing Greenhouse Gas Concentrations at Low Levels: An Assessment of Reduction Strategies and Costs." *Climatic Change.* doi:10.1007/s/10584-006-9172-9.

World Climate Research Programme. 2011. Guide to CMIP5. http://cmip-pcmdi.llnl.gov/cmip5/guide_to_cmip5 .html. Last accessed June 15, 2014.

Supplemental Resources

Dawn J. Wright, ed.; 2015; *Ocean Solutions, Earth Solutions*; http://dx.doi.org/10.17128/9781589483651_d

For the digital content for this chapter, go to the links below, or go to the Esri Press "Book Resources" webpage at esripress.esri.com/bookresources. Then, in the list of Esri Press books, click *Ocean Solutions, Earth Solutions*. On the *Ocean Solutions, Earth Solutions* resource page, click the chapter 5 link to access that webpage and the links to the digital content.

SimCLIM for ArcGIS/Marine

- CLIMsystems/SimCLIM for ArcGIS/Marine, at http://www.climsystems.com /simclimarcgis/marine/

- For additional information on the SimCLIM ArcGIS/Marine add-in and answers to frequently asked questions, see the FAQ, at http://www.climsystems.com/simclimarcgis /marine/faq/

This FAQ is also available on the book resource page.

Downloadable Map Package

To download the digital content explained here, go to the book resource page.

- SimCLIM for ArcGIS/Marine map package

For all the variables presented in this chapter, the map package contains both the baseline and climate-changed situation for 2050 under the assumption of the RCP8.5 emission pathway. This allows for further analysis of the climate change projection as well as usage of the variables in modeling dependent processes and variables. There is no restriction for its usage other than a proper reference to CLIMsystems. All layers have their relevant properties set and use the "cubic convolution (for continuous data)" resampling during display.

CHAPTER 6

A GIS Tool to Compute a Pollutant Exposure Index for the Southern California Bight

Rebecca A. Schaffner, Steven J. Steinberg, and Kenneth C. Schiff

Abstract

Southern California marine ecosystems face a variety of threats to their integrity and health as a result of their proximity to urbanized areas. These threats may be direct, through fishing and resource extraction, or indirect, through exposure to anthropogenic pollutants. Two primary sources of anthropogenic pollutants are treated wastewater, released by publicly owned treatment works through ocean outfalls, and freshwater runoff contained in urban river plumes. We developed a geospatial tool using ArcGIS software to calculate a pollutant exposure index for the Southern California Bight. The pollutant exposure index quantifies long-term exposure to potentially harmful pollutants emanating from these two sources. Recent studies on the dispersal of plumes have resulted in high-quality spatial datasets that predict plume occurrence frequencies as point grids around publicly owned treatment works and river mouths throughout the region. We multiplied the plume frequency values with data on average annual discharge rates, initial dilution factors, and concentration of chemicals in discharges to calculate total exposure to pollutants at each location. Using this approach, we developed maps of the distribution of three important plume constituents: dissolved inorganic nitrogen in the form of nitrate and nitrite, total suspended solids, and copper. A series of Python scripts was created to facilitate geoprocessing of the exposure data and calculate the final pollutant exposure index raster, including (1) creating exposure rasters for each pollutant and source using inverse distance weighting, (2) summing publicly owned treatment works and river plume exposure rasters for each pollutant and normalizing each raster to the maximum exposure value, and (3) creating the pollutant exposure index by summing pollutant exposure rasters and normalizing again to provide values ranging from zero to one. The resulting georeferenced pollutant exposure index raster may be used with other spatial data to examine relative pollution risk for any area of interest within the mapped region. The pollutant exposure index will be incorporated into an ongoing study to examine relative risks posed to marine habitats by pollutants and pressure from fishing.

Dawn J. Wright, ed.; 2015; *Ocean Solutions, Earth Solutions*; http://dx.doi.org/10.17128/9781589483651

Introduction

The Southern California Bight (SCB) is an oceanographically defined region that extends from Point Conception in Santa Barbara County, California, to Cabo Colnett, Mexico, including the Channel Islands (figure 6.1). The SCB is a valuable natural resource that contributes to the regional economy and enhances quality of life for those who work in, live in, or visit the area. Human uses of the coastline and ocean waters of the SCB include recreation, tourism, aesthetic enjoyment, sport and commercial fishing, coastal development, and industry. For example, ocean-dependent tourism contributed approximately $41 billion to the economies of coastal communities surrounding the SCB and supported over eight hundred thousand jobs in 2007 (National Ocean Economics Program 2008).

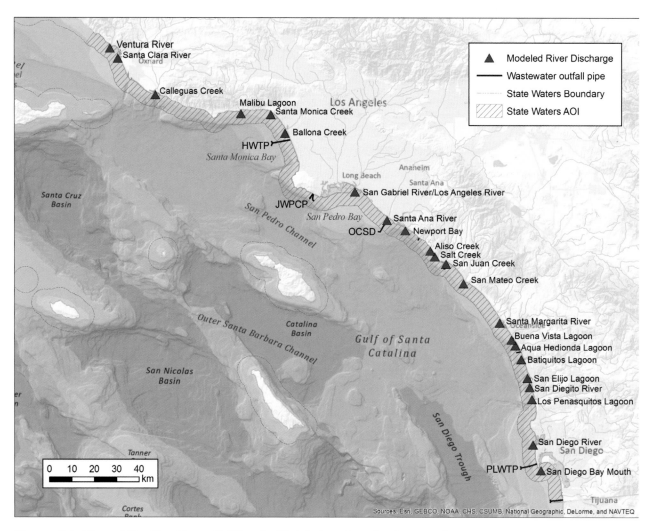

Figure 6.1. Map of the Southern California Bight showing the four major wastewater treatment plant outfalls: City of Los Angeles Hyperion Water Treatment Plant (HWTP), Los Angeles County Sanitation District Joint Water Pollution Control Plant (JWPCP), Orange County Sanitation District (OCSD), City of San Diego Point Loma Water Treatment Plant (PLWTP), and major river mouths draining to the Southern California Bight. Also shown are the state offshore jurisdictional boundary and the boundary encompassing the plumes analyzed in this study. Data sources: Esri, GEBCO, DeLorme, NaturalVue | Sources: Esri, GEBCO, NOAA, CHS, CSUMB, National Geographic, DeLorme, NAVTEQ.

The Southern California Bight a Valuable Resource

The SCB is a unique and highly diverse ecological resource (Schiff et al. 2000). For example, subtidal rocky reefs that support giant kelp *(Macrocystis pyrifera)* are among the most productive habitats in the world. The SCB is a primary stop along the Pacific Flyway, the major migration route for seabirds from North to South America. It also hosts a number of threatened and endangered species, including the blue *(Balaenoptera musculus)* and humpback *(Megaptera novaeangliae)* whales, California least tern *(Sterna antillarum browni),* and least Bell's vireo *(Vireo bellii pusillus).* Altogether, the SCB is home to over five thousand species of invertebrates, 480 species of fish, and 195 species of marine birds (Dailey et al. 1993).

Human Influence in the Region

Throughout the 20th century, significant population growth propelled the coastal community along the SCB from under two hundred thousand in 1900 to over 17 million (US Census Bureau 2010), making the region the largest metropolitan center in the United States. As a result of rapid urbanization, marine resources in the SCB have been placed under extreme pressure, including loss of habitat and discharge of pollutants.

The SCB has a variety of pollutant sources, including publicly owned treatment works (POTWs), urban and agricultural runoff, industrial facilities, power-generating facilities, boating and shipping, dredged material disposal, and atmospheric deposition, among others (Schiff et al. 2000). Historically, the largest of these sources has been POTWs. With increased treatment, pretreatment, and reclamation, both total discharge rates and pollutant inputs from POTWs have decreased over the last four decades, greater than 90% for some individual pollutant types. However, surface runoff generated by storm events that washes off the region's highly developed watersheds delivers large pulses of potential pollutants that enter the ocean without any treatment. Flows and pollutant inputs from surface runoff have generally increased over this same time period as impervious surfaces (e.g., roads and rooftops) have hardened coastal watersheds. As a result, pollutant inputs from runoff now rival those of POTWs.

Of the 17 POTWs in the SCB, four comprise the vast majority of the discharge flow and pollutant inputs (Schiff et al. 2000). These four POTWs are amongst the largest in the United States and include the City of Los Angeles, County of Los Angeles, County of Orange, and City of San Diego (see figure 6.1), which cumulatively discharge over eight hundred million gallons of treated wastewater per day. These inputs remain relatively stable as flows are metered and treatment processes engineered to maintain a consistent effluent quality, but there has been a gradual downward trend in total discharge rates because of improvements in treatment and wastewater reclamation projects (Lyon and Sutula 2011).

The SCB has 23 major watersheds (see figure 6.1). Stormwater discharges are highly variable, with infrequent but intense storms capable of increasing streamflow three to four orders of magnitude in less than one hour (Schiff and Tiefenthaler 2011). Intensive storm water discharge after rainstorms can produce plumes that are easily distinguished from ambient marine waters by their high concentration of suspended matter that changes the color of the ocean surface (Mertes et al. 1998; Sathyendranath 2000; Mertes and Warrick 2001). The movement and persistence of plumes may be influenced by a variety of factors, such as wind and ocean circulation, coastline shape, and bathymetric variability, which may cause resuspension of sediments, as well as the volume, timing, and intensity of storm water discharge (Nezlin et al. 2005).

Study Objective

To date, no study has effectively used a geospatial approach to assess threats to water quality throughout the full extent of the SCB. Prior studies tended to focus on individual sources of pollution such as a single POTW outfall or individual watershed (Bay et al. 2003; Warrick et al. 2004; Ahn et al. 2005). Although these localized studies are valuable, they leave numerous gaps in both spatial and temporal coverage of the SCB, making it impossible to develop a comprehensive understanding of conditions and potential water quality threats. Furthermore, because each study used independent data and methods to assess water quality at a specific time and location, study results are not directly comparable. This variability results in a fragmented understanding of conditions in the SCB, making it difficult to understand how water quality conditions are changing over time as a result of management decisions or regulatory actions intended to protect and improve the water quality of the SCB.

To develop an accurate map of water quality conditions of the entire SCB requires data that is collected in a consistent manner and at an appropriate spatial scale. Such data has only recently become available through the efforts of coordinated regional monitoring programs such as those led by the Southern California Coastal Water Research Project (SCCWRP) (2014) and datasets obtained from regional efforts such as the Southern California Coastal Ocean Observing System (SCCOOS) or satellite imagery data collected by the National Oceanic and Atmospheric Administration (NOAA) and National Aeronautics and Space Administration (NASA).

Our objective was to develop geospatial tools in ArcGIS to model ocean exposure to land-based sources of pollution. Combining sources of newly acquired spatial data for the SCB with information on locations of pollution sources and inputs, we developed a series of Python scripts to calculate a composite pollutant exposure index (PEI) based on pollutant inputs from both POTWs and rivers. Developing tools such as the PEI provides several important benefits. The ArcGIS scripts provide a mechanism to assess pollution impacts throughout the SCB in a consistent and repeatable manner. Second, these scripts provide a documented methodology that may be readily adapted by other researchers. This allows for adjustments that may be useful to incorporate data for additional pollutants or apply these methods in other locations.

For purposes of this study, we focused development of the PEI on three pollutants: dissolved inorganic nitrogen (DIN) in the form of nitrate and nitrite, total suspended solids (TSS), and toxic trace metals (copper). Selection of pollutants for development and testing of the tool was based on the following characteristics:

- The pollutant has known impacts on marine ecosystem health, either through direct toxicity or secondary effects by promoting growth of nuisance algal blooms.
- Datasets are available on the pollutant for both river and wastewater discharge sources.
- The pollutant is discharged frequently and in large enough quantities that it is likely to have significant and ongoing effects on marine ecosystems.
- The pollutant represents a range of sources such as urban, agricultural, or industrial activities.

Nitrate is a critical plant nutrient necessary for the growth of phytoplankton, sea grasses, kelp, and other macroalgae. However, in large concentrations, nitrate can promote the growth of harmful algal species and cause phytoplankton blooms (Anderson et al. 2008). Recent studies have found that treated effluent may have nitrate concentrations equivalent to natural sources such as upwelled seawater (Howard et al. 2014).

TSS can negatively impact nearshore subtidal communities, reefs, and kelp beds through reduction of light availability and, in large quantities, by smothering benthic organisms. Episodic riverine TSS inputs can be quite large in some regions such as the Santa Clara River (up to 5.5×10^7 kg/yr). Additionally, other toxic chemicals adsorb onto the surface of particulate matter, creating a large-scale transport mechanism between the water column and benthic substrates (Hart 1982).

Copper is a trace metal that can cause both acute (lethal) and chronic (impaired growth, reproduction) effects in marine organisms. Copper is a by-product of some industrial activities and a common contaminant found in storm water. Its ubiquity in storm water is because of its use in vehicle brake pads, pesticides, and building materials and from atmospheric deposition (Davis et al. 2001). Copper toxicity to marine plants and invertebrates is so prevalent that it is often used as a biocide in paints used to coat the bottoms of boats and docks. Copper does not transform easily and can settle into ocean sediments, exerting long-term effects (Long et al. 1995).

Methods

We used a regional-scale risk-based approach to creating the PEI. The risk was estimated as a function of pollutant concentration, magnitude, and duration of exposure. Therefore, the primary factors for the index were pollutant loading and pollutant plume frequency. Pollutant loading was estimated for the SCB's two largest sources, POTWs and rivers. Spatial distribution of exposure duration was acquired from multiple sources that utilize advances in plume detection technology. Table 6.1 lists the sources of data used for the water quality index (WQI). Many of the geoprocessing tasks were automated using the ArcGIS ModelBuilder application and Python scripts.

Table 6.1. **Sources of Spatial Data and Pollutant Loading Data for POTWs and River Plumes**

Pollutant Source	Loading Data	Frequency Data	References
Wastewater treatment plants	Discharge monitoring reports; Bight 2008 water quality study	CDOM plume detection method	Lyon and Sutula 2011 Howard et al. 2014 Nezlin et al., in prep.
River plumes	Riverine loading model	HFR-based surface circulation models	Sengupta et al. 2013 Rogowski et al. 2014

Riverine Pollutant Loading

A rational model was developed that uses precipitation data, watershed size, and land cover to predict daily loads of chemicals from watersheds (Ackerman and Schiff 2003; Sengupta et al. 2013). For 23 rivers, we calculated average annual loads of DIN, TSS, and copper based on 11 years of data (2000–2010). Watershed sizes for these rivers range from 25 to 4486 km² and comprise a wide range of predominant land uses, including urbanized (i.e., residential, commercial, and industrial areas), agricultural lands, and open space (i.e., largely undisturbed areas of wetlands, forest, or scrub/shrub vegetation [Sengupta et al. 2013]). The model ignores base flows that result from inland sewage treatment discharges and general dry-weather runoff. For the purposes of this study, we assumed that base river flows are a minor source of pollutants whose effects do not radiate far from river mouths.

Each river mouth was treated as a single point source. Pollutant loads were multiplied by a dilution factor of 1:100 to account for estuarine mixing processes and loss of the pollutant from the plume as it entered the ocean. This factor was selected based on studies showing that during storm events, only about 1% of the TSS originating from riverine inputs is found in the surface plume offshore of river mouths. The majority of sediment settles out in the estuarine mixing zone and within 1 km of the river mouth (Mertes and Warrick 2001; Warrick et al. 2004).

Riverine Plume Mapping

Southern California river plumes have been mapped using several different methods, primarily satellite imagery in combination with in situ sampling of plume tracers (Otero and Siegel 2004; Nezlin et al. 2005; Warrick et al. 2007; Nezlin et al. 2008; Reifel et al. 2009). These methods provide empirical data on plume locations and extents but have several limitations. Generally, they focus on limited areas and individual storm events. Cloud cover can interfere with accurate imaging, preventing visualization of the plume during inclement weather when plumes are forming and dispersing. Remotely sensed plume indicators tend to correlate with the presence of suspended solids in surface waters but may not always agree with concentrations of other dissolved pollutants. One alternative method of mapping plume probability is the use of high-frequency radar (HFR) to model the dispersion of surface water tracers seeded into regions of river discharge. HFR-derived surface currents over a given temporal period are input into a random walk model (Kim et al. 2009; Rogowski et al. 2014). Rogowski et al. (2014) used HFR surface current data to model the probability of plume exposure for 20 river discharges in Southern California, based on two years of hourly surface current data. In the Rogowski et al. model, 50 water tracers were released hourly at each source location (1 km offshore) and independently tracked for three days. A probability density function was used to compute the cumulative sum of water tracers that advected into predefined model grids surrounding each river mouth source location. These grids vary in point density and may overlap (figure 6.2).

Final river plume probability exposure maps used many realizations (with the number varying among the grids) of hourly water tracer trajectory estimates to determine statistical convergence over various time periods. We converted the model output for each river plume into a point feature class representing the centroid of each grid cell, with a river plume exposure probability (RPE) attribute field. The RPE was calculated for each cell based on the maximum number of particles observed at a single point within the grids. Probabilities ranged from 0 to 1, and to reduce noise and eliminate probable nonplume regions of the grids, only cells with a value of ≥ 0.01 were included in the spatial analysis. This value was also used to identify the maximum extent of each plume and create individual plume boundary polygons for clipping interpolated raster layers.

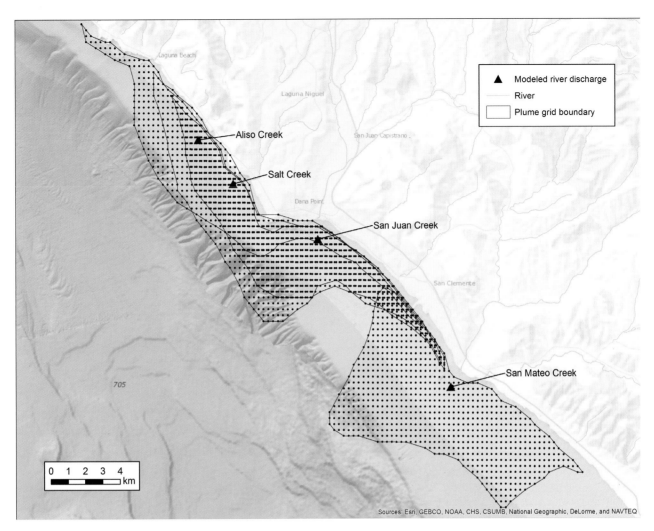

Figure 6.2. Four overlapping river plume grids in Orange County, California. Data from the authors, Esri, GEBCO, NOAA, CHS, CSUMB, National Geographic, DeLorme, NAVTEQ.

To calculate the exposure for each cell, the riverine loading value for each constituent was multiplied by the dilution factor and RPE in equation (6.1):

Exposure = Load × Dilution × Frequency.   (6.1)

We developed Python scripts to automate most of the subsequent geoprocessing tasks (figure 6.3). For each river plume and parameter, exposure values were interpolated across each grid using inverse distance weighting (IDW), an exact interpolator which assumes that the influence of a known point on an unknown location varies inversely with the distance between the points. It is a simple interpolator that ignores all factors other than distance between points. We selected IDW because sufficient datasets are not available to describe the complex variables affecting plume distributions at the scale of the study region and time period necessary to parameterize more sophisticated interpolators such as kriging. The 20 individual river plume exposure grids were interpolated to rasters using a power of 2, a resolution of 250 m², and a maximum search radius of 10 km. The interpolated plumes were clipped by the corresponding plume boundary polygon, "NoData" values were converted to 0, and the spatial extent of the raster was set to the maximum extent of all plumes (river and POTW) to simplify integration of the various layers. The 23 plume rasters were then summed to create a single layer representing the bightwide exposure for each chemical, and a plume boundary polygon that encompassed all 23 grids was created. Finally, the Zonal Statistics tool in ArcMap was used to calculate spatial statistics for the resulting pollutant exposure raster layers, including mean, median, and maximum values, using the bightwide plume boundary polygon as the zone. This polygon exactly encompasses all points in the plume grids and was created using the Aggregate Points tool in the Generalization section of the Cartography toolbox in ArcMap (figure 6.3).

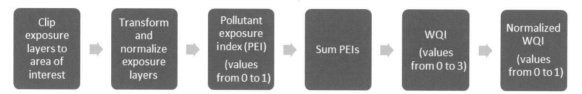

Figure 6.3. **Workflow showing the geoprocessing steps used in the interpolation of pollutant exposure index rasters and calculation of the pollutant exposure index raster.**

Sewage Discharge Pollutant Loading

Monthly discharge flow rates and concentrations of copper and TSS from four major Southern California sewage treatment plants were obtained from Discharge Monitoring Reports that each agency is required to submit to the US Environmental Protection Agency (EPA) (Lyon and Sutula 2011). DIN concentrations in effluent were acquired from a recent study that directly measured nutrients in treated wastewater discharge (Howard et al. 2014). The four POTWs used for this analysis—City of Los Angeles Hyperion Water Treatment Plant (HWTP), Los Angeles County Sanitation District Joint Water Pollution Control Plant (JWPCP), Orange County Sanitation District (OCSD), and the City of San Diego Point Loma Water Treatment Plant (PLWTP)—are shown in figure 6.1. The average annual load of each pollutant from each plant was calculated for the period 2003–9. An instantaneous dilution factor was applied to the loading value to account for immediate dilution of discharged effluent in seawater before dispersal occurs. This dilution factor is calculated individually for each outfall as part of the permitting process, based on empirical data comparing concentrations in the undiluted effluent with concentrations in the seawater immediately adjacent to the outfall (EPA 1985).

Sewage Discharge Plume Mapping

Wastewater outfall plumes are typically modeled by measuring plume tracers such as salinity, ammonia, and bacteria at stations surrounding the outfall. These approaches can be time consuming and costly, so alternative methods of detecting plumes are being assessed. Nezlin et al. (in preparation) used measurements of colored dissolved organic matter (CDOM) to rapidly measure and track plumes emanating from subsurface discharge outlets at the four large sewage treatment plants. Each POTW agency regularly samples water quality at a grid of stations surrounding their outfall pipes (figure 6.4). CDOM datasets were collected at each station on a quarterly basis from 2004 to 2010. For each station, the percentage of sampling events when a plume was observed (at any depth between the outlet and below the mixed layer) was calculated as wastewater plume exposure (WPE). The four individual sampling grids were combined into a single point feature class, and each point was attributed with its associated WPE value.

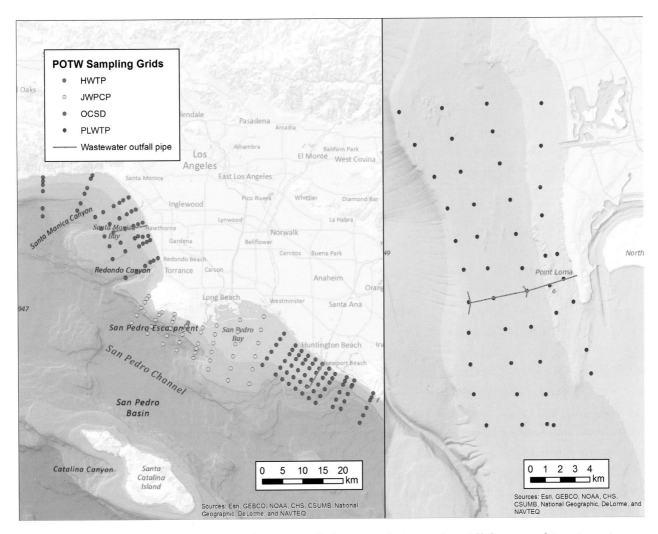

Figure 6.4. Grids of water quality stations sampled quarterly by each publicly owned treatment works (POTW) agency. The distance between stations is variable. The stations are the City of Los Angeles Hyperion Water Treatment Plant (HWTP), Los Angeles County Sanitation District Joint Water Pollution Control Plant (JWPCP), Orange County Sanitation District (OCSD), and City of San Diego Point Loma Water Treatment Plant (PLWTP). Data from Esri, GEBCO, NOAA, CHS, CSUMB, National Geographic, DeLorme, NAVTEQ.

We assumed that plume probability at a given grid point is directly equivalent to the dilution of the treated effluent as it is advected away from the source and that dilution of each chemical within the outfall plume also varies in proportion. Pollutant exposure at each point was calculated using equation (6.1). The exposure values were interpolated across the grid using IDW with a power of 2, a resolution of 250 m², and a maximum search radius of 10 km. A linear shoreline layer was used as a barrier to prevent interpolation of plumes over land areas.

Merging the Pollutant Exposure Layers and Creating the PEI

The geoprocessing steps described in the previous section resulted in six rasters, representing the distribution of each of the three pollutants from both rivers and POTWs within the SCB. For each pollutant, we summed the river plume and wastewater discharge plume rasters to create a single total exposure raster. Each parameter was then normalized to the maximum value before being integrated into the single index. This was a two-step process. First, to reduce skewness for each individual parameter, we calculated the cube root of each raster cell. The cube-root value of each raster cell was divided by the maximum value found in the grid, resulting in a PEI layer with values ranging from 0 to 1 for each chemical. These three normalized PEIs were summed to create a new raster with potential values ranging between 0 and 3. Finally, the cells in this raster were divided by the maximum value in the raster to create the normalized PEI raster, with continuous values ranging between 0 and 1 (see figure 6.3).

Sensitivity Analysis

We conducted a sensitivity analysis to assess variability in riverine pollutant loadings and exposure between wet and dry years. Representative wet and dry years in our 10-year dataset were based on long-term (63 years) rainfall records from three Southern California rain gauge stations (Los Angeles International Airport, Santa Ana Fire Station, and San Diego Lindbergh Field) downloaded from the NOAA National Climatic Data Center (2014). Total rainfall from 2007 (8.2 cm) was less than the 10th percentile of the long-term average. Total rainfall from 2010 (51.3 cm) was greater than the 90th percentile of the long-term average.

For 2007 and 2010, we calculated the total annual loads of each chemical and compared the resulting range of values to the 10-year average annual loading value for each parameter and watershed. Exposure layers were created for both 2007 and 2010 using equation (6.1) described previously, and the total spatial variation in loading was mapped and analyzed. To examine spatial bias, the 20 river plumes assessed were grouped into four regions: Ventura County, Santa Monica Bay (SMB), Orange County, and San Diego County.

Results

Spatial Extents of Plumes

By design, outfalls of wastewater treatment plants are located greater than 8 km offshore and at depths between 60 and 100 m to inhibit effluent from recirculating back into surface and nearshore waters. By contrast, plumes emerging from river mouths are immediately subject to multiple nearshore processes, including estuarine mixing, entrainment in the surf zone, and exposure to surface currents. The resulting differences in the fates of riverine and wastewater plumes are shown in figure 6.5. Wastewater plumes are observed most frequently in the immediate vicinity of the outfalls and carried by prevailing subsurface currents either upcoast or downcoast of the pipe. They rarely reach shallow water, generally occurring within 2 km of shore less than 20% of the time. By contrast, riverine plumes tend to hug the shoreline, with the predominant direction depending on prevailing surface currents. Rapid settling and mixing processes prevent freshwater inputs from extending too far into the ocean, except under extraordinary climatic conditions and flood events. As a result, the portion of river plumes with the greatest concentration of pollutants occurs within 2 km of shore.

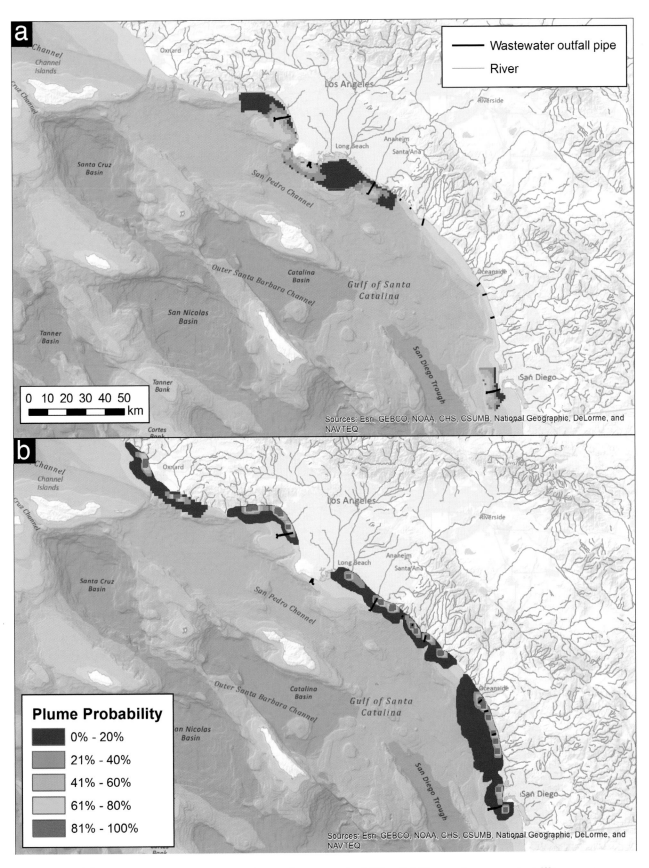

Figure 6.5. **(a) Wastewater treatment plant plume probability. (b) River plume probability.** Data from the authors, Esri, GEBCO, NOAA, CHS, CSUMB, National Geographic, DeLorme, NAVTEQ.

Pollutant Loading and Exposure

The relative loadings of the three pollutants vary between sources, and their spatial distributions within the area affected by plumes vary correspondingly. We calculated exposure statistics for individual and combined pollutants across the entire region, including the annual median, mean, and maximum exposure values (table 6.2). For all pollutants, the median exposure is lower than the mean, and both values are very low compared with the maximum exposure value. This skewness reflects the fact that even within coastal waters impacted by plumes, exposures are generally low over much of the affected area. However, localized areas may be at risk where maxima occur.

Table 6.2. **Spatial Statistics for Individual Pollutants by Source**

Source Parameter	Plume Area (km²)	Average Annual Load (kg/yr)	Exposure (kg/yr)		
			Median	Mean	Max
POTWs					
Copper	1178.9	2.62×10^4	6.5	13.7	81.2
DIN		2.99×10^6	261.8	2813.9	25502.8
TSS		3.39×10^7	11183	171670	89800
Rivers					
Copper	1594.3	1.37×10^4	0.2	0.7	25.4
DIN		9.47×10^5	15.2	49.4	2070
TSS		1.72×10^8	2832	9478	505746
Combined					
Copper	2433.1	4.0×10^4	1.2	7.1	81.2
DIN		1.25×10^6	37.3	1396.9	25502.8
TSS		2.06×10^8	6163	14527	505746

Copper

Large POTWs contribute nearly twice as much copper to the SCB than the rivers considered in this study (see table 6.2). Maximum exposures occur near the HWTP outfall, in Santa Monica Bay. Riverine exposure to copper is highest in the vicinity of the Los Angeles and San Gabriel River mouths, in San Pedro Bay. The Santa Clara River also contributes high loads of copper to the coast of Ventura County.

Dissolved Inorganic Nitrogen

POTWs also contribute three times as much DIN as rivers (see table 6.2). This agrees with recent studies showing that wastewater discharge represents the dominant anthropogenic source of nitrogen to the SCB (Howard et al. 2014). As with copper, highest exposures from wastewater sources are found near the Hyperion outfall. The small but highly urbanized Ballona Creek watershed also contributes relatively high amounts of DIN to Santa Monica Bay (figure 6.6a–b). As with copper, the Los Angeles and Santa Clara Rivers contribute the most riverine DIN overall. The Los Angeles River is predicted to load an average of 219,524 kg/yr, and the Santa Clara River contributes 121,159 kg/yr.

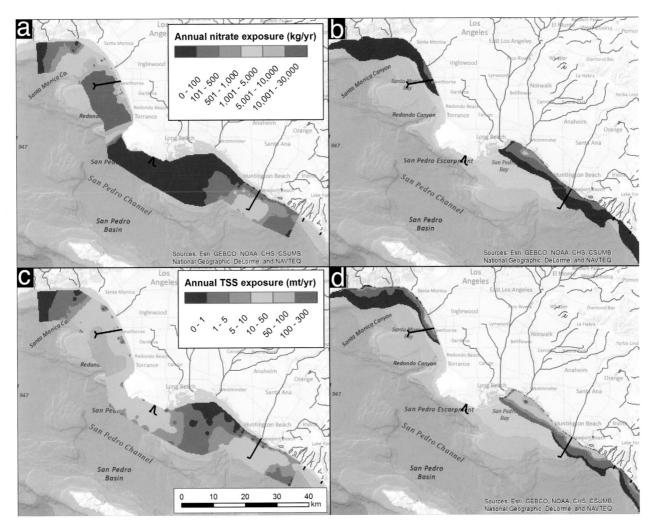

Figure 6.6. Comparison of pollutants by source in Santa Monica Bay/San Pedro Bay: (a) dissolved inorganic nitrogen exposure from publicly owned treatment works, (b) dissolved inorganic nitrogen exposure from rivers, (c) total suspended solids exposure from publicly owned treatment works, and (d) total suspended solids exposure from rivers. Data from Esri, GEBCO, NOAA, CHS, CSUMB, National Geographic, DeLorme, NAVTEQ.

Total Suspended Solids

In contrast to copper and DIN, sediment inputs are dominated by riverine loading, which contributes five times the TSS of POTWs (see table 6.2). Three of the four POTWs employ secondary treatment (activated sludge or biological reactors), which reduces the amount of suspended solids discharged offshore. There are no such controls on the sediments that enter rivers via runoff, especially from watersheds where elevation gradients and land use contribute to high sediment levels. Thus, although offshore exposures from POTW discharges are highest in Santa Monica Bay (figure 6.6c) and offshore of Point Loma, San Diego, they are low in comparison to riverine TSS exposures, particularly from the northernmost watersheds in this study (Ventura River, Santa Clara River, and Calleguas Creek). These watersheds have the highest elevations, most erosion-prone soil types, and greatest rainfall quantities, as well as large areas of agricultural land. As a result, the highest TSS loadings and exposures are found in the nearshore waters of Ventura County. Urbanized areas such as the watersheds emptying into San Pedro Bay also contribute high TSS loads (figure 6.6d).

Pollutant Exposure Index

The PEI developed through this process is designed to be adaptable for use in comparing different polygonal areas of interest within the larger SCB. Figure 6.7 maps the combined PEI score across the bight. For the purpose of this study, the area of interest (AOI) is defined as California state waters inshore of the 3 nm State Seaward boundary and extending from the Mexico border to the northernmost extent of the mapped plumes (hatched area, figure 6.7). Table 6.3 summarizes the percentage area within the AOI found within each range of the normalized pollutant exposure indices, and the normalized PEI. Of the total AOI, over 45% of the AOI had a PEI score of > 0–0.3. Another 24.2% is not covered by any mapped plume. The river plume model includes all major rivers but omits some smaller coastal watersheds and catchments. However, most are relatively undeveloped and unlikely to be major sources of pollutants.

Table 6.3. **Percent Area of California State Waters (inside the 3 nm Limit) within Each Class of the Pollutant Exposure Index for Individual Parameters and the Combined Pollutant Exposure Index (total area = 1986.7 km²)**

	Percent of SCB Area for Each Pollutant Exposure Index Class										
	No Plume	>0–0.1	>0.1–0.2	>0.2–0.3	>0.3–0.4	>0.4–0.5	>0.5–0.6	>0.6–0.7	>0.7–0.8	>0.8–0.9	>0.9–1.0
Copper	24.2	8.2	20.6	15.9	10	8.2	4.6	3.7	3	1.4	0.2
DIN	24.2	28.9	25.5	10.2	3.4	1.2	1.1	1.3	2.4	1.5	0.3
TSS	24.2	4.6	20.2	22.1	19.2	7.3	1.5	0.4	0.2	0.1	0.1
All	24.2	6.5	19.5	19.3	13.4	7.8	3.2	1.8	2.6	1.4	0.3

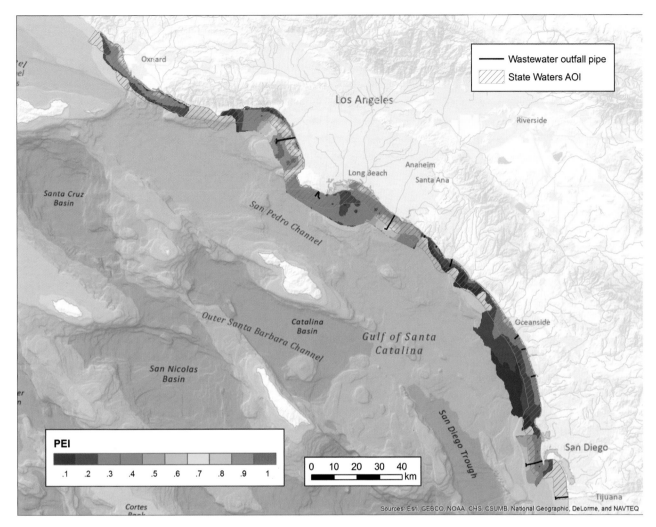

Figure 6.7. Map of pollutant exposure index (PEI) based on combined sources, including state waters areas of interest (AOIs). Data from Esri, GEBCO, NOAA, CHS, CSUMB, National Geographic, DeLorme, NAVTEQ.

Sensitivity Analysis

The interannual variability in precipitation is reflected in total riverine pollutant loading and the spatial extent of plumes (figure 6.8). In dry weather, plumes are ephemeral and restricted to within a few kilometers of river mouths. With increasing precipitation, plumes extend both alongshore and offshore, and plumes from adjacent river mouths may merge. Table 6.4 and figure 6.9 compare the regional mean and maximum exposure values between dry, average, and wet years. In all regions, dry years produce about 20% of the total loads of wet years, whereas average years produce about 45% of the loads of wet years. Such interannual variability can play a large role in the integrity of coastal ecosystems. For example, several dry years in a row may allow the establishment of habitats and communities that may be severely impacted by heavy flows and pollutant inputs in a subsequent wet year.

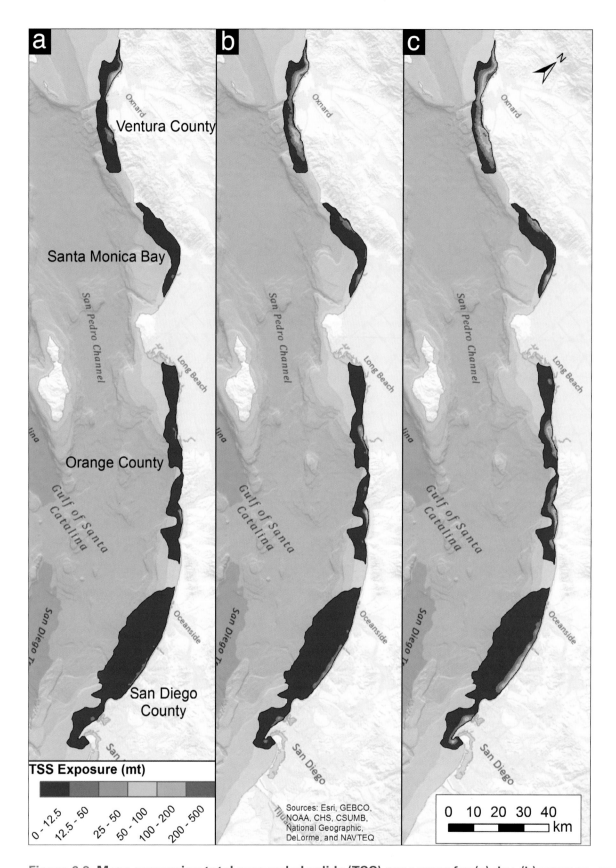

a

Oxnard

Ventura County

Santa Monica Bay

San Pedro Channel

Long Beach

Orange County

Gulf of Santa Catalina

San Diego Tr.

Oceanside

San Diego County

TSS Exposure (mt)

0 - 12.5   12.5 - 50   25 - 50   50 - 100   100 - 200   200 - 500

b

Oxnard

San Pedro Channel

Long Beach

Gulf of Santa Catalina

Oceanside

San Diego

Tijuana

Sources: Esri, GEBCO,
NOAA, CHS, CSUMB,
National Geographic,
DeLorme, and NAVTEQ

c

N

Oxnard

San Pedro Channel

Long Beach

Gulf of Santa Catalina

Oceanside

San Diego

0   10   20   30   40
km

Figure 6.8. Maps comparing total suspended solids (TSS) exposures for (a) dry, (b) average, and (c) wet years. The mapped plumes are grouped into four regions. Data from the authors, Esri, GEBCO, NOAA, CHS, CSUMB, National Geographic, DeLorme, NAVTEQ.

Table 6.4. Sensitivity Analysis: Summary Statistics by Pollutant and Region

	Mean Exposure (kg/yr)			Maximum Exposure (kg/yr)		
	Dry	Average	Wet	Dry	Average	Wet
Copper						
Ventura	0.11	0.32	0.61	1.66	4.13	7.52
SMB	0.04	0.12	0.24	1.03	2.39	5.49
Orange	0.03	0.07	0.18	0.78	1.73	3.93
San Diego	0.16	0.29	0.80	2.84	4.61	14.19
Nitrate						
Ventura	7.2	21.5	40.6	109.7	273.7	498.4
SMB	3.1	9.0	18.2	82.2	190.1	439.9
Orange	1.8	4.6	12.3	52.7	116.8	265.1
San Diego	10.6	18.4	50.8	183.4	297.9	871.3
TSS						
Ventura	7070	21,013	39,643	107,267	267,515	487,234
SMB	1201	4527	7743	18,312	44,679	94,354
Orange	1501	3893	10,739	42,135	95,163	226,030
San Diego	1249	2178	6016	24,564	38,016	102,712

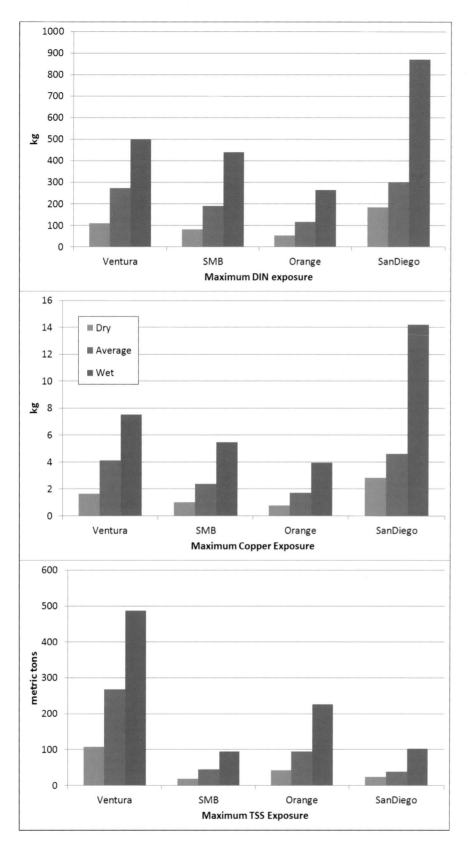

Figure 6.9. Comparison of pollutant exposure (dissolved inorganic nitrogen, copper, and total suspended solids) by region during dry, average, and wet years.

Discussion

As development and population growth continue within watersheds adjacent to the SCB, loads of potentially harmful and toxic chemicals may present a risk to the many valuable resources via both river runoff and treated wastewater. Little is known about the spatial extent of the relative and combined effects of these pollutant sources in different areas of the SCB. With new methods of plume detection, we can now more effectively map and analyze these datasets. Regional-scale spatial analysis of freshwater plumes and pollutant exposures in the marine environment is a relatively new and growing field. To date, few studies have been done to examine regional extents and effects of plumes (Mertes and Warrick 2001; McLaughlin et al. 2003; Álvarez-Romero et al. 2013). As consistent and comparable datasets become more readily available through advances in both in situ sampling instruments and remote-sensing systems, in conjunction with increasing data-processing power, the opportunity to conduct regional spatial analysis and modeling of these datasets is feasible. Satellite imagery allows us to distinguish plumes from ambient water over large areas and longer time scales. Technologies such as HFR aid in the localized prediction of currents and particle movement through surface water. New sensor applications, such as CDOM, enable researchers to capture subsurface water quality information more quickly, consistently, and economically than they could using traditional sampling techniques.

Benefits of Using GIS to Create a Regional PEI

Our development of geoprocessing tools to spatially integrate these datasets with information on pollutant loads to create a regional PEI provides an important next step in mapping and understanding potential threats to water quality in the SCB. This index enables a consistent approach to quantify the spatial extent of exposure to these pollutants and identify critical coastal habitats that may be affected. These tools encapsulate numerous geoprocessing steps into a documented, transferable, and modifiable package. Most of the repetitive steps of this analysis, including the interpolation of individual plume frequency grids, summing of rasters, and generation of spatial statistics, were quickly accomplished by creating models in ModelBuilder or developing Python scripts. These tools can be customized to accept different input variables rather than being hard-coded with set parameters, increasing their flexibility. The separate scripts may then be packaged into an ArcGIS toolbox, which is easily shared with others who wish to perform similar analyses.

GIS also allowed for spatial integration of datasets acquired through different methods and sources. Datasets from multiple studies and sources can be integrated as long as they have sufficient resolution and similar spatial referencing. In this study, we combined plume frequency data from the two primary sources, POTWs and rivers. The POTW plumes were empirically mapped using field sampling techniques (CDOM detectors), while river plume locations were modeled based on interpretations of surface current data. Despite these different approaches, both methods resulted in a simple grid of points attributed with a predicted plume frequency, enabling us to apply similar calculations and interpolation methods for both datasets.

Another advantage of using GIS to model plume exposures is that variables can be easily modified to reflect different environmental conditions and predict different outcomes. For example, the Southern California climate is characterized by high interannual variability in precipitation. This variability has significant effects on the annual loading of pollutants by river plumes. In some years (such as those when El Niño conditions prevail), this can result in massive inputs of sediment, nutrients, and metals; while in other years, the inputs barely rise above baseline. Fortunately, the loading model we used explicitly incorporates precipitation as a variable in predicting loads and provides data on a daily basis, enabling us to subset the data into years that reflect this natural climatic variability. These subsets provide representative end-members for our sensitivity analysis. Other variables that could be modified include the dilution factors used to represent initial loss of pollutants from the plume. Flexibility to adjust these tools to match regional requirements provides a distinct advantage, reducing the need for expensive, long-term empirical field-sampling efforts to capture natural variability.

These GIS-based index development tools have some limitations. As a two-dimensional model, it does not incorporate the dispersal of pollutants with depth, or reflect complex oceanographic mixing processes, and is therefore only applicable to nearshore shallow water. The strength of its predictive capabilities lies in the methods used to integrate plume effects over a large region and long time span, providing for a relative ranking of pollution impacts on coastal areas. Clearly, site-specific or short-duration inputs will require dedicated, localized effort beyond this long-term, regional-scale PEI approach.

Extent of Plumes and Pollutant Exposures

In this study, we identified at least 2,400 km² of nearshore waters and approximately 320 km of coastline between San Diego and Ventura County that could potentially be affected by discharge plumes emanating from either wastewater treatment plants or rivers. About 15% of this area experiences plumes from both sources, making cumulative impacts of anthropogenic influences an important management concern. As expected, areas where plumes overlap also tend to be some of the most highly populated, accessible, and utilized portions of the coastline. Santa Monica and San Pedro Bays receive both treated effluent and riverine runoff from large urbanized areas of the greater Los Angeles Basin, making them two of the most highly impacted areas of the SCB.

The distribution of PEI throughout the SCB indicates that the majority of California state waters within the SCB remain relatively unaffected by anthropogenic pollutants from these sources. Of the 1,986 km² of state waters affected by plumes, 45% have a PEI of < 3, while only 4.2% have a PEI of ≥ 8. However, areas with higher PEI are generally closer to shore and, therefore, more likely to impact both sensitive habitats and areas used for human recreation.

Comparison of Relative Risk from Each Source

The risk from POTWs and rivers varies both temporally and spatially. Treated wastewater discharge from POTWs is a continuous process that discharges pollutants to coastal waters on a more consistent basis. Conversely, the extent and volume of river discharge is directly proportional to precipitation events, resulting in intermittent and somewhat unpredictable plume extent and duration. Subsequently, river plume extent was generally smaller than POTW plume extent. However, river plumes tend to remain inshore of the 30 m isobath, while POTW plumes occur most frequently between the 30 m and 60 m contours. Therefore, pollutants from river plumes are more likely to settle in nearshore sediments or be retained within very shallow waters, potentially causing greater impacts on marine ecosystems (Bay et al. 2003). POTW plumes, though persistent and widespread, emerge from outfalls near the ocean bottom (approximately 60 m) and tend to remain trapped below the upper mixed layer, mixing into larger volumes of water. Ultimately, this decreases the likelihood that POTW discharges will affect shallower habitats. However, wastewater plumes are more likely to impact offshore ecosystems (Howard et al. 2014).

Our sensitivity analysis indicated that in wet years, river plumes can contribute up to five times the pollutant load of dry years and potentially impact 10 times the area, while inputs from sewage discharge remain relatively similar. For this reason, it is important to use multiple years of runoff and loading data (at least 10, as in this study) to capture long-term patterns in pollutant exposure. Although smaller time scales could be modeled using our approach, the accuracy of the model as a predictor of overall exposures would be limited.

Role of PEI in Research and Management

Our development of a spatially referenced PEI for the SCB provides a means to easily and rapidly compare relative anthropogenic threats to a variety of ecosystems. It relies on underlying datasets that model the intrusion and dispersal of freshwater plumes and pollutants into the marine environment. The index provides a simplified, long-term average, and inherently loses some of the accuracy of intensive studies examining plumes over smaller areas and time scales. However, by integrating multiple sources of data, our PEI provides a synoptic look at coastal water quality, providing environmental management agencies the ability to rank and prioritize affected habitats in terms of susceptibility and management opportunities.

The PEI has many implications for mitigation and management practices designed to protect both the integrity of coastal ecosystems and human health. Since enactment of the Clean Water Act in 1972 and subsequent environmental legislation explicitly recognizing the connections between anthropogenic pollutants and ocean health, our understanding of these connections has continually been strengthened by research. This PEI continues the trend by directly linking land-based loads of nutrients, metals, and sediments, arising from multiple individual sources, to ocean circulation patterns. The links can be manipulated to model how upstream changes in land use, flood control and drainage infrastructure, and runoff mitigation attempts might affect loadings on an individual parameter basis. For example, if new regulations require POTWs to change how they process effluent to meet new permitting standards, the downstream effects of those changes can be easily predicted using our tools.

Future Directions

We are aiming to use and modify the PEI in several aspects. The first direction is to examine PEI within a new AOI. There are 27 new or modified marine protected areas (MPAs), comprising approximately 15% of the SCB coastline (including the Channel Islands), that specifically limit fishing and other extractive activities (California Department of Fish and Wildlife 2013). Little is known about the effect of water quality in these MPAs, and the new index is one avenue to ascertain risk and prioritize future resources for characterizing problems in MPA performance.

A second new direction is to examine additional sources of pollutants. Two potentially important sources are small POTWs and runoff from small urbanized coastal catchments. We examined the four largest POTWs in the SCB, but there are seven small POTW outfalls that, despite being only a fraction of the four large POTWs examined herein, may have discrete local impacts. Unfortunately, the plume mapping for these small POTWs is lacking (Jones et al. 2011), and we chose not to extrapolate limited results or large POTW plumes to unmapped small POTW outfalls. Similarly, urban runoff that directly enters the ocean through small creeks and coastal storm drains can be a major localized source of pollution. More storm drain water-quality datasets are becoming available through municipal and volunteer monitoring programs (e.g., Los Angeles Department of Public Health, LA Waterkeeper, and City of San Diego all have regular storm drain monitoring projects). However, there is little available long-term data on the dispersal of these pollutants once they enter the ocean. With the advent of new plume monitoring methods such as gliders, it may become easier and less expensive to map the effects of these sources.

A third new direction is incorporating differential dispersal patterns in the modeled parameters (Reifel et al. 2009). We treated each of the three pollutants as though they were conservative tracers of the plume and, therefore, their concentrations decrease linearly with distance from the plume source. However, it is likely that they are not true conservative tracers of freshwater transport. For example, TSS typically settles more rapidly than nitrate is diluted. The result would be a higher rate of TSS loss from the plume. Prediction of TSS exposure in the nearshore environment is also complicated by processes other than riverine and POTW discharges. In particular, the resuspension of sediments by high wave energy and currents near headlands can result in high TSS exposures that are not accounted for in this study. By adjusting dilution factors for each pollutant to more accurately reflect true oceanographic conditions, accuracy of the model could be further improved for TSS. Likewise, nitrate is typically consumed relatively rapidly by phytoplankton, and these dynamics are not intrinsic to the PEI. To enhance the plume extent portion of the PEI, additional dispersion, advection, or transformation models could be applied. This would allow for more accurate estimates of potential secondary effects, if necessary.

Acknowledgments

The authors would like to thank Nikolay Nezlin and Peter Rogowski for assistance with interpreting their plume grid data; Dan Pondella, Julia Coates, and the members of the Bight '13 MPA Committee for assistance with manuscript development and review; and two anonymous reviewers for their helpful comments and suggestions.

References

Ackerman, D., and K. Schiff. 2003. "Modeling Storm Water Mass Emissions to the Southern California Bight." *Journal of Environmental Engineering* 129 (4): 308–17.

Ahn, J. H., S. B. Grant, C. Q. Surbeck, P. M. DiGiacomo, N. P. Nezlin, and S. Jiang. 2005. "Coastal Water Quality Impact of Stormwater Runoff from an Urban Watershed in Southern California." *Environmental Science & Technology* 39 (16): 5940–53.

Álvarez-Romero, J. G., M. Devlin, E. T. da Silva, C. Petus, N. C. Ban, R. L. Pressey, J. Kool, J. J. Roberts, S. Cerdeira-Estrada, A. S. Wenger, and J. Brodie. 2013. "A Novel Approach to Model Exposure of Coastal-Marine Ecosystems to Riverine Flood Plumes Based on Remote Sensing Techniques." *Journal of Environmental Management* 119:194–207.

Anderson, D. M., J. M. Burkholder, W. P. Cochlan, P. M. Glibert, C. J. Gobler, C. A. Heil, R. M. Kudela, M. L. Parsons, J. E. Rensel, D. W. Townsend, V. L. Trainer, and G. A. Vargo. 2008. "Harmful Algal Blooms and Eutrophication: Examining Linkages from Selected Coastal Regions of the United States." *Harmful Algae* 8 (1): 39–53.

Bay, S., B. H. Jones, K. Schiff, and L. Washburn. 2003. "Water Quality Impacts of Stormwater Discharges to Santa Monica Bay." *Marine Environmental Research* 56 (1–2): 205–23.

California Department of Fish and Wildlife. 2013. Guide to the Southern California Marine Protected Areas. http://www.dfg.ca.gov/marine/mpa/scmpas_list.asp. Last accessed March 16, 2014.

Dailey, M. D., J. W. Anderson, D. J. Reish, and D. S. Gorsline. 1993. "The California Bight: Background and Setting." In *Ecology of the Southern California Bight: A Synthesis and Interpretation*, edited by M. D. Dailey, D. J. Reish, and J. W. Anderson, 1–18. Berkeley, CA: University of California Press.

Davis, A. P., M. Shokouhian, and S. B. Ni. 2001. "Loading Estimates of Lead, Copper, Cadmium, and Zinc in Urban Runoff from Specific Sources." *Chemosphere* 44 (5): 997–1009.

EPA (Environmental Protection Agency). 1985. *Initial Mixing Characteristics of Municipal Ocean Discharges: Volume 1. Procedures and Applications*. Washington, DC: Environmental Protection Agency.

Hart, B. T. 1982. "Uptake of Trace Metals by Sediments and Suspended Particulates: A Review." *Hydrobiologia* 91:299–313.

Howard, M. D. A, M. Sutula, D. A. Caron, Y. Chao, J. D. Farrara, H. Frenzel, B. Jones, G. Robertson, K. McLaughlin, and A. Sengupta. 2014. "Anthropogenic Nutrient Sources Rival Natural Sources on Small Scales in the Coastal Waters of the Southern California Bight." *Limnology and Oceanography* 59 (1): 285–97.

Jones, B. H., M. D. Noble, and G. L. Robertson. 2011. "Huntington Beach: An In-Depth Study of Sources of Coastal Contamination Pathways and Newer Approaches to Effluent Plume to Dispersion." International Symposium on Outfall Systems, Mar del Plata, Argentina. http://www.osmgp.gov.ar/symposium2011/Papers/76_Jones.pdf.

Kim, S. Y., E. J. Terrill, and B. D. Cornuelle. 2009. "Assessing Coastal Plumes in a Region of Multiple Discharges: The US-Mexico Border." *Environmental Science & Technology* 43 (19): 7450–57.

Long, E. R., D. D. Macdonald, S. L. Smith, and F. D. Calder. 1995. "Incidence of Adverse Biological Effects within Ranges of Chemical Concentrations in Marine and Estuarine Sediments." *Environmental Management* 19 (1): 81–97.

Lyon, G. S., and M. A. Sutula. 2011. *Effluent Discharges to the Southern California Bight from Large Municipal Wastewater Treatment Facilities from 2005 to 2009*. Southern California Coastal Water Research Project 2011 Annual Report, 223–36. Costa Mesa, CA: Southern California Coastal Water Research Project.

McLaughlin, C. J., C. A. Smith, R. W. Buddemeier, J. D. Bartley, and B. A. Maxwell. 2003. "Rivers, Runoff, and Reefs." *Global and Planetary Change* 39:191–99.

Mertes, L. A. K, M. Hickman, B. Waltenberger, A. L. Bortman, E. Inlander, C. McKenzie, and J. Dvorsky. 1998. "Synoptic Views of Sediment Plumes and Coastal Geography of the Santa Barbara Channel, California." *Hydrological Processes* 12 (6): 967–79.

Mertes, L. A. K., and J. A. Warrick. 2001. "Measuring Flood Output from 110 Coastal Watersheds in California with Field Measurements and SeaWiFS." *Geology* 29 (7): 659–62.

National Climatic Data Center. 2014. Climate Data Online. http://www.ncdc.noaa.gov/cdo-web. Last accessed June 28, 2014.

National Ocean Economics Program. 2008. National Ocean Economics Program. http://www.oceaneconomics.org/. Last accessed March 16, 2014.

Nezlin, N. P., P. M. DiGiacomo, D. W. Diehl, B. H. Jones, S. C. Johnson, M. J. Mengel, K. M. Reifel, J. A. Warrick, and M. Wang. 2008. "Stormwater Plume Detection by MODIS Imagery in the Southern California Coastal Ocean." *Estuarine Coastal and Shelf Science* 80 (1): 141–52.

Nezlin, N. P., P. M. DiGiacomo, E. D. Stein, and D. Ackerman. 2005. "Stormwater Runoff Plumes Observed by SeaWiFS Radiometer in the Southern California Bight." *Remote Sensing of Environment* 98 (4): 494–510.

Otero, M. P., and D. A. Siegel. 2004. "Spatial and Temporal Characteristics of Sediment Plumes and Phytoplankton Blooms in the Santa Barbara Channel." *Deep-Sea Research Part II: Topical Studies in Oceanography* 51 (10–11): 1129–49.

Reifel, K. M., S. C. Johnson, P. M. DiGiacomo, M. J. Mengel, N. P. Nezlin, J. A. Warrick, and B. H. Jones. 2009. "Impacts of Stormwater Runoff in the Southern California Bight: Relationships among Plume Constituents." *Continental Shelf Research* 29 (15): 1821–35.

Rogowski, P., E. Terrill, L. Hazard, and K. Schiff. 2014. *Assessing Areas of Special Biological Significance Exposure to Stormwater Plumes Using a Surface Transport Model.* La Jolla, CA: Scripps Institution of Oceanography.

Sathyendranath, S., ed. 2000. *Remote Sensing of Ocean Color in Coastal, and Other Optically-Complex, Waters.* Dartmouth, Canada: Reports of the International Ocean Color Coordinating Group.

SCCWRP (Southern California Coastal Water Research Project). 2014. SCCWRP Bight '13 Regional Monitoring. http://www.sccwrp.org/ResearchAreas/HighlightedProjects/Bight13RegionalMonitoring.aspx. Last accessed June 28, 2014.

Schiff, K. C., M. J. Allen, E. Y. Zeng, and S. M. Bay. 2000. "Southern California." *Marine Pollution Bulletin* 41 (1–6): 76–93.

Schiff, K. C., and L. Tiefenthaler. 2011. "Seasonal Flushing of Pollutant Concentrations and Loads in Urban Stormwater." *Journal of the American Water Resources Association* 47:136–42.

Sengupta, A., M. A. Sutula, K. McLaughlin, M. Howard, L. Tiefenthaler, and T. Von Bitner. 2013. *Terrestrial Nutrient Loads and Fluxes to the Southern California Bight, USA.* Southern California Coastal Water Research Project 2013 Annual Report, 245–58. Costa Mesa, CA: Southern California Coastal Water Research Project.

US Census Bureau. 2010. 2010 Census. http://www.census.gov/2010census/. Last accessed June 28, 2014.

Warrick, J. A., P. M. DiGiacomo, S. B. Weisberg, N. P. Nezlin, M. J. Mengel, B. H. Jones, J. C. Ohlmann, L. Washburn, E. J. Terrill, and K. L. Farnsworth. 2007. "River Plume Patterns and Dynamics within the Southern California Bight." *Continental Shelf Research* 27 (19): 2427–48.

Warrick, J. A., L. A. K. Mertes, D. A. Siegel, and C. Mackenzie. 2004. "Estimating Suspended Sediment Concentrations in Turbid Coastal Waters of the Santa Barbara Channel with SeaWiFS." *International Journal of Remote Sensing* 25 (10): 1995–2002.

Supplemental Resources

Dawn J. Wright, ed.; 2015; *Ocean Solutions, Earth Solutions*; http://dx.doi.org/10.17128/9781589483651_d

For the digital content for this chapter, explained below in this section, go to the Esri Press "Book Resources" webpage at esripress.esri.com/bookresources. Then, in the list of Esri Press books, click *Ocean Solutions, Earth Solutions*. On the *Ocean Solutions, Earth Solutions* resource page, click the chapter 6 link to access that webpage and download the digital content.

Plume Interpolation Scripts

Python scripting was indispensable in completing the analyses described in this chapter. We simplified multiple geoprocessing steps using a combination of ArcPy and ModelBuilder and individual geoprocessing tools available in ArcGIS. In particular, we sought to automate the interpolation of pollutant plumes, clipping them to polygonal boundaries, and preparing to sum them together to calculate the pollutant exposure index.

Examples of two of these scripts are provided in the supplemental materials, along with a subset of the data used in our analysis. Both scripts take advantage of a variety of ArcGIS Spatial Analyst functions. Please see the readme.txt file included with the scripts for instructions on saving and running these scripts using the sample data.

The first script, IDW_plume.py, interpolates pollutant load values across the point feature classes that represent individual modeled plume grids. The script uses the same spatial extent and interpolation variables for each grid to ensure they are consistent and can be summed together after additional processing. The second script, Clip_IDW_by_plume_boundaries.py, modifies the output rasters from the first script. This script extracts plume raster values using individual boundary polygons that describe the actual extent of each plume (as determined in the model used to create the plume extents). In this step, all values within the rectangular extent of the raster falling outside the plume boundary are set to NoData by default. Next, it sets NoData values in each raster to zero, so the individual plume rasters may be summed into a single integrated pollutant load raster.

Once these two scripts were run and all interpolated plumes were processed, we used additional geoprocessing tools to sum together the set of plumes for each pollutant and create the pollutant index from these summed plumes. Because these subsequent steps were carried out a limited number of times, we did not develop scripts to automate them.

CHAPTER 7

Development of a Map Viewer for Archipelago de Cabrera National Park, Balearic Islands, Spain

Beatriz Ramos López, Nuria Hermida Jiménez, and Olvido Tello Antón

Abstract

This chapter describes the development of a map viewer (accessible in Spanish at http://www .ideo-cabrera.ieo.es) for the Archipelago de Cabrera National Park, Balearic Islands, Spain. The park is a region with various forms of protection, demonstrating the importance of this area and the need to expand knowledge about it and facilitate the management and monitoring of a wide range of environmental threats and hazards. The development of the map viewer (leveraging ArcGIS Viewer for Flex software) is one of the pioneering initiatives in Spain relating GIS technology to the marine environment. It is the result of extensive interinstitutional collaboration between the EU-US Marine Biodiversity research group of the Instituto Franklin-Universidad de Alcalá, the Instituto Español de Oceanografía, and the Organismo Autónomo de Parques Nacionales. Keywords for this chapter include *GIS*, *web mapping*, *map viewer*, *Visor*, *metadata*, *WMS*, *INSPIRE*, *ArcGIS*, *Geoportal*, *Cabrera*, *parque nacional*, *national park*, *Islas Baleares*, *Balearic Islands*, *España*, *Spain*.

Dawn J. Wright, ed.; 2015; *Ocean Solutions, Earth Solutions*; http://dx.doi.org/10.17128/9781589483651

Part of Archipelago de Cabrera National Park, south of Mallorca, one of the Balearic Islands east of Spain. Photo by Beatriz Ramos López.

Introduction

The merger of GIS and the Internet has led to a revolution of sorts since its inception in 1993 (Fu and Sun 2011), and particularly with the World Wide Web becoming a popular vehicle for the distribution of myriad sources of information. Web-based GIS has quickly evolved accordingly (Shekhar et al. 2001). Indeed, the web has changed the perspective on the role that maps can play (Kraak 2004; Shekhar and Xiong 2008), affecting both the users and developers of GIS, as well as society as a whole. One of the most important advances in GIS is a reliance on web mapping to share and visualize large amounts of data, particularly in the areas of spatial data access, geovisualization and exploration of data, and data processing (Dragicevic 2004). Accordingly, Web 2.0 is a term that has been used to represent the many changes brought forth by software developers and users who interact with the web, helping it to be more interactive, social, and customizable.

Web 3.0 is within reach (Strickland 2008), particularly through the use of more participatory mapping and the recent exchange and flow of data in the cloud. In this vein, web GIS has the potential to make data and information easily accessible to users with limited knowledge of GIS (e.g., Merrifield et al. 2013) and tailor it to specific topics (e.g., wetlands, coastal hazards, and marine protected areas, or MPAs), along with support for environmental decision-making and the proposal of new research projects (Mathiyalagan et al. 2005; Boroushaki and Malczewski 2010).

This chapter seeks to highlight the importance of web GIS technologies as an essential tool in the marine world. Several recent international directives support conservation of the marine environment, and technologies such as GIS are helping to make this possible.

Among the most important is Chapter 17 of the United Nations Agenda 21, which reiterates that the well-being of coasts and oceans is of global concern. In addition, the European Union (EU) Water Framework Directive of December 2000 is regarded as the most important legal stimulus at the EU level for integrated planning, both coastal and inland (Connolly and Cummins 2002). The UN Millennium Declaration of 2000 inspired the design and operation of integrated freshwater and marine management. The World Summit on Sustainable Development in 2002 outlined a broad thematic approach to coastal and ocean management (Vallega 2005). In 2007, the European Union launched an Integrated Maritime Policy, in parallel with the Infrastructure for Spatial Information in the European Community (INSPIRE) Directive (European Commission 2014a) and European Spatial Data Infrastructure (ESDI). In 2008, the Marine Strategy Framework Directive (MSFD) (European Union 2008), another environmental pillar for European maritime policy, was established to greatly improve the environmental status of European seas by 2020, including the creation of a multiresolution digital map from seabed to sea surface (Meiner 2010). In March 2013, the European Commission proposed legislation to create a common framework for maritime spatial planning (MSP) and integrated coastal zone management (European Commission 2012; European Commission 2014a and 2014b).

The preceding directives and initiatives show the continued importance of protecting the oceans and many resources and benefits they provide, as well as the importance of GIS in helping achieve these aims (see also Serral et al. 2009). The creation of the map viewer described in this chapter is a response to the implementation of many of these requirements, with a specific focus on revealing unknown aspects of the Archipelago de Cabrera National Park, Balearic Islands, Spain, along with a geospatial tool for managers to aid in its planning and management.

Spatial Data Infrastructure and Geoportals

Portugal was an early pioneer in terms of European spatial data infrastructure (SDI) conceptual and operational developments, but at present Spain is one of the most active countries on this front. Both countries are actively involved in developing their own national SDIs, focusing, like every EU country, on meeting INSPIRE requirements and national needs (Julião et al. 2009). According to the *Instituto Geográfico Nacional* (IGN), the official definition of an SDI for Spain is an integrated computer system that (1) shows a set of resources (e.g., catalogs, servers, software, data, applications, and websites); (2) is dedicated to managing geographic information (e.g., geographic information maps, orthophotos, satellite images, and place-names); (3) provides resources online while also meeting a number of conditions for interoperability (i.e., standards, specifications, protocols, and interfaces); and (4) allows users of many types and backgrounds to combine various resources according to his or her needs. As such, SDIs aim to catalog and make available to the general public all the information that is often unknown or does not have proper channels for release (Capdevila i Subirana 2004).

The Spanish National Spatial Data Infrastructure (IDEE for *Infraestructura de Datos Espaciales de España*) began in 2002 and was coordinated by the National Geographic High Council, a governmental body composed of representatives of producers of reference and thematic digital geographic data at the national level by agencies such as the National Geographic Institute, Ministry of Agriculture, Cadastre, and so forth; at the regional level by Cartographic Institutes and Regional Services of Cartography; and at the local level. The geoportal of the Spanish national SDI was launched in July 2004 (Julião et al. 2009).

Geoportals such as this organize content and different services such as directories, information resources, data, and applications (Maguire and Longley 2005). One of the earliest marine SDIs in Spain, SIG Pesca, was developed by the *Dirección General de Pesca y Acción Marítima de la Generalitat de Catalunya* in 2008 using the software Miramon (Almazán et al. 2009).

In addition, metadata is one of the most important keys to the discovery of geospatial information as part of an SDI (Ballari et al. 2006; Takken 2008). The benefits of metadata records are many: (1) it provides an inventory of the data; (2) it helps determine and maintain the value of the data; (3) it helps determine the reliability and timeliness of data; (4) it can serve as documents with legal grounds; and (5) it helps maintain and verify accuracy to support good decision-making and cost savings (Esri 2003).

The aim of geoportals is to be interoperable with each other, but a primary obstacle is the lack of standard exchange mechanisms for sharing both data and metadata and facilitating access to the general public (*La Infraestructura de Datos Espaciales de España* 2014; Julião et al. 2009). Recent efforts by the Open Geospatial Consortium (OGC) (2014) have resulted in several specifications to alleviate these problems. The OGC Web Map Service (WMS), as well as Geographic Markup Language (GML), are such standards for developing interoperable web GIS. With the demand for geospatial interoperability and adoption of open standards, GIS is evolving from traditional client-server architecture to web service architecture (Zhang and Li 2005).

The geoportal of the *El Instituto Español de Oceanografía*, or IEO (2014b) (figure 7.1) hosts an important part of Spain's SDI. IEO adopted the INSPIRE Directive as a mandatory component for developing an interoperable SDI, both for Spain and the broader European Union. Within the IEO geoportal, OGC WMS has been created to facilitate access to data and includes three kinds of requests: (1) *Get Capabilities* to return information via Extensible Markup Language (XML); (2) *Get Map* to return a map; and (3) *Get Feature* to return information associated with the map. Certain free GIS programs such as gvSIG are not able to handle WMS 1.3.0, so the IEO geoportal includes WMS 1.1.1 as well. An important feature of the geoportal is that metadata records for the datasets, as well as for WMS services, were created under the INSPIRE Directive.

Geoportal de la Infraestructura de Datos Espaciales
del Instituto Español de Oceanografía

Iniciar sesión   Ayuda   Acerca de

INICIO   BUSCAR   EXAMINAR   ENLACES   VISUALIZADOR DE METADATOS

Inicio

El Instituto Español de Oceanografía (IEO) es un organismo público de investigación (OPI), dependiente del Ministerio de Economía y Competitividad, a través de la Secretaría de Estado de Investigación, dedicado a la investigación en ciencias del mar, especialmente en lo relacionado con el conocimiento científico de los océanos, la sostenibilidad de los recursos pesqueros y el medio ambiente marino. El IEO representa a España en la mayoría de los foros científicos y tecnológicos internacionales relacionados con el mar y sus recursos.

Geoportal ofrece formas fáciles y convenientes de compartir datos geoespaciales.

Figure 7.1. **Screen capture of the El Instituto Español de Oceanografia (IEO) geoportal, based on Esri Geoportal Server software.** Image © 2013 Instituto Español de Oceanografia (Spanish Oceanographic Institute); used by permission.

The Map Viewer

The map viewer for the Archipelago de Cabrera National Park (hereafter referred to as the "Cabrera map viewer") is accessible in Spanish at http://www.ideo-cabrera.ieo.es and located within the IEO geoportal. It seeks to integrate physical and chemical parameters and biological data. It integrates all these parameters to facilitate the understanding of the characteristics of the national park to aid the decision-making of managers and scientists, as well as provide a better knowledge of the park for the general public. The Archipelago de Cabrera National Park (figure 7.2), located south of the island of Mallorca within the Balearic Islands, is one of the 15 national parks of Spain. The park was established in 1991 with an acreage of 10,021 (8,703 terrestrial and 1,318 marine).

Chapter 7   **139**

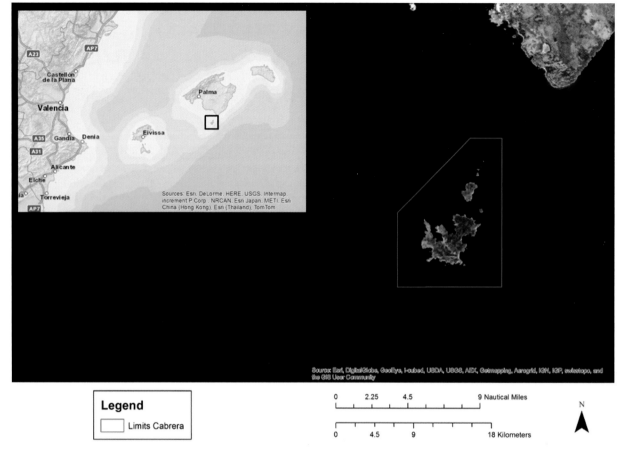

Legend

☐ Limits Cabrera

| 0 | 2.25 | 4.5 | | 9 Nautical Miles |
| 0 | 4.5 | 9 | | 18 Kilometers |

N

Figure 7.2. Location map of the Archipelago de Cabrera National Park. Map data sources: Inset map "World Street Map": Esri, DeLorme, HERE, USGS, Intermap, increment P Corp., NRCAN, Esri Japan, METI, Esri China (Hong Kong), Esri (Thailand), and TomTom. Larger map: "World Imagery" map: Esri, DigitalGlobe, GeoEye, i-cubed, USDA, USGS, AEX, Getmapping, Aerogrid, IGN, IGP, swisstopo, and the GIS User Community.

Map Viewer Benefits

In general, tools such as the map viewer for the Archipelago de Cabrera National Park hold several benefits as a fundamental management tool for MPA management, integrated coastal area management, and marine spatial planning. A map viewer can also offer a general vision of the projects, protection sites, and biological information of Cabrera that is of great utility for stakeholders.

In addition, a map viewer offers an important window into an unknown, relatively unexplored world. Whereas the exploration of outer space is well under way and almost every piece of land on Earth has been discovered and mapped, not much is known about the world's oceans that cover ~70% of the earth's surface, including many biological species (see Vermeulen 2013). With only 5%–10% of the world ocean floors mapped with the resolution of similar studies on land (Sandwell et al. 2003; Wright 2003), marine geomorphology still represents a persistent gap in our knowledge (Wright and Heyman 2008). Despite the high appraisals of the potential of GIS applications in coastal and marine environments, the development of coastal GIS is still challenged by (1) the complexity of highly dynamic coastal and marine systems in a three-dimensional environment; (2) the lack of data worldwide; and (3) the lack of communication between coastal experts and GIS professionals (Green 1995; Hooge et al. 2001; Zeng et al. 2001; Mujabar and Chandrasekar 2010).

Technical Objectives and Approach

Technical objectives of the project included facilitating (1) the search for all available datasets about the park and their subsequent organization into a geodatabase; (2) literature reviews for park resource management and research projects; (3) the creation of maps and WMS services (versions 1.1.1 and 1.3.0) using ArcGIS for Server; (4) the development of metadata in compliance with the INSPIRE Directive; and (5) the deployment of an effective geospatial system for researchers, managers, and the general public. The Cabrera map viewer is based on several technologies, including ArcMap, ArcCatalog, and ArcGIS for Server 10.1 software for backend data development and hosting on the ArcGIS platform; Esri Geoportal Server software for serving metadata and WMS services; Adobe Flash Builder 4.6 for programming; and ArcGIS Viewer 3.3 for Flex for final map viewer deployment.

The first phase of the project was to locate the data that would be displayed in the Cabrera map viewer, such as that of the *Banco de Naturaleza* of *Ministerio de Agricultura, Alimentación y Medio Ambiente* (MAGRAMA) (2014a and 2014b). Other data, such as fisheries and bathymetry, came from services created by the IEO (2014c and 2014d). Still other layers were contributed by nongovernmental organizations (NGOs) such as Oceana.

The next step was the creation of a Cabrera file geodatabase (.gdb), with a geodetic reference system of European Terrestrial Reference System 1989 (ETRS 89), Universal Transverse Mercator (UTM) Zone 31 (with spatial reference European Petroleum Survey Group or EPSG projection 25831). Several groups of feature dataset layers were created with their respective feature classes, grouped into five themes: Protection, Limits, Bathymetry, Biological Data, and Videos. Relationship classes and topology rules were built between the layers.

Next, map services and geoprocessing services were created using ArcGIS for Server. All layers were published together as a representational state transfer (REST) map service, as well as within an ArcMap map document (.mxd) project file. The geodatabase was also mirrored by another server called IDEO2, which was protected against external attacks and in which ArcSDE software was used to manage the Microsoft SQL Server enterprise geodatabase. A REST-style printing service was developed to adapt a new template called printCabreraA3Land, which was created in ArcMap and subsequently exported to the IEO server.

WMS 1.3.0 was created in ArcGIS for Server to enable the WMS capability at the same time we published the map service. As mentioned earlier, WMS 1.1.1 was also created to offer more possibilities to users, thereby expanding the Get Capabilities of both.

INSPIRE-compliant metadata records were created for both the data and WMSs. After creating the metadata in the IEO geoportal, the records were validated on the INSPIRE Geoportal Metadata Validator site (European Commission 2014b).

The next step was to develop the Cabrera map viewer itself, using ArcGIS Viewer 3.3 for Flex. The aim of Adobe Flex programming technology is to offer web developers a tool to quickly and easily build rich Internet applications (RIAs). The Adobe Flash Integrated Development Environment (IDE) was used concurrently while developing in the Flex Viewer. Operational layers of IEO services were configured as map services. Several widgets were configured to provide functionality in the map viewer display, especially to make it easy for users of many skills and backgrounds to interact with the viewer. These widgets (figure 7.3) include:

- *Legend*, showing the symbols assigned to each layer to facilitate their interpretation
- *Print*, allowing for download of a map of the chosen area, transformed into different formats for saving and printing
- *Search* (eSearch), with multiple options for searching between layers in the display
- *Edition* (Draw), allowing the user to draw or measure on the map
- *Links*, allowing access to external resources such as a photo album about Cabrera on Flickr, the IEO geoportal, or to websites for those wishing to visit the Archipelago de Cabrera National Park
- *Bookmarks*, to facilitate navigation among three principal zones: Spain, the Balearic Islands, and Archipelago de Cabrera National Park
- *Location*, to enter coordinates or orient the map in different directions
- *Attribute table*, new in ArcGIS Viewer 3.3 for Flex, allowing display of an attribute table for a selected layer directly in the map viewer
- *Chart*, for charting infauna data (benthic organisms that live buried within sedimentary layers of the seabed)

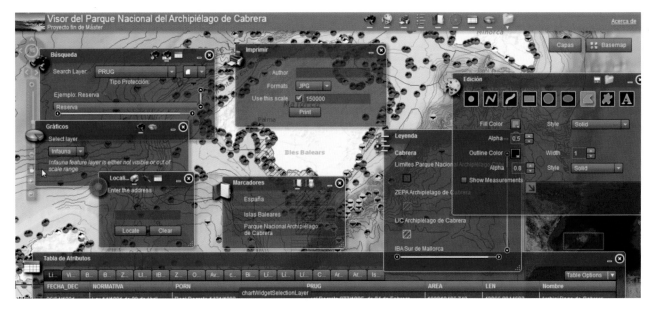

Figure 7.3. Screen capture of the map viewer of Archipelago de Cabrera National Park, showing the various widgets available. Image © 2013 Instituto Español de Oceanografía (Spanish Oceanographic Institute); used by permission.

To offer more map interactivity, pop-up windows with additional detail, including photos of marine mammals and other species, are available when a user selects an object. Once the tools and layers were integrated, the entire project was compiled in Adobe Flash Builder. The final Cabrera map viewer is available at http://www.ideo-cabrera.ieo.es (figure 7.4). Accompanying metadata, in WMS 1.3.0 and 1.1.1, is shown in figures 7.5, 7.6, and 7.7 (IEO 2014b).

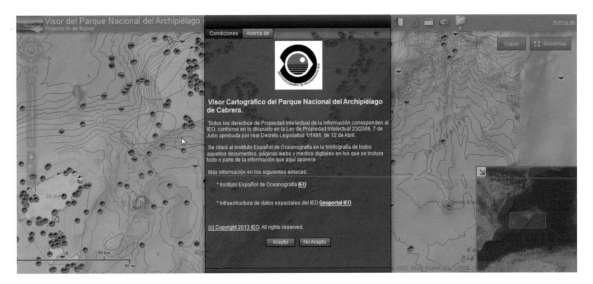

Figure 7.4. **Screen capture of the map viewer of Archipelago de Cabrera National Park.** Image © 2013 Instituto Español de Oceanografía (Spanish Oceanographic Institute); used by permission.

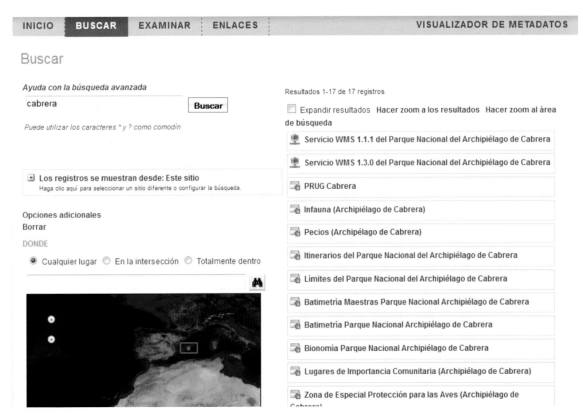

Figure 7.5. **Screen capture of metadata records associated with the map viewer of Archipelago de Cabrera National Park.** Image © 2013 Instituto Español de Oceanografía (Spanish Oceanographic Institute); used by permission.

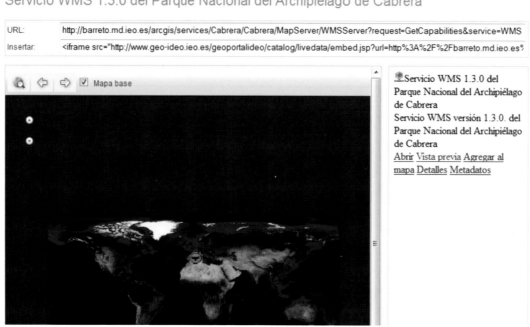

Figure 7.6. **Screen capture of WMS 1.3.0 and metadata associated with the map viewer of Archipelago de Cabrera National Park.** Image © 2013 Instituto Español de Oceanografía (Spanish Oceanographic Institute); used by permission.

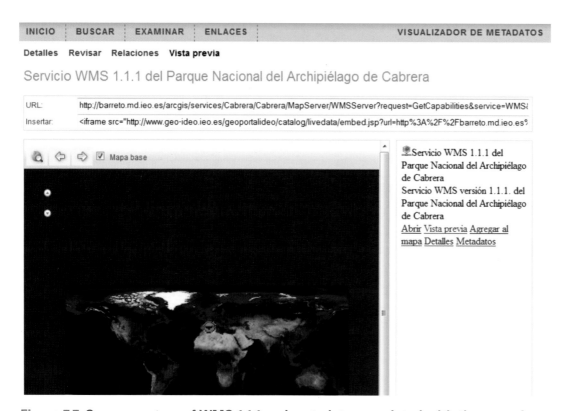

Figure 7.7. **Screen capture of WMS 1.1.1 and metadata associated with the map viewer of Archipelago de Cabrera National Park.** Image © 2013 Instituto Español de Oceanografía (Spanish Oceanographic Institute); used by permission.

The Cabrera map service and geoprocessing service descriptions are shown in figures 7.8 and 7.9 (IEO 2014a and 2014b). The Get Capabilities in WMS 1.3.0 and 1.1.1 are shown in figures 7.10 and 7.11, respectively (IEO 2014e and 2014f).

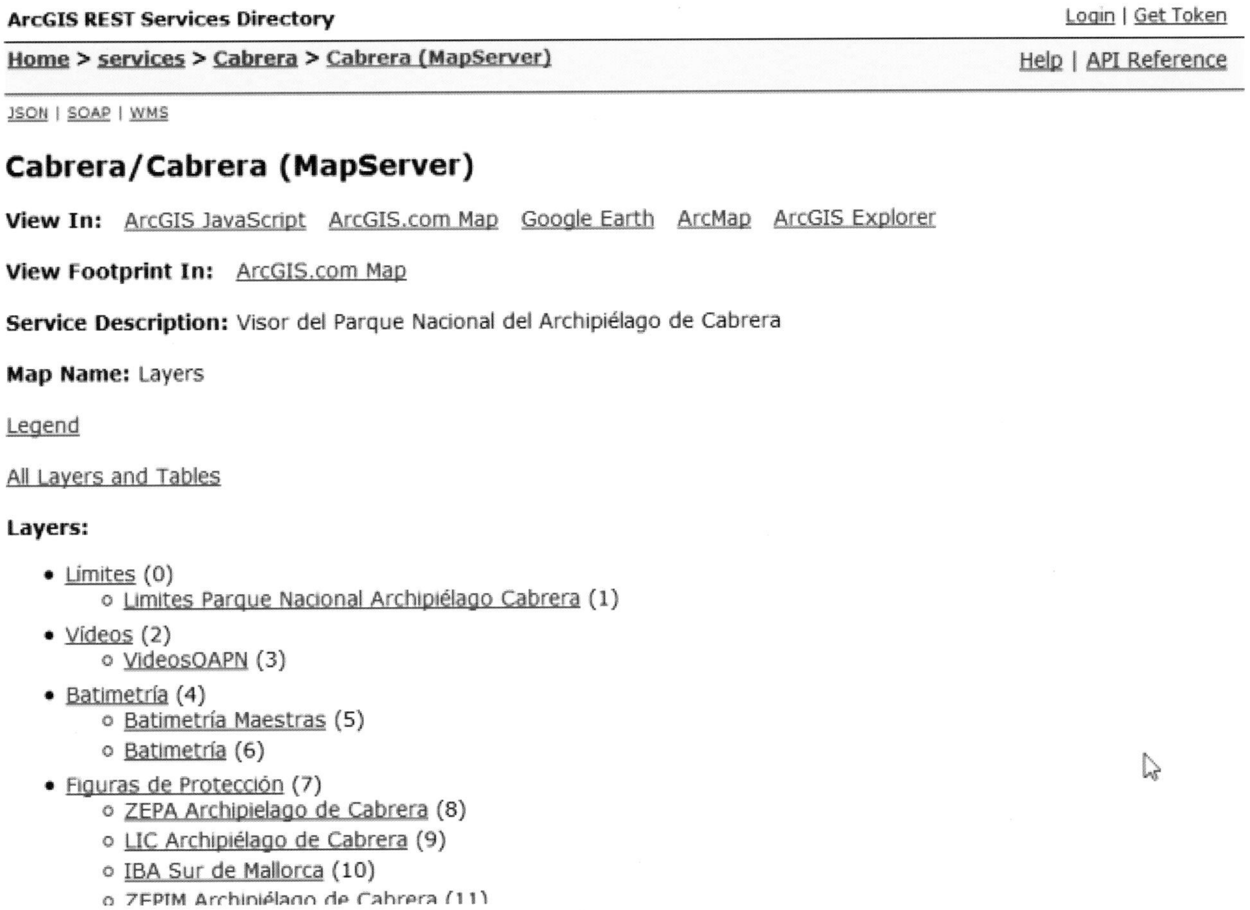

ArcGIS REST Services Directory   Login | Get Token

Home > services > Cabrera > Cabrera (MapServer)   Help | API Reference

JSON | SOAP | WMS

Cabrera/Cabrera (MapServer)

View In: ArcGIS JavaScript   ArcGIS.com Map   Google Earth   ArcMap   ArcGIS Explorer

View Footprint In: ArcGIS.com Map

Service Description: Visor del Parque Nacional del Archipiélago de Cabrera

Map Name: Layers

Legend

All Layers and Tables

Layers:

- Límites (0)
    - Límites Parque Nacional Archipiélago Cabrera (1)
- Vídeos (2)
    - VídeosOAPN (3)
- Batimetría (4)
    - Batimetría Maestras (5)
    - Batimetría (6)
- Figuras de Protección (7)
    - ZEPA Archipiélago de Cabrera (8)
    - LIC Archipiélago de Cabrera (9)
    - IBA Sur de Mallorca (10)
    - ZEPIM Archipiélago de Cabrera (11)

Figure 7.8. Screen capture of the map service description for the map viewer of Archipelago de Cabrera National Park. Image © 2013 Instituto Español de Oceanografía (Spanish Oceanographic Institute); used by permission.

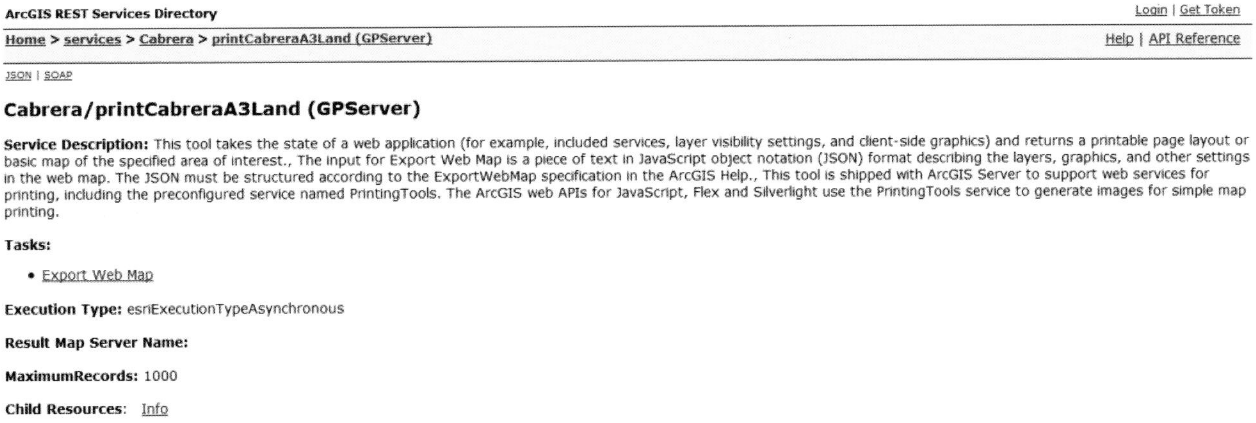

ArcGIS REST Services Directory   Login | Get Token

Home > services > Cabrera > printCabreraA3Land (GPServer)   Help | API Reference

JSON | SOAP

Cabrera/printCabreraA3Land (GPServer)

Service Description: This tool takes the state of a web application (for example, included services, layer visibility settings, and client-side graphics) and returns a printable page layout or basic map of the specified area of interest., The input for Export Web Map is a piece of text in JavaScript object notation (JSON) format describing the layers, graphics, and other settings in the web map. The JSON must be structured according to the ExportWebMap specification in the ArcGIS Help., This tool is shipped with ArcGIS Server to support web services for printing, including the preconfigured service named PrintingTools. The ArcGIS web APIs for JavaScript, Flex and Silverlight use the PrintingTools service to generate images for simple map printing.

Tasks:

- Export Web Map

Execution Type: esriExecutionTypeAsynchronous

Result Map Server Name:

MaximumRecords: 1000

Child Resources: Info

Figure 7.9. Screen capture of the geoprocessing service description for the map viewer of Archipelago de Cabrera National Park. Image © 2013 Instituto Español de Oceanografía (Spanish Oceanographic Institute); used by permission.

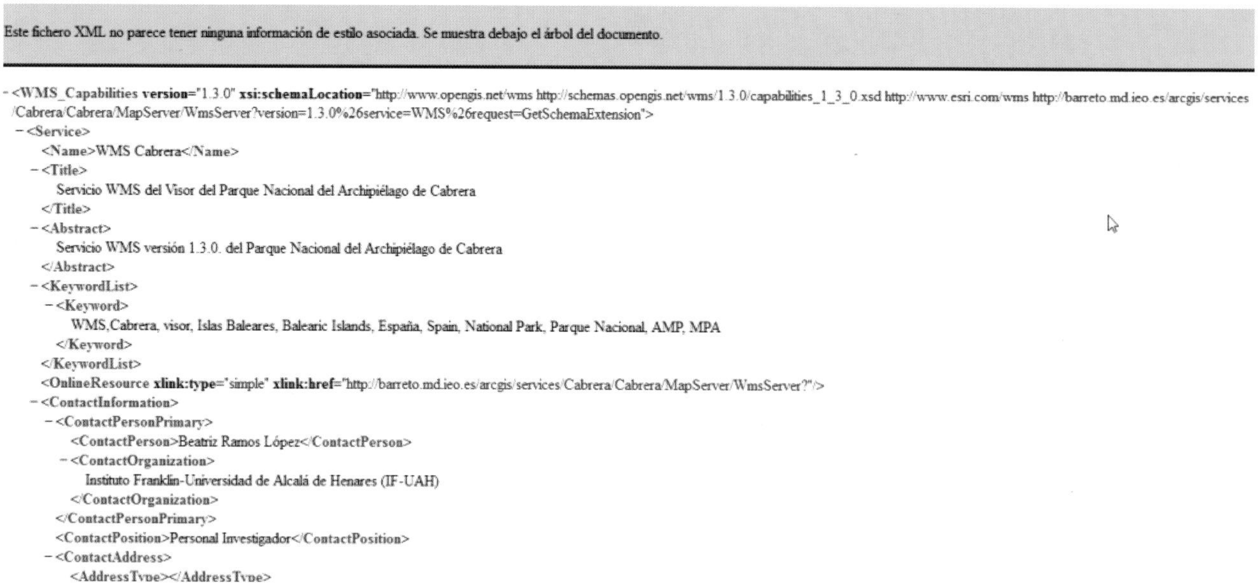

Este fichero XML no parece tener ninguna información de estilo asociada. Se muestra debajo el árbol del documento.

```
-<WMS_Capabilities version="1.3.0" xsi:schemaLocation="http://www.opengis.net/wms http://schemas.opengis.net/wms/1.3.0/capabilities_1_3_0.xsd http://www.esri.com/wms http://barreto.md.ieo.es/arcgis/services
/Cabrera/Cabrera/MapServer/WmsServer?version=1.3.0%26service=WMS%26request=GetSchemaExtension">
  -<Service>
    <Name>WMS Cabrera</Name>
    -<Title>
      Servicio WMS del Visor del Parque Nacional del Archipiélago de Cabrera
    </Title>
    -<Abstract>
      Servicio WMS versión 1.3.0. del Parque Nacional del Archipiélago de Cabrera
    </Abstract>
    -<KeywordList>
      -<Keyword>
        WMS,Cabrera, visor, Islas Baleares, Balearic Islands, España, Spain, National Park, Parque Nacional, AMP, MPA
      </Keyword>
    </KeywordList>
    <OnlineResource xlink:type="simple" xlink:href="http://barreto.md.ieo.es/arcgis/services/Cabrera/Cabrera/MapServer/WmsServer?"/>
    -<ContactInformation>
      -<ContactPersonPrimary>
        <ContactPerson>Beatriz Ramos López</ContactPerson>
        -<ContactOrganization>
          Instituto Franklin-Universidad de Alcalá de Henares (IF-UAH)
        </ContactOrganization>
      </ContactPersonPrimary>
      <ContactPosition>Personal Investigador</ContactPosition>
      -<ContactAddress>
        <AddressType></AddressType>
```

Figure 7.10. Screen capture of WMS 1.3.0 Get Capabilities of the map viewer of Archipelago de Cabrera National Park. Image © 2013 Instituto Español de Oceanografía (Spanish Oceanographic Institute); used by permission.

```
WMSServer.esrimap: Bloc de notas
Archivo  Edición  Formato  Ver  Ayuda
<?xml version="1.0" encoding="UTF-8"?>
<WMT_MS_Capabilities version="1.1.1">
    <Service>
        <Name><![CDATA[WMS Cabrera]]></Name>
        <Title>Servicio WMS del visor del Parque Nacional del Archipiélago de Cabrera</Title>
        <Abstract>Servicio WMS versión 1.1.1. del Parque Nacional del Archipiélago de Cabrera</Abstract>
        <KeywordList>
            <Keyword><![CDATA[WMS]]></Keyword>
            <Keyword><![CDATA[Cabrera]]></Keyword>
            <Keyword><![CDATA[Parque Nacional]]></Keyword>
            <Keyword><![CDATA[National Park]]></Keyword>
            <Keyword><![CDATA[visor]]></Keyword>
            <Keyword><![CDATA[Islas Baleares]]></Keyword>
            <Keyword><![CDATA[Balearic Islands]]></Keyword>
            <Keyword><![CDATA[España]]></Keyword>
            <Keyword><![CDATA[Spain]]></Keyword>
        </KeywordList>
        <OnlineResource xmlns:xlink="http://www.w3.org/1999/xlink" xlink:type="simple" xlink:href="http://barreto.md.ieo.es/arcgis/services/Cabrera/Cabrera/MapS
        <ContactInformation>
            <ContactPersonPrimary>
                <ContactPerson><![CDATA[Beatriz Ramos López]]></ContactPerson>
                <ContactOrganization><![CDATA[Instituto Franklin-Universidad de Alcalá de Henares (IF-UAH)]]></ContactOrganization>
            </ContactPersonPrimary>
            <ContactPosition><![CDATA[Personal Investigador]]></ContactPosition>
            <ContactAddress>
                <AddressType><![CDATA[]]></AddressType>
                <Address><![CDATA[]]></Address>
                <City><![CDATA[]]></City>
                <StateOrProvince><![CDATA[]]></StateOrProvince>
                <PostCode><![CDATA[]]></PostCode>
                <Country><![CDATA[]]></Country>
            </ContactAddress>
            <ContactVoiceTelephone><![CDATA[+34 600 67 41 21]]></ContactVoiceTelephone>
            <ContactFacsimileTelephone><![CDATA[]]></ContactFacsimileTelephone>
            <ContactElectronicMailAddress><![CDATA[beatriz.ramos@institutofranklin.net]]></ContactElectronicMailAddress>
            <ContactElectronicMailAddress><![CDATA[beatriz.ramos1@uah.es]]></ContactElectronicMailAddress>
        </ContactInformation>
        <ContactInformation>
            <ContactPersonPrimary>
                <ContactPerson><![CDATA[Nuria Hermida]]></ContactPerson>
                <ContactOrganization><![CDATA[Instituto Español de Oceanografía (IEO)]]></ContactOrganization>
            </ContactPersonPrimary>
            <ContactPosition><![CDATA[Técnico GIS]]></ContactPosition>
            <ContactAddress>
                <AddressType><![CDATA[]]></AddressType>
                <Address><![CDATA[]]></Address>
                <City><![CDATA[]]></City>
                <StateOrProvince><![CDATA[]]></StateOrProvince>
                <PostCode><![CDATA[]]></PostCode>
                <Country><![CDATA[]]></Country>
            </ContactAddress>
            <ContactVoiceTelephone><![CDATA[]]></ContactVoiceTelephone>
```

Figure 7.11. Screen capture of WMS 1.1.1 Get Capabilities of the map viewer of Archipelago de Cabrera National Park. Image © 2013 Instituto Español de Oceanografía (Spanish Oceanographic Institute); used by permission.

Concluding Remarks

The development of the Cabrera map viewer was an interesting exercise in data integration and map interface design, with an eye toward assisting managers and researchers not only in more fully understanding the resources of the park, but also in envisioning its possible enlargement. The Cabrera viewer seeks to encourage best practices of data management in the context of the INSPIRE Directive, and as new technology breakthroughs aid in handling huge amounts of data (big data), in better visualizing the ocean in three dimensions and managing the ocean via geodesign. The hope is that many new datasets will be made available to the map viewer as the project continues.

In addition, it was gratifying to develop an important dissemination tool to help the general public better understand the richness and importance of this relatively unknown protected area, visit the area, and participate in public events there. It is hoped that with such knowledge of this region, the public will also be motivated to advocate for conservation of other marine environments.

Acknowledgments

We are thankful to several organizations for providing the datasets that were integrated into the Cabrera map viewer, including the *Organismo Autónomo Parques Nacionales* (OAPN), IEO, Oceana, and MAGRAMA. We would like to acknowledge the contribution of Paula Alonso for valuable support in translating this chapter from Spanish to English, and to book editor Dawn Wright for subsequent editing. The comments and suggestions of two anonymous reviewers greatly improved this chapter.

References

Almazán Gárate, J. L., M. C. Palomino Monzón, and A. Verdú Vázquez. 2009. "La Cartografía Marina y los Sistemas de Información Geográfica." *Jornadas Internacionales de Didáctica de las Matemáticas en Ingeniería*, 229–40.

Ballari, D., A. Maganto, J. Nogueras, A. Pascual, and M. Bernabé. 2006. "Experiences in the Use of an ISO19115 Profile within the Framework of the Spanish SDI." In *GSDI-9 Conference Proceedings* (no. 10).

Boroushaki, S., and J. Malczewski. 2010. "Participatory GIS: A Web-Based Collaborative GIS and Multicriteria Decision Analysis." *URISA Journal* 22 (1): 25–32.

Capdevila i Subirana, J. 2004. *Infraestructura de Datos Espaciales (IDE). Definición y Desarrollo Actual en España. Geo Crítica/Scripta Nova. Revista Electrónica de Geografía y Ciencias Sociales* vol. 8 no. 170 (61). Barcelona: Universidad de Barcelona. http://www.ub.edu/geocrit/sn/sn-170-61.htm.

Connolly, N., and V. Cummins. 2002. "Integrated Coastal Zone Management (ICZM) in Ireland, with Particular Reference to the Use of Geographic Information Systems (GIS) and the EU ICZM Demonstration Programme." In *Achievement and Challenge: Rio+10 and Ireland*, edited by F. Convery and J. Feehan. Dublin, Ireland: Environmental Institute, University College Dublin.

Dragicevic, S. 2004. "The Potential of Web-Based GIS." *Journal of Geographical Systems* 6:79–81.

Esri. 2003. "Metadata and GIS: An Esri White Paper." http://www.esri.com/library/whitepapers/pdfs/metadata-and -gis.pdf. Last accessed June 5, 2014.

European Commission. 2012. "*De la Cartografía de los Fondos Marinos a Las Previsiones Oceánicas.*" A European Commission Marine Knowledge 2020 Green Paper. Brussels, Belgium: European Commission. http:// goo.gl/6Ff0iO.

———. 2014a. INSPIRE Geoportal. http://inspire-geoportal.ec.europa.eu. Last accessed June 29, 2014.

———. 2014b. INSPIRE Geoportal Metadata Validator. http://inspire-geoportal.ec.europa.eu/validator2/. Last accessed June 29, 2014.

European Commission Maritime Affairs. 2014. Maritime spatial planning. http://ec.europa.eu/maritimeaffairs/policy /maritime_spatial_planning/index_en.htm. Last accessed June 29, 2014.

European Union. 2007. "INSPIRE EU Directive. 2007. Directive 2007/2/EC of the European Parliament and of the Council of 14 March 2007 Establishing an Infrastructure for Spatial Information in the European Community (INSPIRE)." *Official Journal of the European Union* L 108/1, vol. 50.

———. 2008. "Directive 2008/56/EC of the European Parliament and of the Council of 17 June 2008 Establishing a Framework for Community Action in the Field of Marine Environmental Policy (Marine Strategy Framework Directive)." http://eur-lex.europa.eu/legal-content/en/ALL/?uri=CELEX:32008L0056. Last accessed June 30, 2014.

Fu, P., and J. Sun. 2011. *Web GIS: Principles and Applications.* Redlands, CA: Esri Press.

Green, R. 1995. "User Access to Information: A Priority for Estuary Information Systems." In *Proceedings of Coast GIS '95*, 3550, Cork, Ireland, February 3–5.

Hooge, P. N., W. M. Eichenlaub, and E. K. Solomon. 2001. "Using GIS to Analyze Animal Movements in the Marine Environment." In *Spatial Processes and Management of Marine Populations*, 37–51, Anchorage, AK, October 27–30, 1999. Anchorage, AK: Alaska Sea Grant College Program.

IEO (El Instituto Español de Oceanografía). 2014a. Cabrera/Cabrera (MapServer), ArcGIS REST Services Directory. http://barreto.md.ieo.es/arcgis/rest/services/Cabrera/Cabrera/MapServer. Last accessed June 29, 2014.

———. 2014b. Cabrera/printCabreraA3Land (GPServer), ArcGIS REST Services Directory. http://barreto.md.ieo .es/arcgis/rest/services/Cabrera/printCabreraA3Land/GPServer. Last accessed June 29, 2014.

———. 2014c. Folder: VisorBase, ArcGIS REST Services Directory. http://barreto.md.ieo.es/arcgis/rest/services /visorBase. Last accessed June 29, 2014.

———. 2014d. *Geoportal de la Infraestructura de Datos Espaciales del Instituto Español de Oceanografía.* http://www.geo -ideo.ieo.es/geoportalideo/catalog/main/home.page. Last accessed June 29, 2014.

———. 2014e. Get Capabilities WMS 1.1.1, *del Parque Nacional del Archipiélago de Cabrera.* http://bit.ly/1qp8mQT. Last accessed June 29, 2014.

———. 2014f. Get Capabilities WMS 1.3.0, *del Parque Nacional del Archipiélago de Cabrera.* http://bit.ly/1qI9jl0. Last accessed June 29, 2014.

Julião, R. P., S. Mas, A. Rodriguez, and D. Furtado. 2009. "Portugal and Spain Twin SDIs: From National Projects to an Iberian SDI." In *GSDI-11 Conference Proceedings.* Spatial Data Infrastructure Convergence, Rotterdam, the Netherlands, June 15–19.

Kraak, M.-J. 2004. "The Role of the Map in a Web GIS Environment." *Journal of Geographical Systems* 6:83–93.

La Infraestructura de Datos Espaciales de España. 2014. *Infraestructura de Información Geográfica de España.* http:// www.idee.es. Last accessed June 29, 2014.

MAGRAMA (*Ministerio de Agricultura, Alimentación y Medio Ambiente*). 2014a. *Banco de Datos de la Naturaleza.* http://www.magrama.gob.es/es/biodiversidad/servicios/banco-datos-naturaleza. Last accessed June 29, 2014.

———. 2014b. *Descarga de cartografía de la Red de Parques Nacionales.* http://www.magrama.gob.es/es/red-parques-nacionales/sig. Last accessed June 29, 2014.

Maguire, D., and P. Longley. 2005. "The Emergence of Geoportals and Their Role in Spatial Data Infrastructures." *Computers, Environment and Urban Systems* 29:3–14.

Mathiyalagan, V., S. Grunwald, K. R. Reddy, and S.A. Bloom. 2005. "A WebGIS and Geodatabase for Florida's Wetlands." *Computers and Electronics in Agriculture* 47 (1): 69–75.

Meiner, A. 2010. "Integrated Maritime Policy for the European Union: Consolidating Coastal and Marine Information to Support Maritime Spatial Planning." *Journal of Coastal Conservation* 14 (1): 1–11.

Merrifield, M., W. McClinctock, C. Burt, E. Fox, P. Serpa, C. Steinback, and M. Gleason. 2013. "MarineMap: A Web-Based Platform for Collaborative Marine Protected Area Planning." *Ocean and Coastal Management* 74: 67–76.

Mujabar, P. S., and N. Chandrasekar. 2010. "Web-Based Coastal GIS for Southern Coastal Tamilnadu by Using ArcIMS Server Technology." *International Journal of Geomatics and Geosciences* 1 (3): 649–61.

OGC (Open Geospatial Consortium). 2014. Open Geospatial Consortium. http://www.opengeospatial.org. Last accessed June 29, 2014.

Sandwell, D., S. Gille, J. A. Orcutt, and W. Smith. 2003. "Bathymetry from Space Is Now Possible." *Eos, Transactions of the American Geophysical Union* 84 (5): 37, 44.

Serral, I., X. Pons, R. Jordana, and R. Allué. 2009. "SIG Pesca: An Interoperable GIS Tool for Coastal Knowledge and Management." *Journal of Coastal Research* 56:1587–91.

Shekhar, S., R. R. Vatsavai, N. Sahay, T. E. Burk, and S. Lime. 2001. "WMS and GML-Based Interoperable Web Mapping System." In *Proceedings of the 9th ACM international Symposium on Advances in Geographic Information Systems*, 106–11, Atlanta, GA, November 9–10.

Shekhar, S., and H. Xiong. 2008. *Encyclopedia of GIS.* New York: Springer.

Strickland, J. 2008. How Web 3.0 Will Work. http://bit.ly/1k8YUcH. Last accessed June 29, 2014.

Takken, R. 2008. "Data Sharing from Mash-Up to SDI." *GEOInformatics* 11 (5): 54–55.

Vallega, A. 2005. "From Rio to Johannesburg: The Role of Coastal GIS." *Ocean & Coastal Management* 48 (7): 588–618.

Vermeulen, N. 2013. "From Darwin to the Census of Marine Life: Marine Biology as Big Science." PloS One 8 (1): e54284.

Wright, D. J. 2003. "Introduction." In *Undersea with GIS*, edited by D.J. Wright, xiii–xvi. Redlands, CA: Esri Press.

Wright, D. J., and W. D. Heyman. 2008. "Introduction to the Special Issue: Marine and Coastal GIS for Geomorphology, Habitat Mapping, and Marine Reserves." *Marine Geodesy* 31 (4): 223–30.

Zeng, T. Q., Q. Zhou, P. Cowell, and H. Huang. 2001. "Coastal GIS: Functionality versus Applications." *Journal of Geospatial Engineering* 3 (2): 109–26.

Zhang, C., and W. Li. 2005. "The Roles of Web Feature and Web Map Services in Real-Time Geospatial Data Sharing for Time-Critical Applications." *Cartography and Geographic Information Science* 32 (4): 269–83.

CHAPTER 8

Whale mAPP: Engaging Citizen Scientists to Contribute and Map Marine Mammal Sightings

Lei Lani Stelle and Melodi King

Abstract

Traditional methods of gathering and managing data to map marine mammal distributions requires extensive time and resources so a GIS solution was developed using a volunteered geographic information approach. Whale mAPP consists of mobile and web applications that allow anyone to submit and visualize observations. An Android mobile application, developed with the ArcGIS Runtime software development kit (SDK) for Android application programming interface, uses Global Positioning System–enabled smartphones to record sightings, track and display boat paths, and collect photographs, which are transmitted to a geodatabase through a feature service. The web application, developed using the ArcGIS API for JavaScript application programming interface, allows both the public and researchers to query observations and download data in shapefile format. Curricular materials with conservation themes are designed for middle school students and use ArcGIS Online service to help improve spatial literacy. The additional data collected by users of Whale mAPP will supplement the knowledge base on marine mammals to contribute to research and management efforts.

Understanding Human Impacts on Marine Mammals

The majority of people worldwide live within a short distance of the ocean, yet many are surprisingly unaware of the animals that make our coasts their home. Marine mammals can serve as charismatic megafauna to entice people to make an effort to learn more about marine life and value conservation concerns. Unfortunately, many of the species are endangered or threatened, and baseline data is lacking for most of the populations. Healthy populations of marine mammals serve as a draw to tourists and locals alike (for example, in California millions of people observe whales from boats or

Dawn J. Wright, ed.; 2015; *Ocean Solutions, Earth Solutions*; http://dx.doi.org/10.17128/9781589483651

shore every year [Hoyt 2000]). Photographing and recording videos of the animals often records these encounters, yet this information is rarely shared with anyone beyond the use of social media. Currently, researchers and managers have no way to access this untapped resource. This was the catalyst for the development of Whale mAPP, a citizen science–based web and mobile data collection platform built on the foundation of GIS.

Marine mammals are actually a diverse group of animals and defined as mammals that rely on the marine habitat. They consist of three mammalian orders: *Carnivora* includes seals, sea lions, walruses, polar bears, and otters; *Cetacea* includes whales, dolphins, and porpoises; and *Sirenia* includes the dugongs and manatees. There are approximately 130 extant species of marine mammals worldwide (Committee on Taxonomy 2014).

Marine mammals are found throughout the world's oceans, inhabiting polar, temperate, and tropical regions. Species richness peaks at latitudes of 40, areas of high oceanic productivity (Schipper et al. 2008). Modeling of habitat preference based on surveys of species' presence/absence is increasingly used to identify important predictor variables (Gregr et al. 2013). Common parameters used in these models include bathymetry (depth and slope) and sea surface temperature (SST), both indicators of primary productivity (Kaschner et al. 2006; Becker et al. 2012). Many populations are known to inhabit productive coastal waters along the continental shelf and are thus especially susceptible to threats caused by growing human populations.

There is substantial concern about the conservation of marine mammals, as an estimated 36% of species are considered threatened (Schipper et al. 2008). The exact number of threatened species is difficult to determine because of the challenges associated with surveying animals that spend most of their time out of view and are distributed over a large geographical area, but ranges vary from 23% based only on data-sufficient species but could be as high as 61% if all data-deficient species are also threatened (Schipper et al. 2008). Historically, many populations were decimated by whaling, and some were even extirpated such as the Atlantic gray whale population (Mead and Mitchell 1984). After the International Whaling Commission placed a moratorium on commercial whaling in 1986, many species increased, yet modern populations of most species remain a fraction of their historical size (Roman and Palumbi 2003). Also, both subsistence and commercial harvesting continue currently and threaten 52% of species (Schipper et al. 2008). Today, major threats are incidental mortality from fisheries bycatch and vessel strikes (figure 8.1), pollution (toxins and marine debris, figure 8.2), noise, and climate change (Schipper et al. 2008; Reynolds et al. 2009).

Figure 8.1. Humpback whale injury caused by boat propeller documented off the coast of California. Photo courtesy of Shane Keena Photography.

Figure 8.2. Sea lion entanglement off the coast of Southern California caused by marine debris. Photo courtesy of Shane Keena Photography.

Transforming Data Collection

The lack of reliable data on marine mammal populations makes conservation especially challenging. As the recent Gulf of Mexico oil spill has taught us, it is essential to have baseline data on abundance and distribution just in case of such a disaster so that we can assess and strive to mitigate the damage. As Reynolds et al. (2009, 25) emphasize: "Science does play a crucial role in the conservation of marine mammals and ecosystems." Gregr et al. (2013) describe the importance of modeling habitat use of marine mammals to improve our understanding of species ecology and critical habitat and ability to identify where animals and human threats may overlap. Research needs to focus on core biological questions regarding species distributions, associations, and movement patterns so we can assess the impact of human disturbances.

Gathering this data on marine mammals is time consuming and resource intensive. Regulatory agencies, such as the National Oceanic and Atmospheric Administration (NOAA), collect sightings data from large-scale surveys following predesigned transects (Kinzey et al. 2000). These surveys provide essential data on species distribution and abundance, but because of budget limitations, they have decreased in frequency. In addition, although these studies cover large geographical regions, many areas are still not surveyed seasonally on a regular basis. Taylor et al. (2007) analyzed monitoring data on marine mammal stocks within the United States to calculate that the vast majority of precipitous declines (defined as a 50% decrease in abundance over 15 years) would go undetected. They conclude that increasing survey extent and frequency would improve the ability to detect declines but requires a substantial increase in funding. Researchers have begun to recognize the opportunity provided by the large number of people on whale-watching and other vessels who observe, and often even record, their opportunistic sightings (Palazzo et al. 2004). The difficulty has been in standardizing and sharing this potential wealth of data. With the technological advancements provided by mobile applications, citizen science can greatly expand our knowledge base to support conservation and management. The web and mobile application system Whale mAPP allows anyone equipped with a Global Positioning System (GPS)–enabled Android smartphone (version 4.1 and higher) to record their sightings and transmit that information to a geodatabase that is accessible to researchers and the public. An accompanying website in development (Stelle and King 2014) provides nearly instantaneous data on sighting locations, behavior, group size, and photographs of the species in an interactive format.

Why Citizen Science?

Citizen science, the involvement of interested members of the public in collecting and analyzing data for scientific projects, is not a recent phenomenon. One of the earliest examples is the annual Christmas Bird Count begun by the National Audubon Society, ongoing since 1900. In fact, contributions of bird watchers to the field of ornithology provides one of the greatest examples of citizen science; a recent survey described 10 projects run by Cornell Lab of Ornithology, which contribute over a million records a month (Bonney et al. 2009). Allowing citizens to participate in the scientific inquiry process may bring about awareness, empowerment, and stewardship. Additionally, the inclusion of citizen scientists may help reduce the gaps that have historically divided the public, researchers, and policy makers in environmental management efforts (Connors et al. 2011).

Goodchild (2007) coined the term "volunteered geographic information" (VGI) to describe geographic data provided voluntarily by individuals. The development of Web 2.0, GPS, and the rapid assimilation of mobile technology have made VGI practical. These technologies and the ability to quickly collect, transmit, and submit data is converting citizen scientists into citizen "sensors" (Catlin-Groves 2012). The use of mobile and web applications that utilize a VGI-based strategy in the collection of data in long-term environmental studies is a relatively new field. OakMapper, a working prototype of this data collection method, is a mobile (iPhone) and web-based effort to encourage the public to help monitor the sudden oak death (SOD) of oak trees in California caused by the ramorum leaf blight virus (Geospatial Innovation Facility 2012). The Geyser Notebook 8 application for Android allows users to view information about the Yellowstone geysers and report eruption observations (Glennon 2014). However, there is no peer-reviewed published evidence that this application is used by researchers in understanding geyser activity in Yellowstone National Park. Yet as mobile VGI becomes more common, it is becoming more accepted as useful data by the scientific community. For example, a recent literature search by Catlin-Groves (2012) found over three hundred peer-reviewed publications that included citizen science participation.

Whale mAPP provides a venue for the public, especially students, to become actively involved in valid scientific research efforts. Volunteers are able to contribute sighting records via their mobile application or directly through the website and instantly see their results displayed in comparison to other contributors through the web application. They can easily click a point on the dynamic map and identify details about the sighting, along with additional information on the species. Video interviews and summaries of research efforts educate the user about threats to the species, and a page dedicated to conservation efforts provides additional suggestions of actions that each individual can take to help protect the animals.

Engaging Users

Equally as important as the carefully crafted intention of Whale mAPP's role in citizen science and conservation efforts are its design and execution. User engagement is a key component of attracting data collectors and to keep them coming back. Fundamental to maintaining user engagement is providing an attractive, easy-to-use, and stable application. Web and mobile technologies have already been demonstrated to be successful in a VGI-based study (Connors et al. 2011), and the technology is easy to use and familiar to most users. Applications on these technologies were simultaneously developed to allow users to collect data without the need for an educational background in marine biology. Baseline testing is fundamental to ensuring the apps' stability and is dependent on the profile of the user. However, Whale mAPP is intended for a broad cross section of users in terms of their background in marine biology, technology exposure, and geographic location. Because of this, beta testing with a variety of users has become critical to Whale mAPP's success.

Mobile Application

The Android platform was selected for initial development because of its worldwide popularity compared to its competitors. Specifically, the application was designed to work on versions of the Android operating system greater than or equal to 4.1. The app was built to allow users to contribute track lines to document effort along with marine mammal observations, and visualize both their track line and sightings while their trip is in progress.

When starting a trip from the home page, the user is first asked to fill out a brief profile, including login, boat type, and boat name. Upon entering key trip detail information, the user is presented with a map view (figure 8.3) in which they can visualize their survey as it is under way and record observations. From this view, users can quickly access their camera by clicking the icon and zoom to their current location and explore the surrounding area by panning with their touch screen.

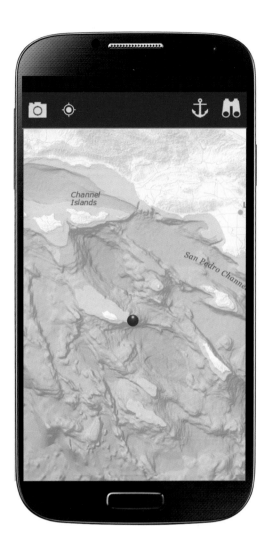

Figure 8.3. **Map view displayed to the user throughout their trip using the Whale mAPP Android application provides them with access to the camera, their current location, and the form for making observations (binoculars icon).** Image by Whale mAPP © 2014. Data sources: Esri, GEBCO, NOAA, National Geographic, De Lorme, HERE, Geonames.org, and other contributors.

The target audience's cellular and GPS reception capabilities exhibit huge variability, especially when on vessels further offshore. Because of this, the Android app was designed to store data locally on the device and operate either in a connected or disconnected map mode. When operating in the connected mode, the app will display the user's track line and observations on top of the Esri Ocean Basemap, but when reception is not available, the user's track line and observations will be displayed on either a blank canvas or basemaps previously downloaded by the user for their particular geographic extent.

When the user chooses to record an observation through the mobile application, they click the binoculars icon and get to a form based on datasheets used in research by both students and Earthwatch Institute volunteers (figure 8.4). The form, which is focused on monitoring the distribution and habitat use of marine mammal species in the Southern California Bight, captures the essential information on the species type, count, behavior, weather, and sea conditions. Additionally, it allows users to associate a photo with the observation and document their confidence in the observation on a scale of one to five (with one star being low confidence and five being high confidence). As the user fills out the form, they can see their responses displayed prior to submitting the form. Giving users the room to fix mistakes in their observation details is an essential component of the user interface design (Norman 1988).

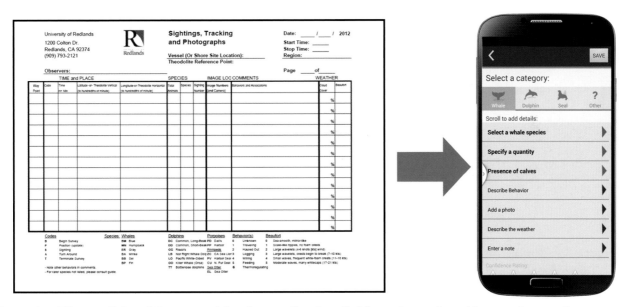

Figure 8.4. The traditional form used for data collection (left) can be replaced by an observation form in a mobile application (right). Image by Whale mAPP © 2014. University of Redlands logo used with permission.

The application is geo-aware, providing users with species lists unique to their location. This is critical as the species present in Southern California, for example, vary in comparison to those in the Baltic Sea. Regions for which unique species lists currently exist include the West Coast of the United States, Baltic Sea, Southeast Alaska/Western Canada, Antarctic Peninsula, Ross Sea in Antarctica, New Zealand, and the East Coast of the United States and Caribbean Sea (figure 8.5). As the user base grows, so will the geography-specific species lists.

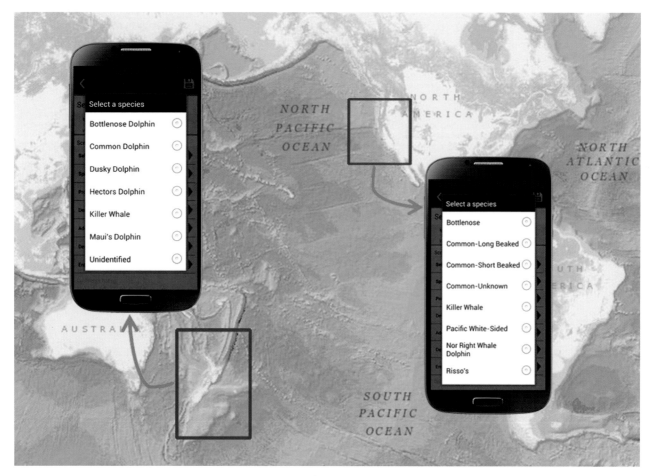

Figure 8.5. **Whale mAPP's geo-specific species view ensures that users are presented with the most accurate species currently known in their region.** Image by Whale mAPP © 2014. Data sources: Esri, GEBCO, NOAA, National Geographic, De Lorme, HERE, Geonames.org, and other contributors.

The My Data View of the application allows users to view basic information about their trip. All the data collected is currently stored in a SQLite database located on the Android device itself. When connectivity is available, trips and observations can be either automatically or manually synced to ArcGIS feature services.

Web Application

A web application has been designed to accompany the Android app. It was important to design the web application with a similar branding technique as the Android application to provide users with a seamless experience. The animated home page provides basic information on the project and describes how to get started (figure 8.6). In addition to the home page, the site includes a learning component and a mAPP component.

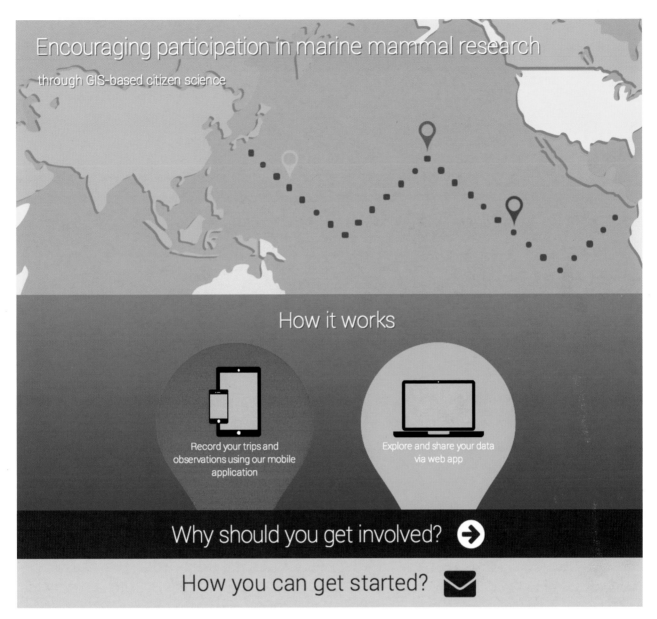

Figure 8.6. **The animated home page of the whalemapp.org website directs the user to additional information on the project and ways to get involved.** Image by Whale mAPP © 2014.

The website's mAPP allows users to visualize and query data. Datasets are presented to the user on a map powered by feature services for both the track lines and observations, and corresponding photos (figure 8.7). Through the mAPP interface, users can choose to view either other contributors' observations or only their own. They can also query by species category and date range. A dashboard on the mAPP shows them how many track lines and observations they have in their current query. The query capabilities and dashboard were designed with the intention of motivating users and providing them with the feeling of belonging to a community of fellow conservationists.

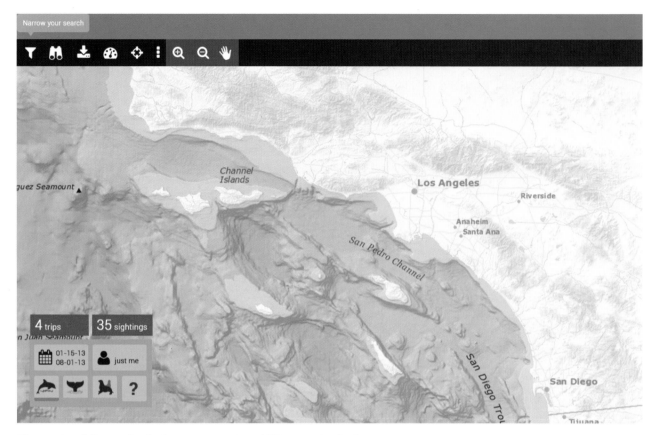

Figure 8.7. View of whalemapp.org's mAPP page includes map controls, querying and printing capability, and a dashboard. Image by Whale mAPP © 2014. Data sources: Esri, GEBCO, NOAA, National Geographic, De Lorme, HERE, Geonames.org, and other contributors.

Another goal of Whale mAPP is to empower users by giving them control of their data. Pop-up windows are used to display observation details and allow users to correct previously recorded observation notes and upload associated photos. Photo upload is essential because photos are often taken with a different device than the phone, such as digital single-lens reflex (DSLR) cameras with zoom lenses. An additional feature of the website's mAPP is the capability to download data in shapefile format. This functionality is targeted at researchers, including those making management decisions, but can also be used in educational settings.

The learning component of the website is meant to engage users in gaining a better understanding of species identification and conservation efforts and provide educational material for use in the classroom. Species identification information is broken up both by individual species and taxonomic group, using common names that are recognizable to a general audience (e.g., whales; dolphins and porpoises; seals, sea lions, and walruses).

Information for each general taxon includes descriptions of the biology of each group, along with images to help users identify species that can be confused such as true seals and sea lions.

Information for individual species includes their common and scientific names, physical characteristics (e.g., length, weight, and coloration), prey, behavioral characteristics, distribution, social system/group size, predators, migration/dispersal patterns, population status, major threats, and "fun facts." Each fact page includes photographs and illustrations (figure 8.8).

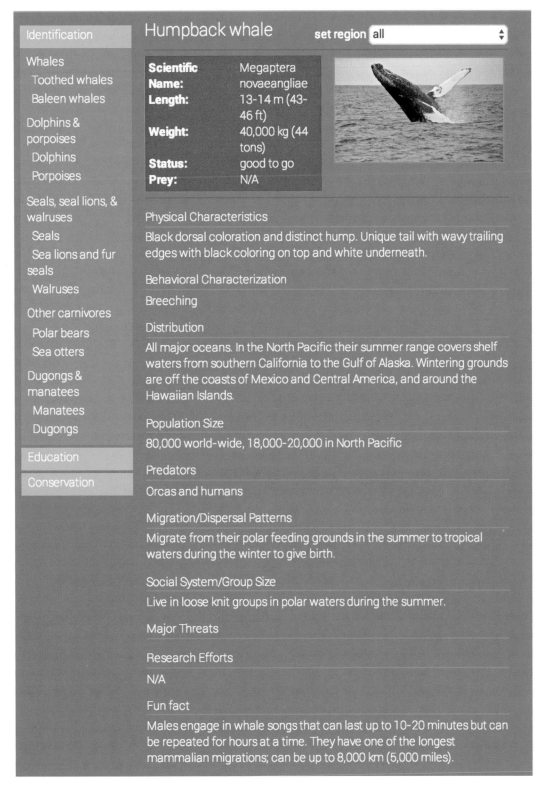

Figure 8.8. Draft view of the learning page of the whalemapp.org website showcasing species identification for humpback whales. Photo courtesy of Shane Keena Photography.

A major goal of the learning component of the website is to increase awareness of conservation issues affecting marine mammals. Threats such as pollution (solid waste, toxins, sound), human interactions (disturbance, entanglements, ship strikes), ecosystem shifts (global climate change, effects of overfishing), and disease are presented with descriptions, images, videos, and case studies. Research efforts are highlighted through the use of project descriptions and video clips of professionals in the field describing their research, and especially the importance of spatial data in their studies.

The learning component of the website also includes educational curricular materials focused on coastal marine mammal species to increase awareness and encourage conservation. GIS-based activities were developed to meet California state standards, specifically the Common Core standards and Next Generation Science standards (California Department of Education, 2014a and 2014b). The materials use ArcGIS Online (arcgis.com), which is freely available to the public, so the only requirements are computers with Internet access. The goal of the educational materials is to help students gain experience actually conducting scientific studies by generating hypotheses and analyzing data. Curricular activities help focus their questions on conservation themes to increase student awareness of threats to marine life and allow them to directly visualize these relationships. Video instructions and frequently asked questions (FAQs) are provided to help teachers get comfortable with the software and labs.

Baseline and Beta Testing

As previously mentioned, because of the variability in the expected user behaviors and skill levels, baseline and beta testing has become a critical component of the app's development. Baseline testing refers to a set of procedures performed on the application prior to its release to beta users. Each step in a procedure has an expected outcome that must be met for the test to be passed. There are baseline procedures for testing the setup menu; syncing functionality, region-specific offline basemaps, and species lists; and recording trips and observations.

Baseline testing helps eliminate errors and identify known issues. Whale mAPP's beta testers vary in geographic location, which impacts cellular and GPS dependability. Additionally, they vary in their intended use of the application. Some beta users are researchers on eight-hour trips in remote areas of the world while others are average citizens on a two-hour whale-watching boat. These two different use cases will have very different expectations of the app and reflect the range of target users. Beta testing has shed light on battery life and connectivity issues that have driven adaptations to the code base to make it more flexible for a wide range of users. For example, beta testing helped demonstrate that incorporating a service that keeps the app as a foreground activity allows users to minimize the app and put their phone to sleep. When minimized, the maps are not being continually updated. Additionally, when connectivity is lost, the maps stop drawing until connectivity is regained rather than continuing to try and draw features. These options have greatly reduced battery usage.

Additionally, beta testing has helped ensure that the accuracy of the locational data being collected is comparable to the data collected scientifically. A beta tester in Dominica (Caribbean) ran the application side by side with their Suunto Ambit GPS watch and contributed their trip data. The horizontal accuracy (~3 ft) of the Suunto watch provided a baseline for the accuracy needed for scientific data collection. The results of this testing (figure 8.9) demonstrate that the horizontal accuracy of the data collected using Whale mAPP is suitable for scientific-grade data collection. However, these results may vary in regions of the world with poor satellite reception.

Data Collection Comparison: Dominica

—— Whale mAPP ——— GPS Watch

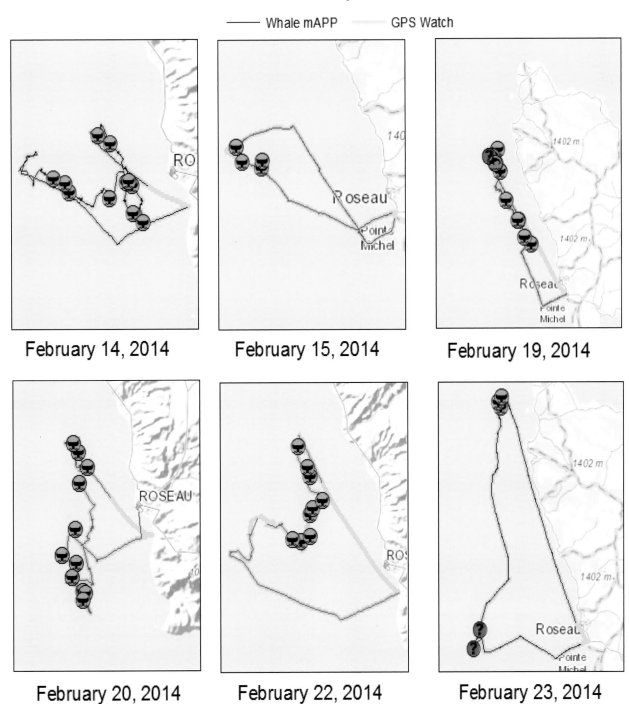

February 14, 2014   February 15, 2014   February 19, 2014

February 20, 2014   February 22, 2014   February 23, 2014

Figure 8.9. Data collected using Whale mAPP (purple) and a Suunto Ambit GPS watch (yellow) in Dominica (Caribbean). Suunto Ambit GPS watch data provided by Ted Cheeseman/Cheeseman's Ecology Safaris. Image by Whale mAPP © 2014. Data sources: Ted Cheeseman/Ted Cheeseman's Ecology Safaris, Esri, HERE, DeLorme, TomTom, Intermap, increment P Corp., GEBCO, USGS, FAO, NPS, NRCAN, GeoBase, IGN, Kadaster NL, Ordnance Survey, Esri Japan, METI, Esri China (Hong Kong), swisstopo, MapmyIndia, © OpenStreetMap contributors, GIS User Community.

Additionally, beta testing has shed light on the stability of the mobile application. Because the app is not available in Google Play Store, it can be difficult to document crashes reported by users. Automated crash reporting for Android (ACRA)/Acralyzer is a light open-source library being used to send crash reports directly to the developer from the user's phone. This has helped identify stability issues with the app without imposing unrealistic demands on the user when or if the app crashes. Finally, beta testing has helped gauge the usability and scientific integrity of the data collected with the app. Beta testers have been key players in identifying errors in species-specific behavior lists and count options on the observation form. Additionally, they have helped identify and resolve unclear steps in the workflow for recording observations.

But Is the Data Good Enough for Science?

Data quality is of serious concern to members of the scientific community. Because of this, several measures have been taken to ensure that Whale mAPP is collecting as high-quality data as possible. As previously mentioned, the observation data collection form in the Android application was based on a survey form used by an expert user and includes standard fields in marine mammal surveys. This ensures that the data collected includes essential information for scientific research.

Additionally, expert and public user accounts (figure 8.10) are both available to select from when users register through the whalemapp.org website. Currently, experts are approved on a case-by-case basis and must provide justification for their credentials, such as an education in marine biology or certification as a whale watch naturalist. This is a key component in making the data useful for researchers as entries can be sorted by user type so that a scientist may choose to include only expert submissions to increase their confidence in the accuracy of the data.

Registration Form View

Novice Users

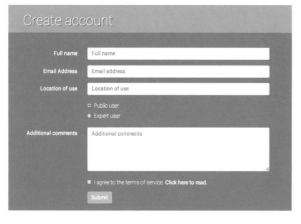

Expert Users

Figure 8.10. **The form used to register as a beta tester changes dynamically, depending on whether the user is a public user (left) or an expert in their field (right)**. Image by Whale mAPP © 2014.

Additionally, options are presented to users as drop-down menus, in the Android application's observation form, rather than requiring them to manually type or guess observation details, which ensures that all the necessary data is recorded. This is particularly useful for fields on the observation form such as counts, behavior, weather, and sea conditions. This also reduces variability in the entry of information submitted. If a user attempts to save their submission before entering all the data, they are alerted with a message to fill in the required fields. Using limited options also eliminates the dependency on user memorization, which is a key component of good user interface design (Norman 1988).

The automatic tracking of each survey is one of the key features that makes data obtained from Whale mAPP useful to scientific investigations. Survey lines provide scientists with information about the effort made in making marine mammal observations. To extrapolate from sightings data to any model of population distribution, information on both presence and absence of the target species is essential.

The last component used for ensuring data quality is the use of the GPS chip on the phone. This reduces the risk of error from transcribing locational information and also allows the app to automatically track surveys and observation locations. The quality of the GPS data can be controlled by the use of time and distance parameters of the location updates being provided by the GPS provider on each device.

Comparison of Traditional Methods to Whale mAPP

To compare the ease, efficiency, and accuracy of the traditional data collection and mapping methods to Whale mAPP, researchers ran both protocols simultaneously for a trip off the coast of Catalina Island, California. Catalina Island is an excellent location for testing because of the island's unreliable cellular reception. The end product using both protocols was a map showing track lines and observations symbolized by species type from the data collected.

Recording observations with the traditional survey form involves writing the time (from a watch), position (from GPS), and species information on a preprinted data sheet. All locations were stored as way points on the handheld GPS (Garmin GPSmap 62s) and the latitude and longitude also written down as a backup. Tracks were stored by clearing the GPS at the start of each survey and saving the survey track at the end of each trip. In the field, the time involved in both forms of data collection was similar, but the traditional method is much easier if done by two people, one to observe and read off the GPS information while the other records, but recording on Whale mAPP requires only one user. The biggest difference in time investment is the additional time required to transform the collected data into a geospatial format. This procedure involves transcribing handwritten data into Microsoft Excel, uploading way points and track lines from the GPS, joining the data in GIS to produce the map, and symbolizing the sightings by species. Processing the data typically requires an experienced user at least one hour for a two- to three-hour survey but obviously varies depending on the number of sightings and GIS experience of the user.

The procedure for the data collected using the Whale mAPP Android application has far fewer steps. Using the application, a trip was started from the home screen, and the track line was automatically recorded and displayed on a map view for the user as the trip was being recorded.

From the map view, an observation form was used to record the details for each sighting. Once the observation was saved, a point symbol was displayed on the map view representing the species category recorded. The app was designed to automatically sync the track lines and point observations to the Whale mAPP feature service. A feature service is an interface between an app and a GIS database that allows the transfer of user data to the database. Because the data is already stored in a geospatial format, the web application makes visualization of the data instantaneous. However, the same map created using the manual data collection method can be created in ArcGIS using a predefined layer to handle symbology based on the species category to allow for additional analyses. The map (figure 8.11) displays data collected using Whale mAPP in comparison to that of Garmin GPSmap 62s (accuracy ~12 ft). This solution decreases the amount of time researchers and volunteers spend in the field and could potentially increase the amount of data collected.

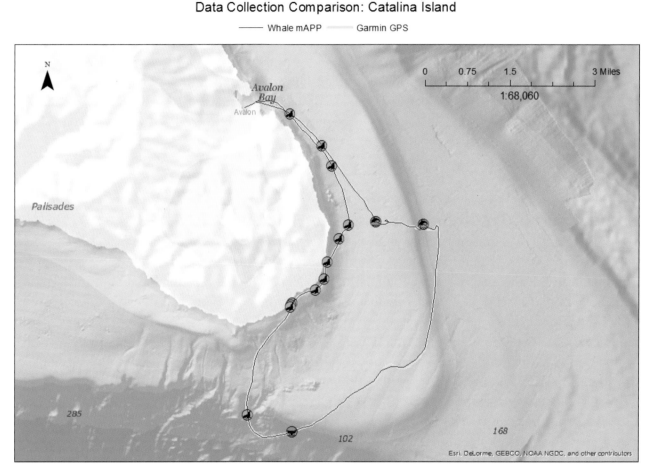

Figure 8.11. **Comparison of track data collected using Whale mAPP and Garmin GPS map 62s.** Image by Whale mAPP © 2014. Data sources: Whale mAPP, Esri, DeLorme, GEBCO, NOAA NGDC, and other contributors.

Future Work

Several efforts are under way to expand the user base while Whale mAPP is still in beta testing.

Expanding and Maintaining the User Base

The first effort in expansion is moving the project as Whale mAPP is currently housed on the University of Redlands campus. However, when the app is released to Google Play Store, it will need a new home because of an expected increase in the number of users. Depending on available funds, this project is perfect for ArcGIS Online. ArcGIS Online could be used to enable editing and image storage through the use of feature services. Additionally, its representational state transfer (REST) end points could be used for data download through the whalemapp.org website. Finally, its built-in user accounts would provide an excellent framework for user management and data sharing. Another important step is to port the application from Android to the iOS platform to expand the potential user base to those with iOS devices.

Collaborative Coding, Collaborative Data Collection

Several efforts are under way to make Whale mAPP a more collaborative project. In addition to syncing data with the database at the University of Redlands, we are eager to extend Whale mAPP to allow users to share and export their data to a commonly used Microsoft Excel comma-separated value (CSV) format directly from their phones or to partner with organizations collecting similar types of data. Additionally, the Whale mAPP project is collaborating with Courtney Hann, a graduate student at Oregon State University who is working with the 501(c)(3) nonprofit Alaska Whale Foundation, coastal Alaskan communities, whale-watching groups, and environmental organizations to evaluate using Whale mAPP as a citizen science mobile application tool for monitoring marine mammals and raising public awareness about important conservation issues in Southeast Alaska. For her research, Hann will be contributing educational components to the Whale mAPP Android application. This social coding endeavor will be performed using a private GitHub repository. Data collected by volunteers using Whale mAPP will be compared to a standardized dataset collected by trained naturalists on National Geographic cruise vessels. Additionally, user surveys will assess the format of the mobile application and its effectiveness in educating participants about important marine mammal topics. Through this collaboration, we will be able to explicitly test the primary hypothesis that Whale mAPP can engage citizen scientists to provide an accurate and low-cost approach to marine mammal monitoring.

Conclusion

Marine mammal species face increasing threats globally because of a growing human population and the associated environmental degradation. Information on each species' distribution and habitat use is essential to management efforts. Yet the vastness of the oceans and lack of funding make such data collection extremely challenging.

Whale mAPP provides an opportunity to capitalize both on the public's fascination with these animals and the increasing number of people observing them in the wild. Mobile technology can greatly expand the cohort of interested citizens while increasing the accuracy of their data. The mobile app and associated website were carefully designed and tested to be user-friendly, engaging, and useful for recording and visualizing essential spatial data.

This global set of tools will continue to evolve with user feedback and collaboration in both the scientific and development communities. Whale mAPP has the capability of serving as a standard for baseline data collection while simultaneously encouraging its users to actively contribute to scientific research and providing them with the education they need to act as stewards of marine life.

Acknowledgments

Funding for the development of Whale mAPP was generously provided by a grant from the California Coastal Commission's Whale Tail License Plate Fund. Additional support was provided by Earthwatch Institute grants and volunteers. The project was initiated with the support of Lei Lani Stelle through a LENS fellowship funded by the Keck Foundation, Melodi King at the University of Redlands GIS master degree's Major Individual Project, Smallmelo Geographic Information Services, and Crown Chimp Design & Development. Assistance was provided by David Smith, GIS and mapping support consultant in the Center for Spatial Studies at the University of Redlands. Invaluable collaborations with Dana Wharf Whale Watch and Cheeseman's Ecology Safaris have provided the opportunity for testing Whale mAPP. Thank you to all the beta testers, especially Ted Cheeseman, Courtney Hann, and Melissa Panfili-Galieti. This manuscript was improved by the constructive critique of two anonymous reviewers.

References

Becker, E. A., D. G. Foley, K. A. Forney, J. Barlow, J. V. Redfern, and C. L. Gentemann. 2012. "Forecasting Cetacean Abundance Patterns to Enhance Management Decisions." *Endangered Species Research* 16 (2): 97–112.

Bonney, R., C. B. Cooper, J. Dickinson, S. Kelling, T. Phillips, K. V. Rosenberg, and J. Shirk. 2009. "Citizen Science: A Developing Tool for Expanding Science Knowledge and Scientific Literacy. *BioScience* 59 (11): 977–84.

California Department of Education. 2014a. Common Core State Standards—Resources. http://www.cde.ca.gov/re/cc. Last accessed July 29, 2014.

———. 2014b. Next Generation Science Standards—Science. http://www.cde.ca.gov/pd/ca/sc/ngssintrod.asp. Last accessed July 29, 2014.

Catlin-Groves, C. L. 2012. "The Citizen Science Landscape: From Volunteers to Citizen Sensors and Beyond." *International Journal of Zoology* 2012:1–14.

Committee on Taxonomy. 2014. Society for Marine Mammalogy: List of Marine Mammal Species and Subspecies. http://bit.ly/1jyhRFR. Last accessed June 28, 2014.

Connors, J. P., S. Lei, and M. Kelly. 2011. "Citizen Science in the Age of Neogeography: Utilizing Volunteered Geographic Information for Environmental Monitoring." *Annals of the Association of American Geographers* 102 (6): 1267–89.

Geospatial Innovation Facility. 2012. OakMapper: Mapping Sudden Oak Death in California. http://www.oakmapper.org. Last accessed June 28, 2014.

Glennon, A. 2014. Geyser Notebook. http://bit.ly/1lZghAN. Last accessed June 29, 2014.

Goodchild, M. F. 2007. "Citizens as Sensors: The World of Volunteered Geography." *GeoJournal* 69 (4): 211–21.

Gregr, E. J., M. F. Baumgartner, K. L. Laidre, and D. M. Palacios. 2013. "Marine Mammal Habitat Models Come of Age: The Emergence of Ecological and Management Relevance." *Endangered Species Research* 22:205–12.

Hoyt, E. 2000. *Whale Watching 2001: Worldwide Tourism Numbers, Expenditures, and Expanding Socioeconomic Benefits.* International Fund for Animal Welfare Technical Report, i-vi. Yarmouth Port, MA: International Fund for Animal Welfare.

Kaschner, K., R. Watson, A. W. Trites, and D. Pauly. 2006. "Mapping World-Wide Distributions of Marine Mammal Species Using a Relative Environmental Suitability (RES) Model." *Marine Ecology Progress Series* 316:285–310.

Kinzey, D., P. Olson, and T. Gerrodette. 2000. *Marine Mammal Data Collection Procedures on Research Ship Line-Transect Surveys by the Southwest Fisheries Science Center.* National Oceanic and Atmospheric Administration Southwest Fisheries Science Center Administrative Report LJ-00-08. La Jolla, CA: NOAA.

Mead, J. G., and E. D. Mitchell. 1984. "Atlantic Gray Whales." In *The Gray Whale, Eschrichtius robustus*, edited by M. L. Jones, S. L. Swartz, and S. Leatherwood, 33–53. Orlando, FL: Academic Press.

Norman, D. 1988. *The Design of Everyday Things.* New York: Doubleday.

Palazzo, J. T., M. A. Iniguez, and M. Hevia. 2004. "A Worldwide Directory of Whale Watching Research." In *Proceedings of the International Whaling Commission Scientific Committee*, Sorrento, Italy, Paper SC/56/WW8. http://bit.ly/1rLB9fP.

Reynolds III, J. E., H. Marsh, and T. J. Ragen. 2009. "Marine Mammal Conservation." *Endangered Species Research* 7:23–28.

Roman, J., and S. R. Palumbi. 2003. "Whales before Whaling in the North Atlantic." *Science* 301 (5632): 508–10.

Schipper, J., J. S. Chanson, F. Chiozza, N. A. Cox, M. Hoffmann, V. Katariya, J. Lamoreux, A. S. L. Rodrigues, S. N. Stuart, H. J. Temple, J. Baillie, L. Boitani, T. E. Lacher, R. A. Mittermeier, A. T. Smith, D. Absolon, J. M. Aguiar, G. Amori, N. Bakkour, R. Baldi, R. J. Berridge, J. Bielby, P. A. Black, J. J. Blanc, T. M. Brooks, J. A. Burton, T. M. Butynski, G. Catullo, R. Chapman, Z. Cokeliss, B. Collen, J. Conroy, J. G. Cooke, G. A. B. da Fonseca, A. E. Derocher, H. T. Dublin, J. W. Duckworth, L. Emmons, R. H. Emslie, M. Festa-Bianchet, M. Foster, S. Foster, D. L. Garshelis, C. Gates, M. Gimenez-Dixon, S. Gonzalez, J. F. Gonzalez-Maya, T. C. Good, G. Hammerson, P. S. Hammond, D. Happold, M. Happold, J. Hare, R. B. Harris, C. E. Hawkins, M. Haywood, L. R. Heaney, S. Hedges, K. M. Helgen, C. Hilton-Taylor, S. A. Hussain, N. Ishii, T. A. Jefferson, R. K. B. Jenkins, C. H. Johnston, M. Keith, J. Kingdon, D. H. Knox, K. M. Kovacs, P. Langhammer, K. Leus, R. Lewison, G. Lichtenstein, L. F. Lowry, Z. Macavoy, G. M. Mace, D. P. Mallon, M. Masi, M. W. McKnight, R. A. Medellín, P. Medici, G. Mills, P. D. Moehlman, S. Molur, A. Mora, K. Nowell, J. F. Oates, W. Olech, W. R. L. Oliver, M. Oprea, B. D. Patterson, W. F. Perrin, B. A. Polidoro, C. Pollock, A. Powel, Y. Protas, P. Racey, J. Ragle, P. Ramani, G. Rathbun, R. R. Reeves, S. B. Reilly, J. E. Reynolds, C. Rondinini, R. G. Rosell-Ambal, M. Rulli, A. B. Rylands, S. Savini, C. J. Schank, W. Sechrest, C. Self-Sullivan, A. Shoemaker, C. Sillero-Zubiri, N. De Silva, D. E. Smith, C. Srinivasulu, P. J. Stephenson, N. van Strien, B. K. Talukdar, B. L. Taylor, R. Timmins, D. G. Tirira, M. F. Tognelli, K. Tsytsulina, L. M. Veiga, J.-C. Vié, E. A. Williamson, S. A. Wyatt, Y. Xie, and B. E. Young. 2008. "The Status of the World's Land and Marine Mammals: Diversity, Threat, and Knowledge." *Science* 322 (5899): 225–30.

Stelle, L., and M. King. 2014. WhalemAPP: Track and Visualize Your Marine Observations. http://www.whalemapp.org. Last accessed June 29, 2014.

Taylor, B. L., M. Martinez, T. Gerrodette, J. Barlow, and Y. N. Hrovat. 2007. "Lessons from Monitoring Trends in Abundance of Marine Mammals." *Marine Mammal Science* 23 (1): 157–75.

CHAPTER 9

Extending Esri Geoportal Server to Meet the Needs of the West Coast Ocean Data Network and Inform Regional Ocean Management

Todd Hallenbeck, Tim Welch, Steven J. Steinberg, and Andy Lanier

Abstract

Lack of accessibility to geospatial data by coastal resource managers has been identified at the regional and national levels as a hurdle to improved ecosystem-based management in the United States. The West Coast Ocean Data Portal (http://portal.westcoastoceans.org) is a project of the West Coast Governors Alliance on Ocean Health to address this problem by working to increase discovery and connectivity of coastal and ocean data users and systems to better inform West Coast ocean health decisions. The West Coast Ocean Data Network is a community of practice connecting West Coast data managers and users to develop best practices for regional data sharing. We demonstrate the West Coast Ocean Data Portal and the strategy for using Esri Geoportal Server software to selectively harvest, curate, and publish access to data resources through a compelling portal experience that is customizable and can integrate novel technologies to inform regional ocean management.

Introduction

The release of the US National Ocean Policy in April 2013 (National Ocean Council 2013) has ushered in an era of ecosystem-based management (EBM), which explicitly recognizes the need for spatial and temporal distribution of ocean resources and activities to minimize conflict between

Dawn J. Wright, ed.; 2015; *Ocean Solutions, Earth Solutions*; http://dx.doi.org/10.17128/9781589483651

human and natural systems (Crowder et al. 2006; Young et al. 2007). As state governments work to address these issues, they are joining together to form regional ocean partnerships that match the geographic extent of the ecosystems being managed (Hershman and Russell 2006). As these partnerships work to inform regional ocean health issues, challenges emerge relating to aggregating and accessing the best available science and geospatial data at this regional scale (WCGA 2006; Hennessey 2011; Obama 2013).

The West Coast Ocean Data Portal (WCODP) (WCGA 2014) is a project of the West Coast Governors Alliance on Ocean Health (WCGA) designed to address data access and aggregation issues by working to improve discovery and connectivity of coastal and ocean data and people. The portal is working to better inform West Coast ocean resource management, policy development, and planning. WCODP is a web application built on Esri Geoportal Server software (Esri 2014) to selectively harvest, curate, and visualize data resources from the West Coast Ocean Data Network (WCODN).

The West Coast Ocean Data Network

WCODN began in 2012 as a grassroots group of data managers and users from state and federal agencies, tribal governments, nongovernmental organizations (NGOs), and universities. Members of the network are dedicated to sharing best practices appropriate to increasing the accessibility of West Coast ocean and coastal data for regional ocean management, policy development, and planning. WCODN is organized into three working groups: data standards and priorities, information technology and software, and outreach and communication (figure 9.1). WCODN members contribute to and guide the development of WCODP and share lessons learned from their respective data management efforts.

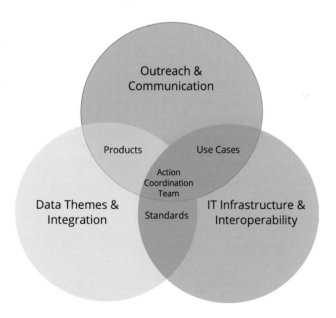

Figure 9.1. West Coast Ocean Data Network structure and function. Three working groups, data, information technology, and outreach, support the development of the West Coast Ocean Data Portal and develop and share data best practices.

The underlying design strategy of WCODP was to avoid duplication of existing efforts of network members in the region. To accomplish this, WCODP connects data catalogs maintained by members of WCODN and selectively harvests and curates the best available data resources from each contributor (figure 9.2). These datasets are then published through a modern, lightweight portal experience designed to meet the expressed needs of WCGA for data organization and discovery. The actual datasets themselves are not harvested by WCODP. Instead, standardized metadata is harvested, which contains service links to access the data in all its various forms, whether it is a file or web service. The benefit of this approach is that the dataset resides with the original provider, who is able to control what is published, and WCODN ensures that members can communicate their changes.

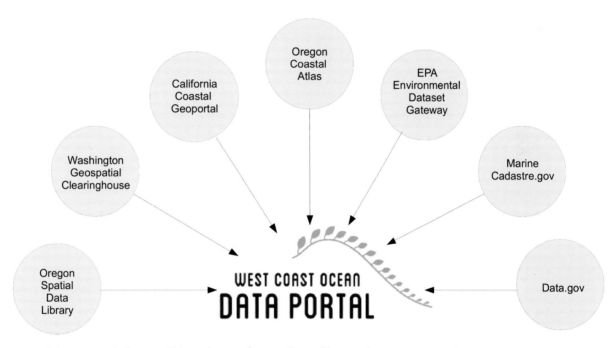

Figure 9.2. Diagram of diverse West Coast Ocean Data Network partner catalogs.

Members of WCODN contribute data to WCODP because they see value in collaborating with regional partners to advance their agency, organization, or tribal mission or mandates for coastal management. WCODN members also benefit from wider advertisement and accessibility of data holdings to new user groups and stakeholders, technical capacity building through WCODN working groups and training, demonstration of the regional collaboration and sharing often required by grant funding agencies, and inclusion of data in decision-making for regional issues such as sea-level rise adaptation, ocean acidification, and marine planning.

The challenge to this networked approach is that WCODN members are both geographically and technologically diverse with respect to their individual data-sharing infrastructure and practices. For example, WCODN members currently use at least seven different data catalog products, including Esri Geoportal Server software; the Comprehensive Knowledge Archive Network (CKAN) (The Open Knowledge Foundation 2014); GeoNetwork; pycsw (Karlidis et al. 2014); Thematic Real-Time Environmental Distributed Data Services (THREDDS) Data Server (Unidata Program Center 2014); Mercury (Oak Ridge National Laboratory 2013); and ArcGIS Online service (arcgis.com).

Within these catalogs, data resources are represented by at least four different metadata standards, including Dublin Core, Federal Geospatial Data Committee (FGDC), Content Standard for Digital Geospatial Metadata (CSDGM), International Organization for Standardization (ISO) 19115, and ISO 19115-2. In order to succeed, WCODP had to accommodate the diversity of technology and standards used by WCODN partners.

Data Catalog Requirements and Testing

In late 2012, four different data catalog products were thoroughly assessed and considered for the West Coast Ocean Data Portal, including Esri Geoportal Server, GeoNetwork, CKAN, and pycsw (Welch et al. 2012). The primary requirements assessed included that the product:

- possess an open-source license, well documented, with a responsive development team;
- provide a full-featured application programming interface (API) allowing the catalog to be readily integrated into a larger portal solution with its own distinct user interface (UI);
- provide faceted search capability, allowing users to drill down or filter their search by theme, geographic area, source, service type, or other characteristics;
- support selective metadata harvesting from data catalogs using the standard Catalog Services for the Web (CSW), as well as CSW server capabilities so that others could harvest WCODP itself;
- support alternative or nonstandard metadata harvesting when needed (e.g., from web-accessible folders [WAF], THREDDS, and ArcGIS Online);
- support a variety of metadata standards, including ISO 19115, ISO 19115-2, FGDC CSDGM, and Dublin Core; and
- allow metadata records to be curated by organizing into one or more groups, effectively supporting the creation of a simple taxonomy for West Coast–specific categories and issues.

As of January 2013, testing showed that only two data catalog products met most of the requirements. It came down to Esri Geoportal Server, which supported all the metadata standards but not faceted search, and CKAN, which supported faceted search but did not provide full CSW capabilities or support all the required metadata standards. Conversations with both development teams determined that CKAN was in the process of covering some, but not all, the identified feature gaps through enhancements being developed for the Data.gov platform. However, it would not be ready in time for this project. The Esri Geoportal team demonstrated that they were working on support for faceted search and would be able to provide early access to those enhancements. For this reason, Esri Geoportal Server was selected as the WCODP data catalog solution, and over the months to follow, the WCODP team and Point 97 (2014), an ocean technology solution company in Portland, Oregon, worked closely with the Esri Geoportal development team to help shape and test these enhancements.

West Coast Ocean Data Portal Architecture

In July 2013, Point 97 released a beta version of WCODP that incorporated Esri Geoportal Server with a PostgreSQL back end, leveraging a newly released Geoportal functionality called the Geoportal Facets Customization (GFC) (figure 9.3). The GFC component pulls metadata documents from the backend Esri Geoportal Server database and passes them through to Apache Solr (The Apache Software Foundation 2012), a search platform that provides full-text, faceted, and geospatial search capabilities accessible through a robust API. The WCODP UI uses this Solr API to provide users with the ability to rapidly search and filter through the available datasets by keyword, category, location, or source (figure 9.4a). Discovery of resources through faceted search is often used in modern e-commerce websites, allowing visitors to quickly drill down and find their product of interest. In the case of WCODP, instead of filtering by product price and feature, the GFC component creates search facets tuned for discovery of spatial datasets. Once a user finds a dataset they are interested in, they are provided with a range of additional information provided through the metadata, including the title, abstract, date published, creator, publisher, contact, use constraints, and all the available links for accessing the resource (figure 9.4b).

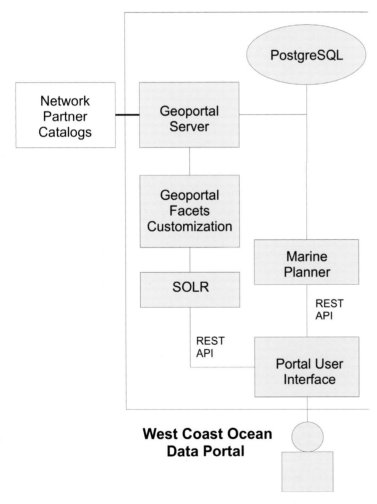

Figure 9.3. Diagram of West Coast Ocean Data Portal system architecture. Network partners contribute metadata in standards-based XML documents to Esri Geoportal Server. The Geoportal Facets Customization pulls discovery metadata fields from the Esri Geoportal Server and pushes them into Solr, which provides a robust, faceted search API that drives the portal user interface.

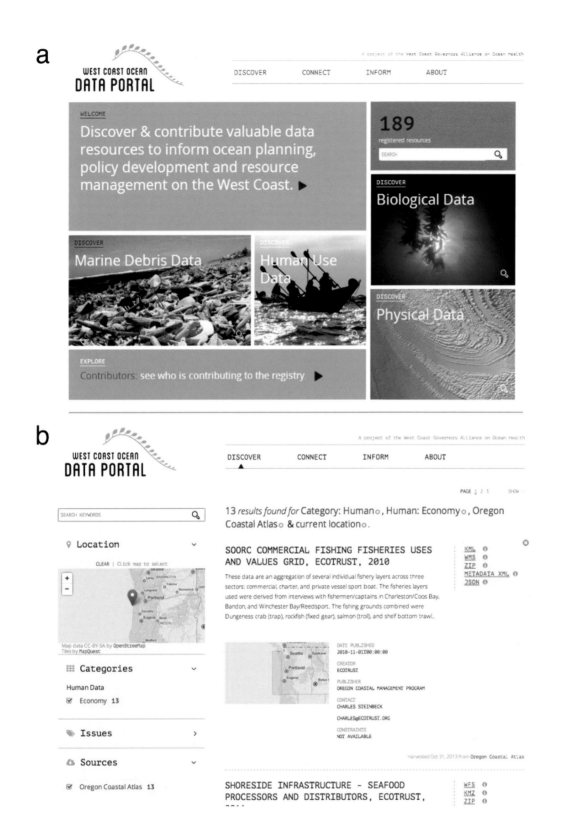

Figure 9.4. The (a) custom home page and (b) search interface of the West Coast Ocean Data Portal. The search interface allows users to drill down and discover datasets using a variety of search facets, including location, category, issue, and source. Images courtesy of the West Coast Governors Alliance; location map data CC-by-SA by OpenStreetMap; titles by MapQuest; "SOORC Commercial Fishing Fisheries Uses and Values Grid, Ecotrust, 2010" courtesy of Oregon Coastal Management Program, harvested October 31, 2013, from Oregon Coastal Atlas.

This detailed information, in many cases, is queried directly from Esri Geoportal Server, using its provided representational state transfer (REST) API, rather than Solr. The strategy, in this case, is not to push everything through Solr, but to provide it with only the metadata fields needed for search and discovery of datasets and their service links. More detailed metadata can then be queried directly from Esri Geoportal Server. Point 97 developed an attractive, custom front end to consume the Geoportal's REST and Solr JavaScript Object Notation (JSON) APIs. This API-driven approach allows a variety of software components, which each do a separate job, to be integrated into a unified website through customizable user experience.

In January 2014, the West Coast Ocean Data Portal was formally released to the public with an initial catalog of over 180 data resources for the West Coast contributed by seven network partners, including state and federal agencies. The strategy is quality over quantity, and the catalog will continue to expand as new data resources come online and additional network partners are connected.

Prioritizing Portal Expansion for Marine Debris

Early in 2013, WCGA formally recognized marine debris, an issue that agencies and organizations have mobilized for decades to address through both cleanup and prevention strategies, as one of the highest priority issues for the West Coast region. WCGA recognized that there was of a lack of data at the West Coast scale that could answer questions such as where debris was occurring (sink), where it was coming from (source), how much debris there was (density), what type of debris it was (plastic, metal, wood), and whether it was changing over time—information that could be used by these groups to better target their activities and measure their effectiveness.

Plans were already under way to expand WCODP with data visualization and map-viewing capabilities in 2014. The decision was made to focus these efforts on the marine debris issue, organize the information needed, and synthesize it into a form that could be readily used by cleanup coordinators, state regulators, and policy advocates. The goal was to expand the data portal in a general-purpose or repeatable manner so that it could scale to support the other priority issues of the region, such as sea-level rise and ocean acidification. In November 2013, this work began by convening experts working in all areas of marine debris to look at how the West Coast Ocean Data Portal and the resources it harvests could support them in their work. The discussion evolved around two particular debris-related data resources. First is the West Coast Marine Debris Database (WCGA 2012), which aggregates beach cleanup and derelict fishing gear removal data from organizations working up and down the West Coast (figure 9.5). The WCGA Marine Debris Action Coordination Team and Point 97 developed the database. The second resource is the National Oceanic and Atmospheric Administration (NOAA) Marine Debris Monitoring and Assessment Project (NOAA 2013), an initiative to collect baseline data on the amount and type of debris in the environment from an increasing number of sites up and down the West Coast.

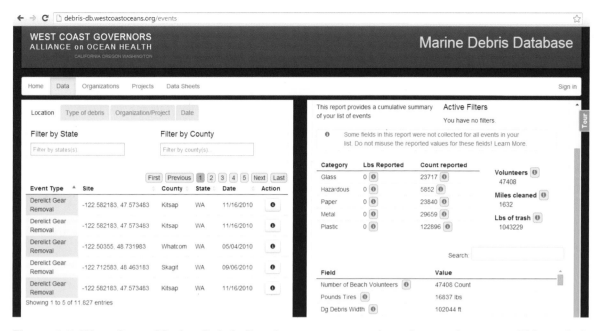

Figure 9.5. West Coast Marine Debris Database event search and reporting page. This website provides access to beach cleanup and derelict fishing gear event data. The West Coast Ocean Data Portal plans to visualize web services from this database, along with services from other datasets in portal, to inform the issue of marine debris preventions and cleanup on the West Coast. Courtesy of West Coast Governors Alliance on Ocean Health.

The decision was made to start by incorporating the West Coast Marine Debris Database and summarizing the information it contained in more useful ways, including visualizing it on a map. One example is aggregating derelict gear removal events from the site level up to larger geographic units and coloring these units based on the frequency of removals (figure 9.6).

Figure 9.6. Design wireframe of the West Coast Ocean Data Portal map viewer using Marine Planner technology. Shown with derelict gear data layer with real-time filtering by category and time.
Courtesy of West Coast Ocean Data Portal; data courtesy of Esri, GEBCO, NOAA, National Geographic, DeLorme, NAVTEQ, Geonames, and others.

Rather than displaying a snapshot of the data, this map view would be created in real time by querying the most recent data from the West Coast Marine Debris Database, using its REST API. Furthermore, users would be able to select a specific time range of interest, as well as the specific type of debris they wanted to see, and the map view would update in real time to show that specific subset of data. Other data layers pertaining to source and sensitive habitats could be added to the map, using web mapping services hosted and published through partner organizations. This functionality will be added and built into WCODP using Marine Planner (see figure 9.6), a platform developed by Point 97 in partnership with the Nature Conservancy to provide robust map visualization capabilities as well as a wide variety of scenario and sketch planning tools (figure 9.7).

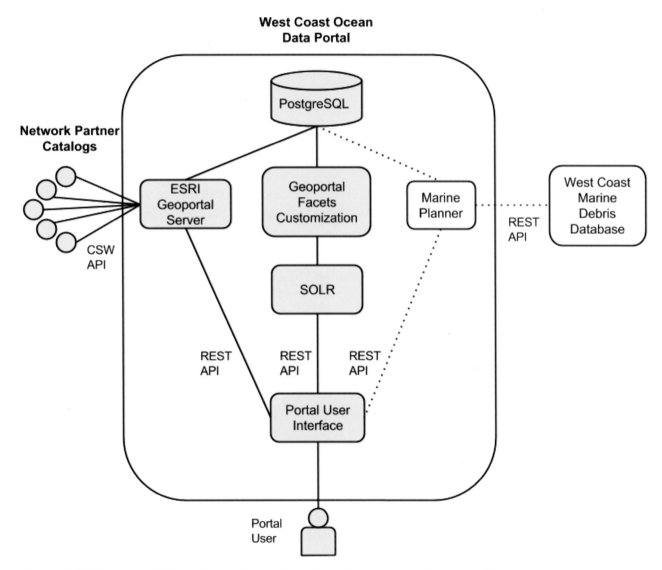

Figure 9.7. Diagram of West Coast Ocean Data Portal system architecture. Network partners contribute metadata in standards-based XML documents to Esri Geoportal Server. The Geoportal Facets Customization parses metadata from Esri Geoportal Server and passes it through to the ocean data portal user interface. The user interface also provides users access to the Marine Planner data visualization tool.

The latest version of WCODP was released with the new map viewer and marine debris datasets in mid-2014, complemented by a variety of new datasets and map layers, including:

- beach access points,
- boat launches,
- population centers,
- river and sewer outfalls,
- sensitive marine areas,
- watershed boundaries,
- biological inventories,
- nearshore habitat characteristics (shore zone),
- oceanographic data (wind, surface currents), and
- essential fish habitat.

Marine Planner technology will further allow users to quickly create map views, print them, and share them.

Best Practices for Regional Data Sharing

The West Coast Ocean Data Portal demonstrates how metadata, combined with web technologies, is changing the way we discover and use geospatial data for regional ocean management. Traditionally, metadata has been used primarily to document the science behind a particular dataset (Qin et al. 2012). Although that role is essential in providing user context and the certainty necessary to select and use a dataset properly, metadata is now an important driver in the discovery process. Additionally, as web services become the norm for data sharing and web applications, metadata that is packaged with web address links for direct download, web services, and other formats is being increasingly used to advertise agency, institutional, or tribal data assets (Tsou 2002). Because the technology still relies on the underlying quality of the metadata, the challenge for data providers is to adapt their metadata practices to serve both needs: ensuring that their data is documented properly and that it can also be readily indexed by catalog systems to facilitate discovery.

There has been a growing shift over the last decade from metadata repositories being used as static text documents to new, dynamic pathways for data access, discovery, and use (Nopgueras-Iso et al. 2005). This shift is important because it reinforces the need for high-quality metadata and highlights standards such as ISO 19115-2 as being critical to documentation of data in multiple formats within one metadata record. In the past, metadata was an afterthought, but with the rise of metadata catalogs such as the California Coastal Atlas, Data.gov, and WCODP, metadata records have become the face of data, and sometimes are the first data component seen. UI design that styles the metadata and leverages the information it contains to offer discovery features such as facets, temporal search, and constraining by geography is important to connecting users seamlessly to data through the metadata.

The distributed framework for sharing via WCODP required the adoption and promotion of data-sharing best practices to ensure that partner data catalogs would be interoperable with the data portal. The WCODN data and information technology working groups started developing practices and policies related to registering data in the portal; development of metadata, web services, and catalogs; and the implementation of Open Data policies within partner organizations. Part of this effort was also developing criteria for inclusion of data in the portal. For example, the data must have documentation in an FGDC or ISO metadata standard format, must be state or regional in scale, and must meet one or more priority needs of WCGA. It is recommended that the record be accessible through a standards-compliant, CSW-enabled catalog; provide links to web services (both Open Geographic Consortium and vendor-specific) or direct download; and make use of existing keywords from established taxonomies. On this last point, WCODN has promoted the use of ISO keywords as well as building and publishing a West Coast taxonomy that can be included in metadata records to facilitate easy search and discovery. Additional best practices are aimed at reducing redundancy and confusion around published records by encouraging data providers to publish only data that they develop and maintain. This helps reduce uncertainty about the "freshness" and authoritativeness of data. WCODP does not republish records to other catalogs unless the data is originally developed by WCGA. Development of best practices is helping promote interoperability and widespread data sharing, which has implications not only for how data is discovered, but also how datasets are used in regional ocean ecosystem management.

Relevance to Ecosystem-Based Management and Regional Decision-Making

Lack of accessibility to geospatial data has been identified at the state, regional, and national levels as a hurdle to improved EBM in the United States (WCGA 2006; Hennessey 2011; Obama 2013). For regional EBM, data needs are cross-disciplinary and cross-jurisdictional, compounding the challenge to acquire and use data in public policy processes. The need to aggregate data from multiple agencies and subject domains places a high premium on the development of frameworks that bring together geodata from multiple sources in an easy-to-use portal. Technologies that take advantage of standards-based metadata formats and web services allow for a distributed network of data providers to contribute to common systems that can better inform regional decision-making.

However, despite availability of required technological infrastructure to support distributed sharing, the success of these systems relies on a strong network of partners to develop, adopt, and promote common standards. Fostering communication within and between members of the network is essential to maintaining a strong foundation and accountability for contribution of data. Furthermore, it is imperative to formalize relationships between partners in the network to ensure coordination and foster communication mechanisms. Without human networks such as WCODN, the success of distributed and dispersed data sharing would not be possible.

The development of the West Coast Ocean Data Portal is an important first phase in development of a regional system for cataloging data, connecting people, and informing ocean issues through web mapping applications. These systems are increasingly being developed and used across the country as marine spatial planning and EBM take place at regional scales. Within regions including the Northeast, mid-Atlantic, and Caribbean, regional planning bodies are forming to tackle the issue of marine spatial planning. These policy-making bodies are relying on quick and intuitive information delivery systems to understand the conflicts and opportunities in their ocean space. Web-based applications such as WCODP provide a source of authoritative science and make it transparent and accessible to policy makers and stakeholders alike.

WCGA has identified regional ocean priorities related to marine debris, sea-level rise, and ocean acidification. WCODP will develop data and visualization tools around these issues to inform coastal managers, policy makers, planners, and stakeholders about the trends and impacts of these issues on coastal ecosystems and communities. Furthermore, WCODN will continue to act as a forum for the discussion and coordination of data-related issues to facilitate the growing interoperability and function of systems and data on the West Coast.

Conclusion

The West Coast Ocean Data Portal is already providing benefits to the discovery and connectivity of coastal and ocean data and people. This network of information and expertise better informs West Coast ocean resource management, policy development, and planning. WCGA and Point 97 successfully customized an instance of Esri Geoportal Server to harvest, curate, and publish geospatial resources from a diverse set of WCODN partner catalogs and information nodes. By building on Esri Geoportal Server's powerful back end to harvest and publish diverse metadata resources, Point 97 was able to customize the front end with a visually appealing and intuitive UI to enhance data discovery. Rather than attempting to aggregate data in a single portal location, WCGA opted for a federated data portal approach as a model for developing an effective, decentralized regional data portal. A primary advantage of this approach is that data and its associated metadata remain close to the authoritative source, which helps to ensure that the data is maintained by the entities directly responsible for its creation and upkeep. As visualization tools are developed and new partner organizations are registered, WCODP and WCODN will continue to grow to better inform management and planning decisions at a regional scale.

Acknowledgments

The development of the portal was funded by an NOAA Regional Ocean Partnership grant, which is gratefully acknowledged. We also thank the West Coast Regional Data Framework Action Coordination Team, including the information technology and data working groups therein, for their guidance and support. The comments and suggestions of two anonymous reviewers greatly improved this chapter.

References

The Apache Software Foundation. 2012. Apache Lucene—Apache Solr. http://lucene.apache.org/solr. Last accessed July 6, 2014.

Crowder, L. B., G. Osherenko, O. R. Young, S. Airame, E. A. Norse, N. Baron, J. C. Day, F. Douvere, C. N. Ehler, B. S. Halpern, S. J. Langdon, K. L. Mcleod, J. C. Ogden, R. E. Peach, A. A. Rosenberg, and J. A. Wilson. 2006. "Resolving Mismatches in US Ocean Governance." *Science* 313:617–18.

Esri. 2014. Esri/Geoportal Server—GitHub. https://github.com/Esri/geoportal-server. Last accessed July 6, 2014.

Hennessey, J. 2011. *Marine Spatial Planning in Washington: Final Report and Recommendations of the State Ocean Caucus to the Washington State Legislature.* Olympia, WA: Washington Department of Ecology.

Hershman, M., and C. W. Russell. 2006. "Regional Ocean Governance in the United States: Concept and Reality." *Duke Environmental Law and Policy Forum* 16 (227): 227–65.

Karlidis, T., A. Tzotsos, and R. Clark. 2014. pycsw: Metadata Publishing Just Got Easier. http://pycsw.org. Last accessed July 6, 2014.

National Ocean Council. 2013. *The National Ocean Policy Implementation Plan.* http://www.whitehouse.gov//sites /default/files/national_ocean_policy_implementation_plan.pdf. Last accessed June 8, 2014.

NOAA (National Oceanic and Atmospheric Administration). 2013. NOAA Marine Debris Monitoring and Assessment Project. http://marinedebris.noaa.gov/research/marine-debris-monitoring-and-assessment-project. Last accessed June 8, 2014.

Nopgueras-Iso, J., F. J. Zarazaga-Sonia, P. J. Alvarez, and P. R. Muro-Medrano. 2005. "OGC Catalog Services: A Key Element for the Development of Spatial Data Infrastructure." *Geospatial Research in Europe* 31 (2): 199–209.

Oak Ridge National Laboratory. 2013. MERCURY: Distributed Metadata Management, Data Discovery and Access System. http://mercury.ornl.gov. Last accessed July 6, 2014.

Obama (President Barack Obama). 2013. *Memorandum on Open Data Policy: Managing Information as an Asset* (May 9, 2013). http://www.whitehouse.gov/sites/default/files/omb/memoranda/2013/m-13-13.pdf. Last accessed June 8, 2014.

The Open Knowledge Foundation. 2014. CKAN: The Open Source Data Portal Software. http://ckan.org. Last accessed July 6, 2014.

Point 97. 2014. Point 97: Technology Solutions for Ocean Management. http://www.pointnineseven.com. Last accessed July 6, 2014.

Qin, J., A. Ball, and J. Greenberg. 2012. "Functional and Architectural Requirements for Metadata: Supporting Discovery and Management of Scientific Data." In *Proceedings of the 12th International Conference on Dublin Core and Metadata Applications*, 62–71. Kuching, Malaysia, September 37.

Tsou, M.-H. 2002. "An Operational Metadata Framework for Searching, Indexing, and Retrieving Distributed Geographic Information Services on the Internet." In *Geographic Information Science*, edited by M. J. Egenhofer and D. M. Mark, 313–32. Berlin: Springer.

Unidata Program Center. 2014. Unidata: THREDDS Data Server (TDS). http://www.unidata.ucar.edu/software /thredds/current/tds. Last accessed July 6, 2014.

Welch, T., R. Hodges, and T. Haddad. 2012. *WCGA RDF Data Registry Design Assessment.* Technical Report, West Coast Governors Alliance on Ocean Health Regional Data Framework Action Coordination Team. Portland, OR: Ecotrust. http://www.westcoastoceans.org/media/Data_Network_ACT/WCGA_RDF_Data_Registry _Design_Assessment_2013.pdf.

WCGA (West Coast Governors Alliance on Ocean Health). 2006. *West Coast Governors Agreement Action Plan.* http:// www.westcoastoceans.org/media/WCGA_ActionPlan_lowest-resolution.pdf. Last accessed June 8, 2014.

———. 2012. WCGA Marine Debris Database. http://debris-db.westcoastoceans.org. Last accessed June 8, 2014.

———. 2014. West Coast Ocean Data Portal. http://portal.westcoastoceans.org. Last accessed July 6, 2014.

Young, O. R., G. Osherenko, J. Ekstrom, L. B. Crowder, J. C. Ogden, J. A. Wilson, J. Day, F. Douvere, C. N. Ehler, K. L. McLeod, B. S. Halpern, and R. Peach. 2007. "Solving the Crisis in Ocean Governance: Place-Based Management of Marine Ecosystems." *Environment* 49 (4): 21–32.

Supplemental Resources

Dawn J. Wright, ed.; 2015; *Ocean Solutions, Earth Solutions*; http://dx.doi.org/10.17128/9781589483651_d

For the digital content for this chapter, go to the Esri Press "Book Resources" webpage at esripress .esri.com/bookresources. Then, in the list of Esri Press books, click *Ocean Solutions, Earth Solutions*. On the *Ocean Solutions, Earth Solutions* resource page, click the chapter 9 link to access that webpage and the links to the digital content.

Ocean and Coastal Resources

The resource page includes the following resources:
- WCGA "West Coast Ocean Data Sharing Best Practices and Policies" PDF
- "WCGA RDF (Regional Data Framework) Data Registry Design Assessment" PDF
- OOI Cables Metadata XML
- Solr Schema Metadata XML

Additional Ocean Resources

For additional ocean resources, go to the following links:
- West Coast Ocean Data Portal, at http://portal.westcoastoceans.org

- Action Coordination Team website, at http://www.westcoastoceans.org/wcodp

- GitHub Repository, at https://github.com/Ecotrust/wc-data-registry

- Geoportal Facets Customization GitHub Repository, at https://github.com/Esri /geoportal-server/wiki/Geoportal-Facets

- Apache Solr, at http://lucene.apache.org/solr

Hyperlinks to these websites are available on the book resource webpage for chapter 9.

CHAPTER 10

Linking Landscape Condition Impacts to Coral Reef Ecosystem Composition for the East End of Saint Croix

Daniel S. Dorfman, Simon J. Pittman, Sarah D. Hile, Christopher F. G. Jeffrey, Alicia Clarke, and Chris Caldow

Abstract

In this land-sea characterization, we endeavored to map spatial patterns of the connections between actions on land and impacts at sea. Specifically, we analyzed 2007 land-cover data, evaluated land-use patterns, and applied a Landscape Development Intensity Index for watersheds adjacent to the East End Marine Park of Saint Croix, US Virgin Islands. We then correlated the distribution of benthic species and coral reef habitats within 300 m buffer watershed impact zones of the landscape development intensity index to identify and explore potential linkages between land-use patterns and ecological impacts on coral reefs. We compared the benthic habitat composition of watershed impact zones within classes of anticipated impacts from land-based sources of pollution. This was done using benthic habitat data, both from benthic habitat maps and in-water surveys. The benthic habitat maps indicated a positive correlation between the Landscape Development Intensity Index and seagrass presence and a negative correlation between the index and coral cover. The in situ surveys revealed higher coral cover in medium-impact classes compared with high and low impact. Although the results from comparing benthic habitat maps and in situ surveys are inconsistent, we anticipate that this could be because of the low number and uneven distribution of the in-water surveys. Additionally, we identified watersheds where species known to be susceptible to land-based sources of pollution are located. The process described here is intended to evaluate potential linkages between landscape condition and marine ecosystem condition. We expect that the methods described here could be employed to track the impacts of land-based sources of pollution on benthic habitats and species composition in the nearshore environment.

Dawn J. Wright, ed.; 2015; *Ocean Solutions, Earth Solutions*; http://dx.doi.org/10.17128/9781589483651

Introduction

Coral reefs are among the most diverse and productive ecosystems on Earth. In many locations they are the economic engine of the marine environment, providing a wide range of ecosystem goods and services, including food, as well as supporting recreation and tourism activities, and coastal protection from storm and wave action. It is estimated that the total annual economic value of coral reef services to the United States is US$3.4 billion (Brander and Van Beukering 2013), with coral reefs of the US Virgin Islands valued at approximately US$187 million annually (Van Beukering et al. 2011).

The United States and its territories are home to thousands of acres of coral reefs that stretch from remote, relatively untouched areas in the Pacific Ocean to densely populated regions in the Caribbean Sea and Atlantic Ocean. With their vibrant colors and array of fish, sea grass, and invertebrates, coral reef ecosystems are the foundation of many unique and special places throughout the country. In the US Virgin Islands, there are two coral reef national monuments, a national park, and several territorial marine protected areas, including the East End Marine Park on the island of Saint Croix, to enhance protection of coral reef ecosystems.

Despite their economic, biological, and cultural significance, coral reefs in the US Virgin Islands and those across the globe face a myriad of threats and stressors, such as storms, thermal stress from elevated water temperature, diseases, and land-based sources of pollution (Rogers and Beets 2001). Research indicates that multiple threats and stressors interact to negatively impact coral reefs, and poor water quality from runoff reduces the resilience of corals to disturbance from thermal stress (Carilli et al. 2009; Ban et al. 2014).

Coastal development and land-based sources of pollution, such as runoff, are a major concern to marine resource managers in the US Virgin Islands and other small islands, where steep terrain results in direct and rapid transport of runoff into coastal waters after rainfall events (Smith et al. 2008; Waddell and Clarke 2008). In these environments, unpaved "dirt" roads in the watershed were identified as a major source of sediment onto reefs (Begin et al. 2014).

These land-sea connections are of particular interest to coastal resource managers of the Saint Croix East End Marine Park, where the upland development of watersheds poses a primary threat to the health of adjacent marine ecosystems. This concern is documented in the management plan for the East End Marine Park (The Nature Conservancy 2002). The Saint Croix East End Marine Park, established in 2003 as the first multiuse marine park managed by the US Virgin Islands Department of Planning and Natural Resources, encompasses nearly 39,000 acres of underwater habitats and protects the largest island barrier reef system in the Caribbean. Geographical patterns of predicted erosion suggest that the eastern half of Saint Croix is more seriously impacted than the west. The eastern half is also where considerable long-term investments were made to protect coral reef ecosystems through establishment of marine protected areas by federal and territorial governments supported by nongovernmental organizations and community groups.

At the time of the park's establishment, there were substantial data gaps in knowledge about the living marine resources, watersheds, and environmental interactions that influence the park's marine resources. To address these data gaps, the Government of the US Virgin Islands requested the support of the National Oceanic and Atmospheric Administration (NOAA) National Centers for Coastal Ocean Science (NCCOS) to characterize

- the landscape and adjacent seascape condition relevant to threats to coral reef ecosystem condition; and
- the marine communities within the Saint Croix East End Marine Park boundaries to increase local knowledge of resources exposed to various stressors.

This chapter provides a synopsis of the land-sea characterization of the Saint Croix East End Marine Park. It describes the calculation and mapping of the landscape development intensity index (LDII) as a spatial proxy for evaluating the threat from runoff and analyses of biotic communities within watershed impact zones (WIZs) associated with each watershed. The focus is on examining and identifying locations where vulnerable species were exposed to threats from land-based sources of pollution and providing a baseline for tracking potential changes in habitat composition. Pittman et al. (2013) make a comprehensive report on the land-sea characterization of the Saint Croix East End Marine Park.

Background

Alteration of the natural landscape for development, road construction, or agriculture can have adverse impacts on coral reefs through increased delivery of sediment and pollution to coastal waters. The threat associated with land clearing is higher in areas of steep relief, intense precipitation, and where soils are erosive in nature (World Resources Institute and NOAA 2006). Our land-sea characterization seeks to highlight potential linkages between land-based sources of pollution (LBSP) and the condition of coral reef ecosystems for the East End Marine Park by building on previous efforts to help managers prioritize actions and guide local action strategies. Specifically, we quantified and mapped watershed condition in the landscapes in closest proximity to the Saint Croix East End Marine Park, and then examined spatial distribution patterns of marine communities and potentially vulnerable species to determine relative coral reef condition.

Previous studies have characterized spatial patterns in stressors and threats to coral reef ecosystems, but only a few studies have examined landscape condition as a proxy for threats to coral reefs and linked assessment of ecological condition of coral reef ecosystems to adjacent watersheds. Burke et al. (2011) analyzed coastal development, watershed-based pollution, marine pollution and damage, and overfishing to characterize factors affecting coral reef ecosystems and categorized the US Virgin Islands as experiencing "high" levels of exposure to threats from local human activities. Exposure to threats increased even further when projected climate change stressors (sea temperature and acidification) were included in the risk assessment. Another study analyzed the relative erosion potential of watersheds as well as estimated erosion from roads, and found that the majority of the eastern end of Saint Croix had high vulnerability to land erosion (World Resources Institute and NOAA 2006).

To address increasing concerns about watershed condition, the NOAA Coral Reef Conservation Program commissioned two projects by Horsley Witten Group to assess threats from land-based sources of pollution to watersheds adjacent to the Saint Croix East End Marine Park. The first document, *St. Croix East End Watersheds Existing Conditions Report* (Horsley Witten Group 2011a), identified basic watershed characteristics such as soils, rainfall, land use, and infrastructure. It also examined individual sites, the potential for reduction in land-based sources of pollution, and feasibility of implementing restoration projects. Several restoration projects were highlighted as priorities for implementation. Building on watershed restoration activities associated with the American Recovery and Reinvestment Act of 2009, the NOAA Coral Reef Conservation Program, US Virgin Islands Department of Planning and Natural Resources, US Department of Agriculture, and The Nature Conservancy are coordinating comprehensive watershed restoration plans for six watersheds surrounding the park, some of which are 303(d)-listed impaired water bodies. In the second report, the *St. Croix East End Watersheds Management Plan* (Horsley Witten Group 2011b), one objective was to protect marine resources by reducing the negative impacts of land-based sources of pollution through reducing sediment and nutrient loads. The Watersheds Management Plan focuses on actions that can be taken on land to reduce negative impacts to marine ecosystems and natural resources.

In 2011, a team of scientists with the US Environmental Protection Agency applied an index of human disturbance to correlate landscape development with nearshore marine ecosystem conditions in Saint Croix (Oliver et al. 2011). Their analysis focused on the relationship between impervious surfaces and coral reef assemblage metrics such as stony coral colony density, taxa richness, coral colony size, and total coral cover. The LDII, an index of human disturbance, quantified impervious surfaces, agriculture, and other land-cover types to create a composite value for each watershed and represented a spatial proxy of the potential threat to water bodies from runoff. Oliver et al. (2011) found that the LDII was more robust than other indicators of human activity and correlated negatively with stony coral colony density, taxa richness, colony size, and total coral cover. The LDII was also an effective indicator of human impacts to corals, and highlighted the link between land-based human activity and marine ecosystems.

For this study, we adapted and applied the LDII techniques developed by Oliver et al. (2011). However, we expanded the LDII to include the additional variable of "dirt roads." Furthermore, we correlated the LDII to coral-cover metrics to highlight areas where human-impacted terrestrial conditions are proximal to high-priority marine species, habitats, and biodiversity hot spots.

Methods

The US Virgin Islands recognizes six watersheds, encompassing a total area of 3,145 ha, that make up the land adjacent to the park. These watersheds range in size from 807 ha for Great Pond Bay to 281 ha for Turner Hole, and were too coarse for the scale of this analysis. To examine potential linkages between land-based sources of pollution and ecological condition of coral reef ecosystems, we used the island's stream network and aerial photographs to aggregate 121 catchment basins identified from the US Geological Survey (USGS) National Hydrography Dataset (NHD) into 42 watershed units.

One of the primary contributors to land-based sources of pollution is erosion from land-cover conversion. To analyze the potential contribution to sedimentation for each analytical unit, we analyzed the land-cover data developed in 2007 by the NOAA Office for Coastal Management (formerly Coastal Services Center) Coastal Change Analysis Program (C-CAP) (figure 10.1). Our analysis applied the LDII developed by the Center for Environmental Policy of the University of

Florida (Brown and Vivas 2005) (table 10.1). The index is intended to serve as a measure of human disturbance. It is calculated based on the level or density of human activity in a given area based on land-cover properties. The LDII quantifies impervious surfaces, agriculture, and other land-cover types to create a composite value for each watershed, forming a spatial proxy of potential threat to nearshore waters from runoff. Oliver et al. (2011) first linked the landscape development intensity index to coral reef condition along the coast of Saint Croix. The researchers found that the LDII was more robust than other indicators of human activity, exhibiting negative correlations with stony coral colony density, taxa richness, colony size, and total coral cover. The LDII provided an effective indicator of human impacts to corals, highlighting the link between land-based human activity and marine ecosystems. Using a more recent land-cover product and finer scale watershed units focused only on Saint Croix East End Marine Park, we modified the approach of Oliver et al. (2011) by incorporating dirt road density into the LDII. Our approach quantified and ranked landscape development intensity for each watershed unit, and then identified areas where high landscape development intensity is likely to threaten sensitive marine species and habitats.

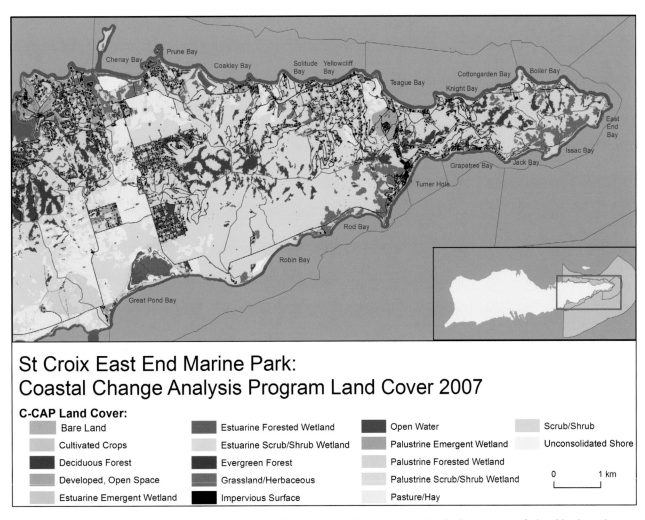

Figure 10.1. Land-cover map developed by the Coastal Change Analysis Program of the National Oceanic and Atmospheric Administration Office for Coastal Management, formerly the Coastal Services Center. Map by authors; data source: NOAA NOS Office for Coastal Management (formerly Coastal Services Center) Coastal Change Analysis Program.

Table 10.1. Landscape Development Intensity Index Coefficients

Land Cover Class	Landscape Development Intensity Index Coefficient
Impervious Surface	8.28
Pasture/Hay	3.03
Grassland/Herbaceous	2.06
Scrub/Shrub	2.06
Bare Land	1.85
Developed, Open Space	1.85
Deciduous Forest	1.00
Palustrine Forested Wetland	1.00
Palustrine Scrub/Shrub Wetland	1.00
Palustrine Emergent Wetland	1.00
Estuarine Forested Wetland	1.00
Estuarine Scrub/Shrub Wetland	1.00
Estuarine Emergent Wetland	1.00
Unconsolidated Shore	1.00
Open Water	1.00

Table by authors; data source: EPA.

We created WIZs adjacent to the watersheds of the Saint Croix East End Marine Park (figure 10.2). The WIZs were constructed using a 300 m buffer along the coast of the park that was then subdivided into smaller areas immediately adjacent to the watershed units. The zones were used as replicate sample units to determine potential effects of land-based sources of pollution on marine biota. A buffer of 300 m was selected based on the likelihood that impacts from land-based sources of pollution to marine biota are detectable within 300 m of the land-sea interface. We acknowledge that land-based sources of pollution can influence a broader geographical extent because local water movements could disperse terrigenous materials offshore and alongshore beyond 300 m. However, impacts from land-based sources of pollution on marine biota are likely to decrease with increasing distance from shore because of dilution. Therefore, our WIZs represent conservative regions of impact, within which marine biota most likely will be exposed to land-based sources of pollution. In addition, the intensity of a threat from land-based sources of pollution was included by weighting the WIZs with LDII values to account for spatial variability in land development within the study area.

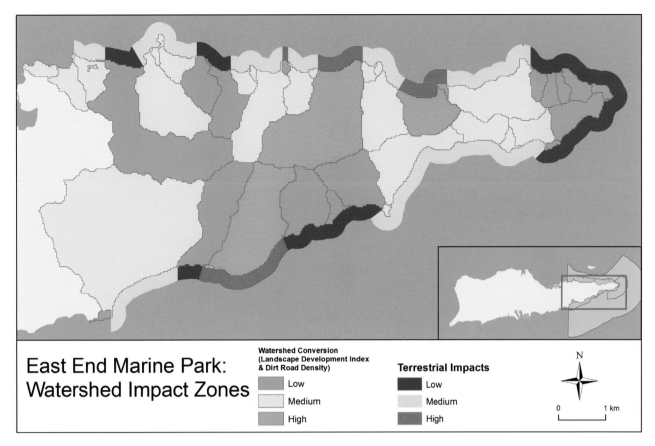

Figure 10.2. **Terrestrial impacts and watershed impact zones.** Map by authors; data source: NOAA National Centers for Coastal Ocean Science.

LDII provides one measure of the expected threat to coastal ecosystems from watershed-based pollution. Another expected threat is soil erosion from dirt roads. This threat was not captured by the LDII, so it was characterized independently. The area of dirt roads mapped by Horsley Witten Group was quantified for each analytical unit and used as a proxy for soil erosion from dirt roads. Horsley Witten Group interpolated 2009 aerial orthophotos and parcel boundaries as part of the Saint Croix East End Watershed Planning effort funded by the NOAA Coral Reef Conservation Program (Horsley Witten Group 2011a and 2011b). Road type was verified during January 2011 field assessment work (Horsley Witten Group 2011a). The LDII and dirt road variable were combined to create a single metric that represented potential negative impacts to nearshore habitats (see figure 10.2). Figure 10.3 summarizes the process used to map and link land-based stressors to adjacent WIZs.

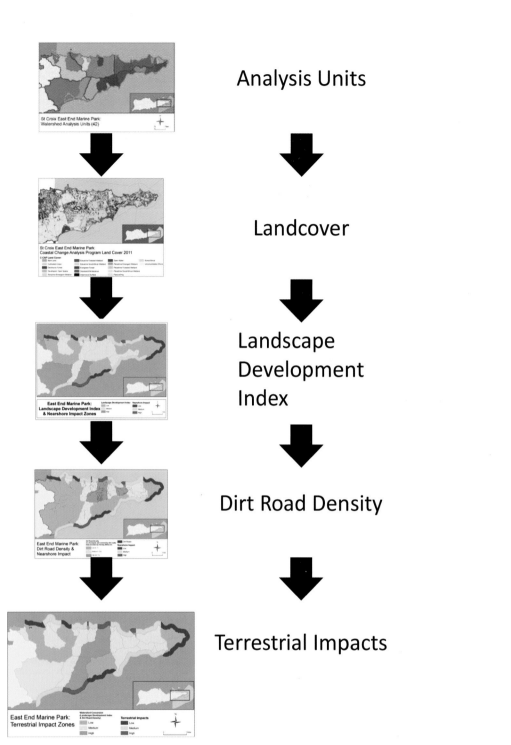

Figure 10.3. **Analytical process for deriving terrestrial impacts to watershed impact zones.** Map by authors; data source: NOAA NOS Office for Coastal Management (formerly the Coastal Services Center) Coastal Change Analysis Program and Horsley Witten Group.

Characterization of Marine Benthic Communities in Watershed Impact Zones

Benthic habitats were evaluated using two distinct datasets: (1) benthic habitat maps developed by the NOAA Coral Reef Conservation Program (Kendall et al. 2002), and (2) direct observation conducted by underwater surveys. Information from NOAA benthic habitat maps was analyzed to quantify the composition and area of habitat types (e.g., sea grasses, coral reef, sand) in each zone. This data provides managers with information on the proportions of each habitat type within each management unit. Benthic habitats were digitized from high-resolution aerial photographs, and their accuracy was assessed using underwater validation surveys (Kendall et al. 2002).

In addition to the information collected from benthic maps, we used a 10-year (2002–2010) dataset of in situ, field-based observations of benthic composition from direct observation, using 251 point locations to describe benthic communities within each WIZ, as well as management zones within the park. In situ variables analyzed from this study were the percentage cover of scleractinian corals, hydrocorals, algae gorgonians, sponges, and sea grasses. Datasets were collected with a stratified random sampling design to comprehensively assess faunal populations and benthic communities around Buck Island Coral Reef National Monument and the Saint Croix East End Marine Park (Menza et al. 2006; Pittman et al. 2008). Point locations were selected randomly from two strata based on benthic habitat maps of the area (Kendall et al. 2002). The "hard" stratum comprised bedrock, pavement, rubble, and coral reefs. The "soft" stratum comprised sand, sea grasses, and macroalgal beds. A subsample of the data was analyzed by selecting only sites that were located within the WIZs of the park (*n* = 190; figure 10.4).

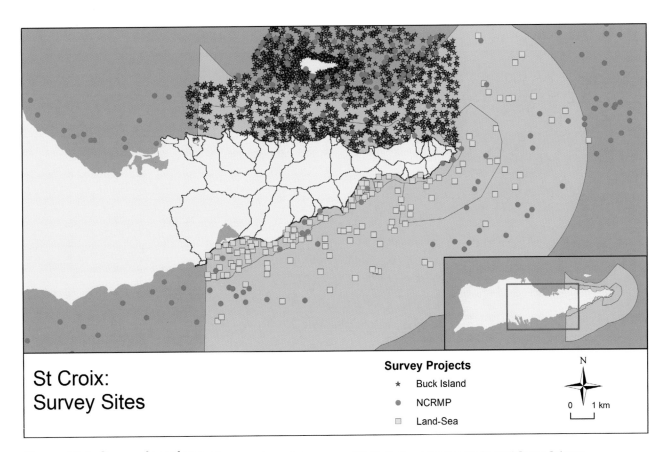

Figure 10.4. Survey locations. Map by authors; data sources: NOAA National Centers for Coastal Ocean Science.

Coral species that are considered sensitive to land-based sources of pollution were selected based on published literature (Bak and Elgershuizen 1976; Rogers 1983; Pastorok and Bilyard 1985; Gleason 1998; Nemeth and Nowlis 2001; Pait et al. 2007). Although observed seagrass species were not considered sensitive to contamination or sedimentation for this study, when exposed to nutrient enrichment or eutrophication, sea grasses experience changes in growth rates and relative dominance among species (Fourqurean et al. 1995; Ferdie and Fourqurean 2004; Burkholder et al. 2007).

Results

Marine Communities in the Watershed Impact Zones
Benthic Communities
Analysis of the benthic habitat maps indicates that coral areas are concentrated along the eastern end of the park, while sea grasses and macroalgae dominate the western side (figure 10.5). We analyzed benthic habitat composition as reported from the benthic habitat maps for differences among the impact classes (Kendall et al. 2002) (figure 10.6). We found significant differences between high-, medium-, and low-impact classes for the distribution of coral reef and colonized hard bottom and submerged aquatic vegetation (SAV) habitats. For coral reef and colonized hard bottom, there was a clear trend of greater coral cover in areas with lower anticipated impacts. For SAV, high-impact areas had greater amounts of SAV and low-impact areas had lower levels.

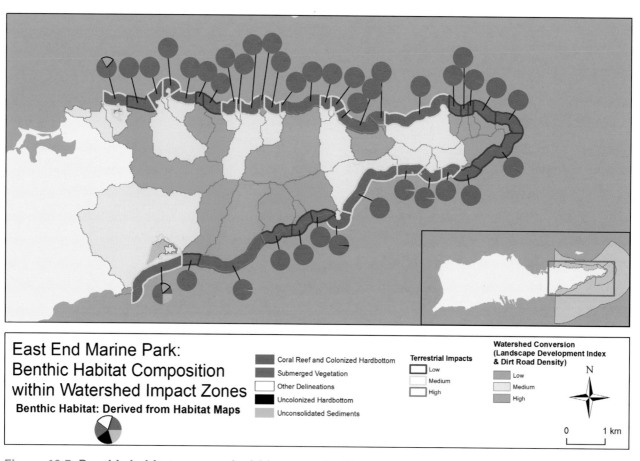

Figure 10.5. **Benthic habitats mapped within watershed impact zones.** Map by authors; data source: NOAA National Centers for Coastal Ocean Science.

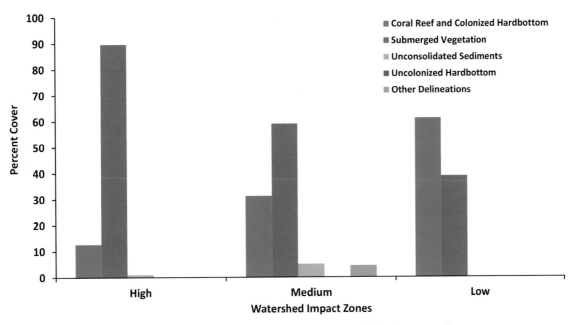

Figure 10.6. **Graph of the distribution of habitat types within impact classes.** Chart by authors; data source: NOAA National Centers for Coastal Ocean Science.

Based on the underwater surveys, the 300 m buffer contained primarily algae and sea grasses. Very little coral reef was found within the WIZs. The highest coral cover is found in Teague and Boiler Bays on the northeast end of the park. Figure 10.7 shows the composition of benthic habitat types for each of the WIZs, based on information collected through underwater surveys. Figure 10.8 shows benthic habitat composition within each impact class.

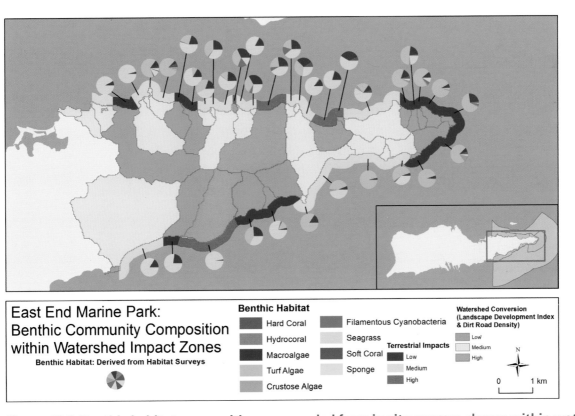

Figure 10.7. **Benthic habitat composition as recorded from in situ surveys shown within watershed impact zones.** Map by authors; data source: NOAA National Centers for Coastal Ocean Science.

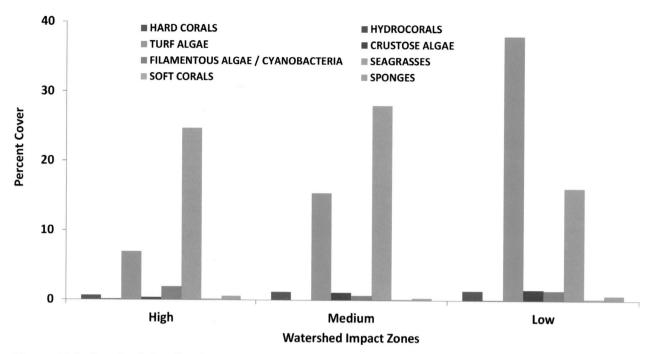

Figure 10.8. Graph of the distribution of habitat types within exposure classes of watershed impact zones based on National Oceanic and Atmospheric Administration underwater surveys. Chart by authors; data source: NOAA National Centers for Coastal Ocean Science.

Significant differences were detected among impact classes for three habitat types: hard corals occurring on hard substrate, turf algae on hard substrate, and hard corals in areas dominated by soft substrate. For hard corals occurring on hard substrate, the difference between medium impact and low impact was most significant, with hard coral habitats more abundant in areas classified as having medium terrestrial impact. There was also a difference in abundance between medium- and high-impact classes, with lower abundance in high-impact areas. A single site drove both of these results with nearly 33% coral cover occurring in an area classified as medium impact. This site is found in Boiler Bay. For turf algae on hard substrate, differences were significant between high and low classes and between high and medium classes. Turf algae distribution was greater in low and medium classes than in high. In areas dominated by soft-bottom substrate, significant differences were found between hard coral classes. Specifically, the difference was significant between high and low classes and between medium and low classes. For high-impact zones, the difference was driven largely by a single site with nearly 2% coral reef and colonized hardbottom cover.

Coral Reef Community

To assess the potential threat to coral species known to be sensitive to impacts from land-based sources of pollution, coral locations were mapped within the WIZs (300 m coastal buffer; figure 10.9). Elkhorn coral *(Acropora palmata)*, a listed species in the Endangered Species Act (ESA), exists within the WIZs, primarily concentrated on the eastern tip of the island where watersheds are least developed. Staghorn corals *(Acropora cervicornis)* were not sighted within any of the WIZs.

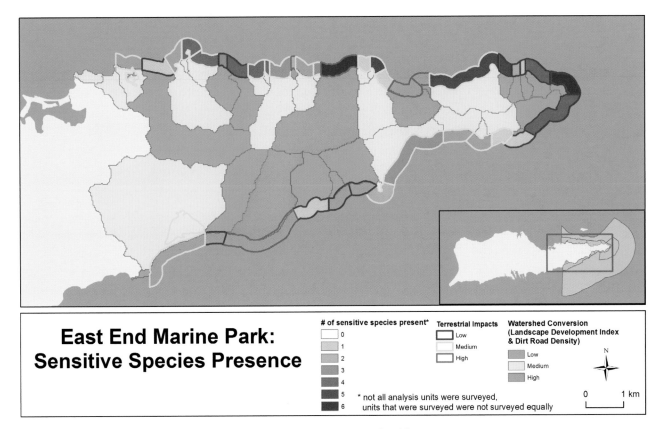

East End Marine Park: Sensitive Species Presence

# of sensitive species present*
0
1
2
3
4
5
6

Terrestrial Impacts
Low
Medium
High

Watershed Conversion
(Landscape Development Index
& Dirt Road Density)
Low
Medium
High

N

* not all analysis units were surveyed,
units that were surveyed were not surveyed equally

0   1 km

Figure 10.9. Sensitive species distributions within watershed impact zones. Map by authors; data source: NOAA National Centers for Coastal Ocean Science.

Brain coral (*Pseudodiploria strigosa*) was observed in 20 of 38 WIZs. Golfball coral (*Favia fragum*) was observed in 12 impact zone units along the north shore, and boulder star coral (*Orbicella annularis complex*) was observed in six units along the north shore. Yellow pencil coral (*Madracis auretenra*) was not observed within the impact zone. Finger coral (*Porites porites*) and mustard hill coral (*Porites astreoides*) were observed within 20 m and 17 m of the surveyed impact zone units, respectively. No significant differences were detected between threat classes when data on the distribution of sensitive coral species was combined (*Acropora palmat, Pseudodiploria strigosa, Favia fragum, Orbicella annularis* complex, *Madracis auretenra, Porites porites,* and *Porites astreoides*).

Many of the WIZs contained species that are known to be sensitive to sedimentation and land-based sources of pollution. Highest species richness for coral sensitive to sedimentation and pollution is found on the north shore of the park with varied exposure to threat (see figure 10.9). The Yellowcliff Bay watershed, containing six sensitive species, is estimated to be high threat, and several medium-threat impact zones along the northeast shore contain at least five species.

Sea grasses
Shoalgrass (*Halodule wrightii*) was observed in three WIZs. Paddle grass (*Halophila decipiens)* is known to be sensitive to land-based sources of pollution and was not observed in any of the analysis units. Manatee grass (*Syringodium filiforme*) was observed in 27 analysis units. Turtle grass (*Thalassia testudinum*) was widely distributed within the WIZs.

Discussion and Conclusion

On Caribbean islands, land-use patterns are an important concern to the conservation of coral reef ecosystems. Modifications to watersheds such as urbanization and agriculture have the potential to greatly increase erosion and sediment delivery to coastal waters. Many coral species are highly sensitive to declines in water quality caused by runoff of material from land. In response, best management practices in land development are being introduced to reduce erosion and runoff of land-based sources of pollution into coastal waters. The land-sea characterization presented here provides a technique to prioritize areas for management attention where potential threat to sensitive coral species is greatest.

We recognize that activities on land can influence the health of nearshore marine ecosystems. In particular, we understand that sedimentation and pollution delivered from land to coral reef ecosystems can kill corals and, by extension, negatively affect entire ecosystems. There are particular species of corals that are particularly susceptible to pollution and sedimentation. In this study, we highlight areas within the Saint Croix East End Marine Park where some of these species are present. The recognition of areas where sensitive species are present can assist coral reef managers in reaching their conservation objectives by highlighting the need for terrestrial management upstream of sensitive coral species. By reducing land-cover conversion and/or road development upstream of sensitive species, we can improve survival conditions for coral species of concern.

This study does not address the issue of how pollution or sediments are transferred from land to sea. It also does not address the complicated issue of exactly where benthic communities will be influenced by upstream watershed conditions. This study identifies areas that are expected to produce high levels of pollution and sediment and evaluates the composition of benthic communities associated with the areas expected to be impacted. When the Saint Croix East End Marine Park was established, it was recognized that the health of the ecosystems that the park was designed to protect would depend on both marine ecosystem management and management of the land adjacent to the park. Here, we hope to have identified one mechanism by which we can associate land-based management with its effects on nearshore marine ecosystems.

In this study, we found inconsistent results between our analysis of habitats from benthic maps and in-water surveys. From the mapped benthic habitats, we found coral cover decreasing where terrestrial impacts are greater and sea grasses increasing with greater terrestrial impacts. However, we did not find the same trend when we evaluated the information collected through underwater surveys. Here, we found greater coral cover in areas with medium terrestrial impacts. We expect that this discrepancy is a result of the limited number and uneven geographic distribution of surveys. The results of the survey information analysis appear to be driven by a single record where coral cover was recorded as 33%. This value is significantly different from the average value of 1.18% coral cover across the 190 surveys. We expect that greater survey effort and a more even geographic distribution of surveys might enable us to better track the impacts of terrestrial influence to direct observations of the benthic community.

Acknowledgments

We gratefully acknowledge the NOAA Coral Reef Conservation Program for funding this research and the US Virgin Islands Department of Planning and Natural Resources for its input and guidance. We also wish to thank The Nature Conservancy and National Park Service for their assistance in conducting the underwater surveys. The comments and suggestions of two anonymous reviewers greatly improved this chapter.

References

Bak, R. P. M., and J. H. B. W. Elgershuizen. 1976. "Patterns of Oil-Sediment Rejection in Corals." *Marine Biology* 37:105–13.

Ban, S. S., N. A. Graham, and S. R. Connolly, 2014. "Evidence for Multiple Stressor Interactions and Effects on Coral Reefs." *Global Change Biology* 20 (3): 681–97.

Begin, C., G. Brooks, R. A. Larson, S. Dragicevic, C. E. R. Scharron, and I. M. Cote. 2014. "Increased Sediment Loads over Coral Reefs in Saint Lucia in Relation to Land Use Change in Contributing Watersheds." *Ocean & Coastal Management* 95:35–45.

Brander, L. M., and P. van Beukering. 2013. *The Total Economic Value of US Coral Reefs.* Coral Reef Conservation Program Technical Report. Silver Spring, MD: NOAA. http://bit.ly/1lULkhU.

Brown M. T., and M. B. Vivas. 2005. "Landscape Development Intensity Index." *Environmental Monitoring and Assessment* 101:289–309.

Burke, L., K. Reytar, M. Spalding, and A. Perry. 2011. *Reefs at Risk Revisited.* Washington, DC: World Resources Institute.

Burkholder, J. M., D. A. Tomasko, and B. W. Touchette. 2007. "Seagrasses and Eutrophication." *Journal of Experimental Marine Biology and Ecology* 350:46–72.

Carilli, J. E., R. D. Norris, B. A. Black, S. M. Walsh, and M. McField. 2009. "Local Stressors Reduce Coral Resilience to Bleaching." *PLoS One* 4 (7): e6324.

Ferdie, M., and J. W. Fourqurean. 2004. "Responses of Seagrass Communities to Fertilization along a Gradient of Relative Availability of Nitrogen and Phosphorus in a Carbonate Environment." *Limnology and Oceanography* 49:2082–94.

Fourqurean, J. W., G. V. N. Powell, W. J. Kenworthy, and J. C. Zieman. 1995. "The Effects of Long-Term Manipulations of Nutrient Supply on Competition between the Seagrasses *Thalassia testudinum* and *Halodule wrightii* in Florida Bay." *Oikos* 72:349–58.

Gleason, D. F. 1998. "Sedimentation and Distribution of Green and Brown Morphs of the Caribbean Coral *Porites astreoides lamarck.*" *Journal of Experimental Marine Biology and Ecology* 230:73–89.

Horsley Witten Group. 2011a. *St. Croix East End Watersheds Existing Conditions Report.* National Oceanic and Atmospheric Administration Coral Reef Conservation Program Technical Report. Sandwich, MA: Horsley Witten Group. http://www.horsleywitten.com/stx-east-end-watersheds/pubs/final/01_Intro_110719.pdf.

Horsley Witten Group. 2011b. *St. Croix East End Watersheds Management Plan*. National Oceanic and Atmospheric Administration Coral Reef Conservation Program Technical Report. Sandwich, MA: Horsley Witten Group. ftp://ftp.nodc.noaa.gov/pub/data.nodc/coris/library/NOAA/CRCP/project/20471/Final-STXEEMP-WMP .pdf.

Kendall, M. S., C. R. Kruer, K. R. Buja, J. D. Christensen, M. Finkbeiner, R. Warner, and M. E. Monaco. 2002. *Methods Used to Map the Benthic Habitats of Puerto Rico and the US Virgin Islands*. Technical Memorandum 152. Silver Spring, MD: NOAA. http://ccma.nos.noaa.gov/products/biogeography/usvi_pr_mapping/manual.pdf.

Menza, C., J. Ault, J. Beets, C. Bohnsack, C. Caldow, J. Christensen, A. Friedlander, C. Jeffrey, M. Kendall, J. Luo, M. E. Monaco, S. Smith, and K. Woody. 2006. *A Guide to Monitoring Reef Fish in the National Park Service's South Florida/Caribbean Network*. Technical Memorandum NOS NCCOS 39. Silver Spring, MD: NOAA. http://www.ccma.nos.noaa.gov/publications/Reefmonitoringguide.pdf.

The Nature Conservancy. 2002. *St. Croix East End Marine Park Management Plan*. Technical Report. St. Thomas, Virgin Islands: University of the Virgin Islands and Department of Planning and Natural Resources.

Nemeth, R. S., and J. S. Nowlis. 2001. "Monitoring the Effects of Land Development on the Near-Shore Reef Environment of St. Thomas, USVI." *Bulletin of Marine Science* 69 (2): 759–75.

Oliver, L. M., J. C. Lehrter, and W. S. Fisher. 2011. "Relating Landscape Development Intensity to Coral Reef Condition in the Watersheds of St. Croix, US Virgin Islands." *Marine Ecology Progress Series* 427:293–302.

Pait, A. S., D. R. Whitall, C. F. G. Jeffrey, C. Caldow, A. L. Mason, J. D. Christensen, M. E. Monaco, and J. Ramirez. 2007. *An Assessment of Chemical Contaminants in the Marine Sediments of Southwest Puerto Rico*. Technical Memorandum NOS NCCOS 52. Silver Spring, MD: NOAA.

Pastorok, R. A., and G. R. Bilyard. 1985. "Effects of Sewage Pollution on Coral-Reef Communities." *Marine Ecology Progress Series* 21:175–89.

Pittman, S. J., D. Dorfman, S. D. Hile, C. F. G. Jeffrey, M. A. Edwards, and C. Caldow. 2013. *Land-Sea Characterization of the St. Croix East End Marine Park, US Virgin Islands*. Technical Memorandum NOS NCCOS 170. Silver Spring, MD: NOAA.

Pittman, S. J., S. D. Hile, C. F. G. Jeffrey, C. Caldow, M. S. Kendall, M. E. Monaco, and Z. Hillis-Starr. 2008. *Fish Assemblages and Benthic Habitats of Buck Island Reef National Monument (St. Croix, US Virgin Islands) and the Surrounding Seascape: A Characterization of Spatial and Temporal Patterns*. Technical Memorandum NOS NCCOS 71. Silver Spring, MD: NOAA. http://1.usa.gov/1sZftBq.

Rogers, C. S. 1983. "Sublethal and Lethal Effects of Sediments Applied to Common Caribbean Reef Coral in the Field." *Marine Pollution Bulletin* 14 (10): 378–82.

Rogers, C., and J. Beets. 2001. "Degradation of Marine Ecosystems and Decline of Fishery Resources in Marine Protected Areas in the US Virgin Islands." *Environmental Conservation* 24 (4): 312–22.

Smith, T. B., R. S. Nemeth, J. Blondeau, J. M. Calnan, E. Kadison, and S. Herzlieb. 2008. "Assessing Coral Reef Health across Onshore to Offshore Stress Gradients in the US Virgin Islands." *Marine Pollution Bulletin* 56 (12): 1983–91.

Van Beukering, P., L. Brander, B. van Zanten, E. Verbrugge, and K. Lems. 2011. *The Economic Value of the Coral Reef Ecosystems of the United States Virgin Islands*. Institute for Environmental Studies Report R-11/06. Amsterdam: Instituut voor Milieuvraagstukken (IVM). http://bit.ly/1mLv1iU.

Waddell, J. E., and A. M. Clarke, eds. 2008. *The State of Coral Reef Ecosystems of the United States and Pacific Freely Associated States*. Technical Memorandum NOS NCCOS 73. Silver Spring, MD: NOAA. http://ccma.nos.noaa.gov/ecosystems/coralreef/coral2008.

World Resources Institute and NOAA. 2006. *Land-Based Sources of Threat to Coral Reefs in the US Virgin Islands*. Washington, DC: World Resources Institute. http://pdf.wri.org/usvi_atlas_web.pdf.

Supplemental Resources

Dawn J. Wright, ed.; 2015; *Ocean Solutions, Earth Solutions*; http://dx.doi.org/10.17128/9781589483651_d

For the digital content for this chapter, listed in this section, go to the Esri Press "Book Resources" webpage at esripress.esri.com/bookresources. Then, in the list of Esri Press books, click *Ocean Solutions, Earth Solutions*. On the *Ocean Solutions, Earth Solutions* resource page, click the chapter 10 link to access that webpage and view photos of sea life in the East End Marine Park, Saint Croix, US Virgin Islands. The photos were taken between 2007 and 2012 and are courtesy of the NOAA NCCOS Biogeography Branch. The photos include:

- Blackbar soldierfish (*Myripristis Jacobus*)
- Blue tang (*Acanthurus coeruleus*) in dense gorgonian habitat
- Coral within the park
- Elkhorn coral (*Acropora palmata*), an endangered species
- Finger coral (*Porites porites*)
- Large school of adult and juvenile grunts (family *Haemulidae*)
- Red hind (*Epinephelus guttatus*)
- Scuba divers collecting fish and benthic composition data

CHAPTER 11

Using GIS Tools to Develop a Collaborative Essential Fish Habitat Proposal

Sophie de Beukelaer, Karen F. Grimmer, Jennifer A. Brown, and Chad King

Abstract

In 2012, the Monterey Bay National Marine Sanctuary brought together stakeholders representing different interests (e.g., fishing and conservation) to work toward a collaborative proposal to the Pacific Fishery Management Council for the design of spatial modifications to current Pacific coast groundfish essential fish habitat conservation areas. This was done by adding new areas containing sensitive resources in need of protection from bottom-trawl fishing gear and reopening some areas to gain access to historical fishing grounds of economic importance to bottom-trawl fishermen. The collaborative proposal was submitted in July 2013 after 11 months of work among stakeholders. This accomplishment was largely because of the suite of GIS tools that were employed, including GIS projects and products such as a customized ArcGIS Online map and spatial analysis tools, which helped inform discussions. GIS was used to provide a clear understanding of the current spatial management, distribution of sensitive resources, and distribution of historical and current fishing efforts. In addition, marine Sanctuary staff used desktop GIS maps during group meetings to manage discussions, enabling the group to move forward even during times of tension and disparity. Data layers were added to help elucidate the topic of conversation or focus the map on a different area to refocus the discussion. At the time of this writing, the proposal was still under consideration by the Pacific Fishery Management Council for implementation. If accepted and approved, the comprehensive essential fish habitat proposal will increase habitat protection for hundreds of known coral and sponge specimen locations, as well as reopen historically important bottom-trawling areas in the Monterey Bay National Marine Sanctuary. The collaborating entities that were part of the completed proposal included the Sanctuary, Alliance of Communities for Sustainable Fisheries, Monterey Bay trawl fishermen, the City of Monterey, Oceana, the Natural Resources Defense Council, Ocean Conservancy, The Nature Conservancy, California Risk Pool, and the Environmental Defense Fund.

Dawn J. Wright, ed.; 2015; *Ocean Solutions, Earth Solutions*; http://dx.doi.org/10.17128/9781589483651

Introduction

The Monterey Bay National Marine Sanctuary (MBNMS) is a federally protected area offshore California's central coast. Its boundaries encompass 6,094 sq mi of remarkably productive offshore waters from Cambria to north of San Francisco (figure 11.1). MBNMS is one of 13 National Marine Sanctuaries in the United States. The goal of the National Marine Sanctuaries Act is to conserve pristine ocean areas for future generations, just as the nation has protected national parks and forests since the 19th century. The National Marine Sanctuaries Act mandates the MBNMS to "maintain for future generations the habitat and ecological services of the natural assemblages of living resources that inhabit" the Sanctuary. In addition to resource protection and conservation, MBNMS was established for the purpose of research, education, and compatible public use.

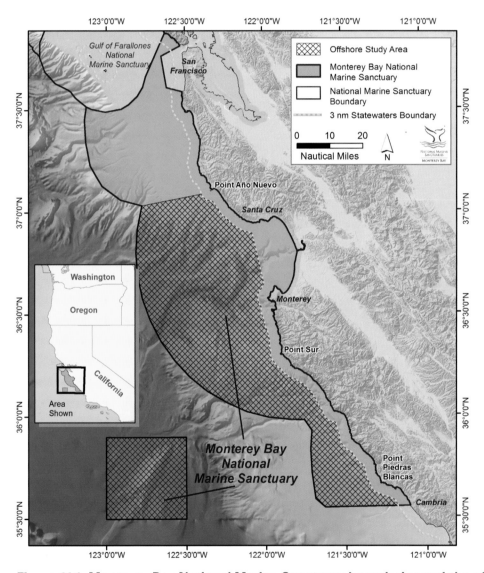

Figure 11.1. Monterey Bay National Marine Sanctuary boundaries and the offshore study area boundary. The offshore study area includes all the Sanctuary south of Point Año Nuevo and west of the 3 nm state waters boundary line. The state waters boundary represents the seaward boundary of the coastal states of the United States. Courtesy of NOAA Monterey Bay National Marine Sanctuary. Data from NOAA, California Department of Fish and Wildlife, Monterey Bay National Marine Sanctuary, Esri; logo courtesy of Monterey Bay National Marine Sanctuary.

Human uses of the sanctuary include both commercial and recreational activities, such as fishing, whale watching, sailing, diving, kayaking, and surfing. Some human activities that threaten the health of sanctuary resources, such as oil drilling, ocean dumping, and seabed mining, are not allowed within the sanctuary. Other human uses that have the potential to impact sensitive resources in the sanctuary are allowed but managed either directly by MBNMS or other state and federal resource management agencies.

Building partnerships with other resource management agencies and fostering strong public involvement are key elements in helping MBNMS attain its management plan goals. For example, the goal of the *Bottom Trawling Effects on Benthic Habitats* action plan, a component of the MBNMS management plan (Office of National Marine Sanctuaries 2008), is to maintain the natural biological communities and ecological processes in MBNMS by evaluating and minimizing adverse impacts of bottom trawling in benthic habitats while allowing the long-term continuation of sustainable local fisheries in MBNMS. Strategies to address that goal include developing partnerships with fishermen, assessing trawling activity, identifying habitats vulnerable to trawling, and identifying and implementing potential ecosystem protection measures. MBNMS can only attain those types of goals by working with stakeholders and engaging with partners in natural resource management processes.

MBNMS does not regulate fisheries, but fisheries do have an impact on the habitats and ecological services that MBNMS is mandated to protect. As such, MBNMS engages as a partner agency and stakeholder in the various fishery management processes led by state and federal fisheries management agencies. For example, the National Marine Fisheries Service (NMFS) and Pacific Fishery Management Council (PFMC) use a variety of avenues, such as quotas, trip limits, gear restrictions, and permanent and time-area closures, to regulate fishing in the United States, and there are a variety of management processes to identify and implement these management strategies. One type of permanent area closure is Pacific coast groundfish essential fish habitat (EFH). MBNMS staff and partners engaged in the first EFH review process after EFH was established in 2006.

NMFS and PFMC established EFH boundaries for groundfish and habitat areas of particular concern (HAPCs) in 2006 in Amendment 19 to fulfill the requirements for EFH established by the Magnuson-Stevens Act (Office of the Federal Register 2006). EFH is defined as those waters and substrate necessary to fish for spawning, breeding, feeding, or growth to maturity. The entire US West Coast continental margin was designated as EFH, but this designation alone did not include any fishing restrictions. Rocky reefs, for example, were designated as HAPCs because of their importance to groundfish, their rarity in the region, and their sensitivity to impacts, but that designation alone did not provide protection either. EFH conservation areas, however, are areas permanently closed to certain types of bottom-contact gear to protect groundfish EFH and HAPCs. In 2006, NMFS and PFMC established 51 conservation areas offshore the US West Coast, including three areas either wholly or partially in MBNMS (figure 11.2). However, key stakeholders were not at the table when these large conservation areas were designed.

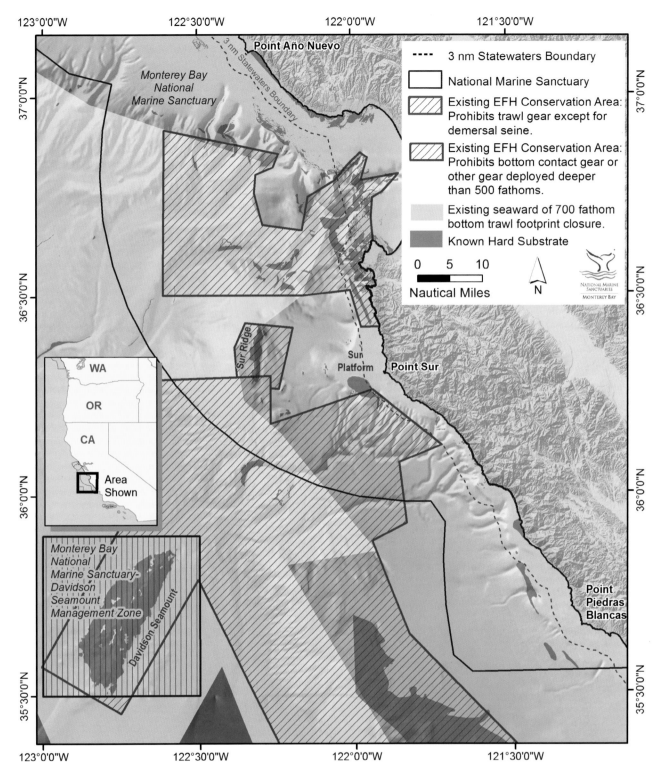

Figure 11.2. **Essential fish habitat (EFH) conservation areas within the Monterey Bay National Marine Sanctuary.** Courtesy of NOAA Monterey Bay National Marine Sanctuary. Data from NOAA National Marine Fisheries Service, California Department of Fish and Wildlife, Monterey Bay National Marine Sanctuary, Esri, Moss Landing Marine Laboratories.

The EFH review process offered an opportunity for stakeholders to provide input on the boundaries of current EFH conservation areas, HAPCs, and other suggestions (e.g., research and gear options) by submitting a proposal to PFMC. As the regional fishery management council that has jurisdiction over the exclusive economic zone off Washington, Oregon, and California, PFMC recommends management measures to be implemented by NMFS. As a local stakeholder, MBNMS was interested in engaging in the EFH review process. Others such as trawl fishermen and nonprofit conservation organizations were also interested in engaging because each entity recognized that EFH regulations are part of a valuable suite of spatial management tools. The trawl fishermen wanted more fishing opportunities while the conservation groups were interested in protecting more areas. MBNMS was interested in both of these goals, and thus brought representatives of the different groups together to facilitate the development of the EFH proposal. This meant the local stakeholders most affected by the outcomes were able to participate in the review process for evaluating existing groundfish EFH regulations.

Prior to engaging in the EFH review process, MBNMS had to determine whether there were areas in the sanctuary containing habitats important to groundfish that were also vulnerable to bottom trawling. Groundfish are those fish that live on or near the bottom of the ocean and include rockfish (e.g., bocaccio, cowcod, yelloweye), flatfish (e.g., petrale sole, Dover sole), roundfish (e.g., lingcod; figure 11.3), sharks, and skates. Many managed fish species are dependent on hard-substrate habitats during some portion of their life cycle. MBNMS staff analyzed available data on hard-substrate habitats, including rocky reefs, soft substrate, and biogenic habitat, in the sanctuary. Benthic organisms attached to the substrate, such as kelp, corals, and sponges, can play an important role in increasing the structural complexity of a bottom habitat, and are therefore considered biogenic habitat. Identifying the location and presence of hard substrate, corals, and sponges was paramount because these habitats are important to groundfish. These habitats provide important physical structure and are also very sensitive to impacts from bottom trawling and are slow to recover from these impacts. Bottom trawling is a fishing technique in which a net is dragged across the seafloor to catch groundfish, but other organisms living on the seafloor, such as corals and sponges, could also be caught by the net. By analyzing new biological, geological, and human-use data in GIS, MBNMS staff were able to illustrate that there were areas within MBNMS important to groundfish but also vulnerable to bottom trawling.

Figure 11.3. Lingcod (*Ophiodon elongatus*) and Cnidaria, including anemones and gorgonians.
Courtesy of NOAA Monterey Bay National Marine Sanctuary.

As part of the EFH review process, MBNMS submitted a proposal in July 2013 to PFMC with anticipated benefits for all three stakeholder groups: trawling, conservation, and resource management. The main goal of the collaborative proposal was to balance conservation and sustainable uses. Specifically, the proposal aimed to balance the protection of EFH for groundfish with opportunities for groundfish bottom-trawl fishing within MBNMS.

MBNMS offered several key ingredients: a venue, GIS technical support, the expertise for developing a sound and complete proposal, and mediation between the stakeholders. The success of this collaborative endeavor is owed in large part to the suite of GIS tools that were employed, including GIS projects and products that helped inform discussions. MBNMS used GIS to provide a clear understanding of the current spatial management, distribution of sensitive resources, and distribution of historical and current fishing efforts. Using GIS interactively during the group process (e.g., quickly turning data layers on or off within an ArcGIS project shown on a screen to everyone in the room) proved valuable, as it enabled stakeholders to see the trade-offs of various proposals in real time. MBNMS staff not only displayed proposed boundary changes to EFH conservation areas in GIS, but also used GIS to manage discussions (e.g., switching to different areas or the whole region to enable the group to move forward, even during times of tension and disparity). GIS helped members of the stakeholder group commit to staying involved, as it helped them envision a great variety of potential future spatial management scenarios. As such, the process was not unlike similar activities elsewhere where collaborative, participatory planning and decision-making activities among various stakeholders were guided effectively by the real-time use of GIS (e.g., Gleason et al. 2010; Lanier et al. 2011; Paul et al. 2012; Gass et al. 2013; Stelzenmuller et al. 2013).

Methods

Study Area
The proposal's study area was established as solely within the federal waters (i.e., > 3 nm from shore) of MBNMS south of Point Año Nuevo (see figure 11.1). State waters were excluded from the study area because trawling is currently not allowed within California's state waters, and California recently implemented a network of marine protected areas (MPAs) using an extensive stakeholder process. The study area also included the Davidson Seamount Management Zone, a noncontiguous offshore portion of MBNMS. The study area covers a total of 3,992 sq mi.

Identifying Stakeholders and Data Needs
The proposal submitted to PFMC was the product of an 11-month collaborative stakeholder process that began in August 2012 with meetings of MBNMS and fishing industry stakeholders. These stakeholders included MBNMS Advisory Council representatives (harbors and commercial fishing), the Alliance of Communities for Sustainable Fisheries, Monterey Bay trawl fishermen, and the City of Monterey, California. Meeting with fishing interests was an essential first step in the process, as it allowed MBNMS and fishing industry representatives to discuss their perspectives on EFH process and management implementations. It also allowed MBNMS to understand where the important historically trawled areas were located. Last but not least, it allowed MBNMS to showcase the GIS datasets to be used in this process, especially as the fishing interests did not have GIS tools readily

available to them. In June 2013, MBNMS invited additional stakeholders, those representing conservation interests, to be part of the proposal development, as it was beneficial to all involved to discuss options for EFH modifications with all groups of different interests. The conservation representatives were from Oceana, the Natural Resources Defense Council, Ocean Conservancy, The Nature Conservancy, California Risk Pool, and Environmental Defense Fund.

In early discussions, group members recognized that the focus of the collaborative proposal would be on spatial modifications to EFH conservation areas by adding new areas containing sensitive resources in need of protection from bottom-trawl fishing gear, while also reopening some areas to gain access to historically fished areas of economic importance to the bottom trawling community. This strategy was settled on because the representatives were satisfied, for the most part, with the spatial management implementations of EFH. But they also wanted a few small changes to certain areas to maximize outcomes that were important to them. Knowing that spatial modifications would be the focus of most of the discussions, the group had to familiarize themselves with various data, including:

1. Benthic habitat needed to identify which areas were consistent with groundfish EFH within MBNMS but were not protected by current EFH management measures. This included habitats that are unique, rare, or ecologically sensitive, either because of geologic features or that contain known biogenic habitat (i.e., deep-sea corals and sponges), and habitats that could be vulnerable to impacts of fishing activities for groundfish managed by PFMC.

2. Data to identify where new EFH HAPCs and new conservation areas that minimize adverse fishing impacts on groundfish EFH could be proposed, and further the protection of both biogenic and physical habitat for groundfish through prohibitions on bottom trawling.

3. Data to identify areas for proposing the reopening of certain portions of existing EFH conservation areas that will return valuable historical fishing grounds to the bottom trawling fleet.

Regional Data Acquisition

MBNMS GIS staff became well acquainted with the regional data and associated metadata made available through the *5-Year Review of Pacific Coast Groundfish EFH, Consolidated Geographic Information Data Catalog and Online Registry* (Pacific Coast Ocean Observing System 2014). The datasets available in the data catalog were fully documented in the *Pacific Coast Groundfish 5-Year Review of Essential Fish Habitat Report* (PFMC 2012), the EFH synthesis report (NMFS 2013b), and the appendix to the EFH synthesis report (NMFS 2013a). The data portal allowed for easy downloading of data and maps of known physical habitats of the US West Coast, trawling effort, presence of biogenic species, groundfish occurrence models, MPAs, and nonfishing pressures. MBNMS GIS staff delivered presentations to other MBNMS staff and stakeholders showcasing the most relevant regional data for the collaborative proposal within the MBNMS study area, and conveying important caveats with regard to how the datasets were collected, portrayed, and analyzed.

MBNMS staff integrated the relevant regional data into a desktop GIS project to use during the stakeholder meetings. The stakeholder group used data layers largely provided by the National Oceanic and Atmospheric Administration (NOAA) Deep Sea Coral Research and Technology

Program (DSCRTP), which indicated the locations where sponges, corals, and pennatulids (sea pens) have been observed in situ or collected in NMFS's bottom-trawl surveys. These datasets helped all members of the group understand where biogenic habitat was located while the stakeholders discussed boundary changes during stakeholder meetings. Additionally, bathymetry, physical habitat, and current spatial management layers were always useful during discussions. MBNMS staff also incorporated the occurrence models for six groundfish species that were chosen to represent the suite of groundfish on the US West Coast as modeled by the NOAA Northwest Fisheries Science Center (NWFSC) and NOAA National Centers for Coastal Ocean Science (NCCOS). The models were based on NMFS bottom-trawl surveys and some in situ observations.

Local Data Acquisition

The regional datasets provided by NMFS spanned the entire US West Coast, so MBNMS GIS staff augmented it with local data from the Monterey Bay Aquarium Research Institute (MBARI) and California State University, Monterey Bay (CSUMB). Both institutions have developed databases with in situ observations of fish and invertebrates, which MBNMS staff used to corroborate the NMFS and NCCOS models and better understand groundfish distributions.

MBARI used high-resolution video equipment to record observations of the marine environment during more than three hundred remotely operated vehicle (ROV) dives per year. In MBARI's 25 years of ROV operations (1989 to the present), well over 16,000 hours of videotapes have been archived, annotated, and maintained as a centralized institutional resource. This video library contains detailed footage of the biological, chemical, geological, and physical aspects of the Monterey Bay submarine canyon and other areas of the US West Coast. MBARI has developed the Video Annotation and Reference System (VARS), consisting of software and hardware to facilitate the creation, storage, and retrieval of video annotation records from the ROV dive tapes. The public can run queries on VARS, but the most recent data is embargoed for two years to allow MBARI scientists the first opportunity to publish findings using the ROV data. The groundfish query results used for the collaborative proposal were run internally by MBARI staff, and thus include the most recent data.

The MBARI VARS query was run for 10 species and one complex (table 11.1). Species selection was based on the groundfish species modeled by both NMFS and NCCOS and the additional species modeled only by NCCOS (NMFS 2013a and 2013b). Each query was exported as a text file, and then imported to Microsoft Excel where headings were modified for better compatibility with ArcGIS software. The files were converted to an Esri shapefile, and hyperlinks to images were activated.

Table 11.1. Local Database Queries for Select Groundfish and Sessile Invertebrates

Common Name	Scientific Name	MBARI VARS Query	CSUMB Query
Dark blotched rockfish	*Sebastes crameri*	x	None in database
Yelloweye rockfish	*Sebastes ruberrimus*	x	x
Sablefish	*Anoplopoma fimbra*	x	x
Shortspine thornyhead	*Sebastolobus alascanus*	x	None identified to species yet
Longspine thornyhead	*Sebastolobus altivelis*	x	None identified to species yet
Thornyhead complex	*Sebastolobus complex*		None in MBNMS EFH proposed areas
Greenstriped rockfish	*Sebastes elongatus*	x	x
Petrale sole	*Eopsetta jordani*	x	x
Dover sole	*Microstomus pacificus*	x	x
Lingcod	*Ophiodon elongatus*	x	x
Pacific ocean perch	*Sebastes alascanus*	None in database	None in database
Chilipepper rockfish	*Sebastes goodei*	x	x
Bocaccio	*Sebastes paucispinis*	Not run	x
Pacific halibut	*Hippoglossus stenolepis*	Not run	x
Canary/Vermillion/Yelloweye rockfish complex		Not run	x
Canary rockfish	*Sebastes pinniger*	Not run	x
Young of the Year (YOY) rockfish		Not run	x
Widow rockfish	*Sebastes entomelas*	Not run	None identified to species yet
Corals		Included in DSCRTP Database	x
Sea pens and sea whips		Included in DSCRTP Database	x
Sponges (>10 cm)		Included in DSCRTP Database	x

x = records returned from the query.

A similar process was used to query the CSUMB database for demersal fish and invertebrate species of interest. CSUMB staff built a Microsoft Access database containing habitat and fauna observational data from approximately 45 hours of ROV video and 150 hours of towed camera sled video collected in MBNMS. ROV datasets were from 2009, 2010, and 2012. The towed camera sled datasets were from 2007 to 2011. Additional video was taken from the camera sled in 2006, but those datasets were analyzed separately for the EFH proposal and are not yet in the database. Organisms were identified to the lowest taxonomic level possible, which in many cases was just to genus or species complex rather than species. This was because of many factors, including visibility at depth, vehicle lighting, and organism position relative to the video camera. Database query results included the organism or complex of interest, count, and associated latitude and longitude coordinates. Each query was exported as a Microsoft Excel file and converted to an Esri shapefile.

CSUMB database queries were run for invertebrates in addition to fish (see table 11.1). CSUMB's invertebrate datasets were not included in the regional DSCRTP database available on the EFH catalog, so those CSUMB queries helped augment the in situ data from DSCRTP.

Finally, the proposal was enhanced by local knowledge and information from fishermen. This anecdotal information was of a larger cartographic scale than the trawling effort data that was provided in the EFH data catalog. Because of the confidentiality of fishing effort data, only areas fished by three or more vessels can be made public through the EFH data catalog. However, by working with the trawl fishermen, MBNMS staff were able to understand which local areas were important to each individual fisherman. MBNMS GIS staff helped the fishermen capture these areas within GIS so that their datasets were available in the same medium as the biological, geological, and management data. During the first few meetings, MBNMS GIS staff digitized their areas of interest in a live editing session using ArcGIS 10.1 for Desktop software (figure 11.4). The trawl fishermen provided site-specific socioeconomic information on the locations of historically productive bottom-trawl fishing grounds that were then captured as attributes in GIS. Trawl fishermen also informed MBNMS staff and other stakeholders about the operational logistics that aid in an economically viable bottom-trawl fishery within the study area. These datasets provided vital information that helped the stakeholder group develop a collaborative proposal.

Figure 11.4. **Trawl fisherman pointing to an area on an ArcGIS for Desktop map projected on a screen, which enabled MBNMS staff to capture spatial comments.** Courtesy of NOAA Monterey Bay National Marine Sanctuary.

GIS Tools

Relevant US West Coast and local data was compiled into an ArcGIS Online web map (see a link to the EFH Story Map in the "Supplemental Resources" section at the end of this chapter). The datasets that were referred to most often, aside from the base layers that oriented the user, were the biogenic habitat data layers and associated image links. Whenever possible and appropriate, MBNMS GIS staff converted regional US West Coast raster datasets to vector form and clipped them to the study area (as vector files use less service credits). The ArcGIS Online web map allowed all staff and stakeholders to explore the dataset and develop strategic input for meetings based on these datasets. The nontechnical staff of MBNMS also used the ArcGIS Online service while writing the introduction and discussion sections of the proposal. ArcGIS Online allowed them to quickly access biological and geological characteristics of each individual area and select ROV-captured images to include in the proposal to best illustrate the habitats of these areas.

In addition, an Esri Map Book app was developed for individual areas using data-driven pages to showcase the biogenic, management, bathymetric, and geological data for each area. The Data Driven Pages toolbar was enabled, and the EFH boundary modification layer proposed for MBNMS was selected as the index layer. The map book was an asset because it provided a quick reference to larger-scale maps of each of the proposal areas, which focused on the locations of biogenic habitat from regional and local sources.

Calculations

The EFH modification polygons in the proposal were developed using an iterative process, requiring constant modification and refinement in GIS on the fly during the numerous meetings with stakeholders. The MBNMS GIS team also provided summary statistics for each boundary iteration. In all GIS processes, ArcGIS 10.1 for Desktop was used, often with the ArcGIS Spatial Analyst extension. MBNMS developed a matrix in Microsoft Excel to store important information (e.g., total area of various substrate types, number of corals seen in situ), most of it calculated using GIS, for each of the modified polygon iterations, making it easy to compare and contrast the data. The most relevant data was transferred from the matrix to the attribute tables of the polygons in both the ArcGIS desktop and online projects.

MBNMS GIS staff added a variety of useful habitat data as attributes. For example, the ArcToolbox function Zonal Statistics as Table (Spatial Analyst extension) was used to summarize raster values of depth and percent slope by each depth zone (continental shelf: 30–100 m; continental shelf break: 100–300 m; continental slope 1: 300–800 m; continental slope 2: 800–3,000 m) for each of the proposed areas. This tool summarizes the values of a raster within the zones of another raster or feature layer and reports the results to a table. The summary attributes included the mean, min, max, median, and standard deviation of the values of the raster grids per zonal area. Other important attributes summarized included physical habitat type (hard or soft) and whether an area was classified as being in a canyon or not. Areas were then calculated in ArcGIS using the Calculate Geometry tool. The resulting shapefile was exported as a Microsoft Excel table, and a variety of pivot tables were generated so the data could be analyzed from different perspectives.

Within Excel, the Shannon-Wiener diversity index, H' (Shannon 1948), was calculated as a way to examine habitat diversity within each proposed area. There were 11 possible habitat classes, which combined the depth zone, substrate type, and canyon habitat. Higher diversity values indicated a balanced proportion of habitat classes.

Results

The final proposal contains 10 areas to add to EFH conservation areas and five areas to reopen (figure 11.5). The proposal considered these modifications as a comprehensive package that will minimize, to the extent practicable, the adverse effects of bottom-trawl gear on groundfish EFH within the study area. In-depth and overview GIS analyses of the individual areas and the whole package were incorporated in the proposal. For example, the proposal adds 260 sq mi to EFH conservation areas, of which 84% is thought to be soft and 16% is hard bottom. In addition, 153 sq mi, almost all of which is estimated to be soft bottom, is proposed to be reopened to bottom trawling.

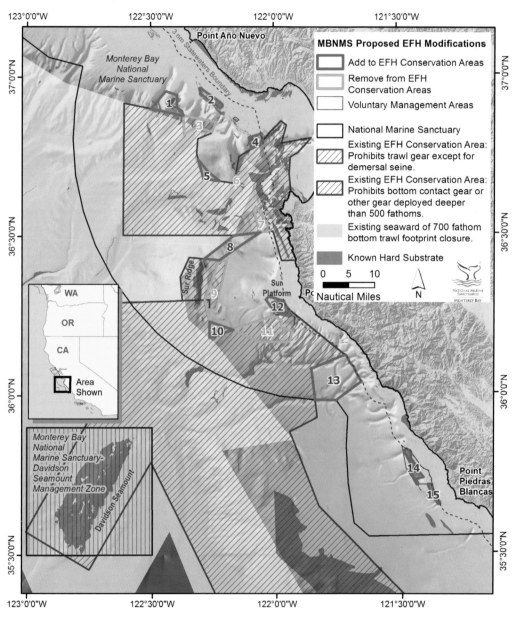

Figure 11.5. Essential fish habitat (EFH) conservation area modifications within the Monterey Bay National Marine Sanctuary study area as proposed by the collaborative proposal. Modifications include adding (in green) or removing (in orange) areas from the existing conservation areas. Courtesy of NOAA Monterey Bay National Marine Sanctuary. Data from NOAA National Marine Fisheries Service, California Department of Fish and Wildlife, Monterey Bay National Marine Sanctuary, Esri, Moss Landing Marine Laboratories.

Since MBNMS GIS staff had integrated a variety of layers in both the ArcGIS for Desktop project and ArcGIS Online web map, seafloor imagery could be used to validate substrate data. One such example pertained to Area 12 at Point Sur Platform, where the habitat layer obtained from the EFH data catalog delineated a smaller area on the platform as hard substrate compared with two independent sources of substrate data provided by the US Geological Survey (USGS) and CSUMB. This proved to be vital hard-substrate habitat and led us to proposing Area 12 as an additional EFH conservation area.

Many maps were included in the proposal to illustrate the spatial relationships between various data. Figure 11.6 is a map book example page and showcases biogenic data in Area 12 and the lack thereof in Area 11, Sur Canyon, which was an area fishermen wanted to reopen to trawling for deep continental slope fishes. This map shows the biogenic data (e.g., sponges) from the DSCRTP and CSUMB databases. Area 12 is an area that is a proposed addition to EFH conservation areas. As an additional map portfolio example, MBNMS GIS staff combined the NMFS groundfish occurrence models with the local in situ observations of those groundfish, and then overlaid the proposed EFH conservation area boundaries to have a better understanding of the local applicability of the models, in addition to which species were captured within the proposed boundaries. For example, figure 11.7 shows the NMFS modeled data for longspine thornyhead occurrence overlaid by MBARI observations and effort data. The longspine thornyhead preference for soft continental slope habitats is clearly captured by the model. MBARI in situ observations and effort data suggest that many fish are also found in slightly shallower waters than the model predicts.

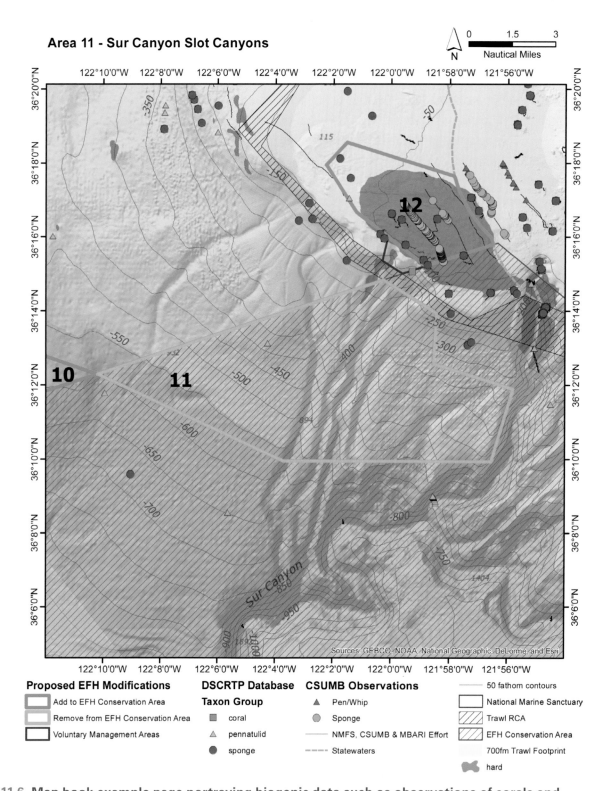

Area 11 - Sur Canyon Slot Canyons

Figure 11.6. **Map book example page portraying biogenic data such as observations of corals and sponges and proposed essential fish habitat conservation area modifications, including Areas 11 and 12.** Courtesy of NOAA Monterey Bay National Marine Sanctuary. Data from NOAA National Marine Fisheries Service, GEBCO, National Geographic, DeLorme, and Esri.

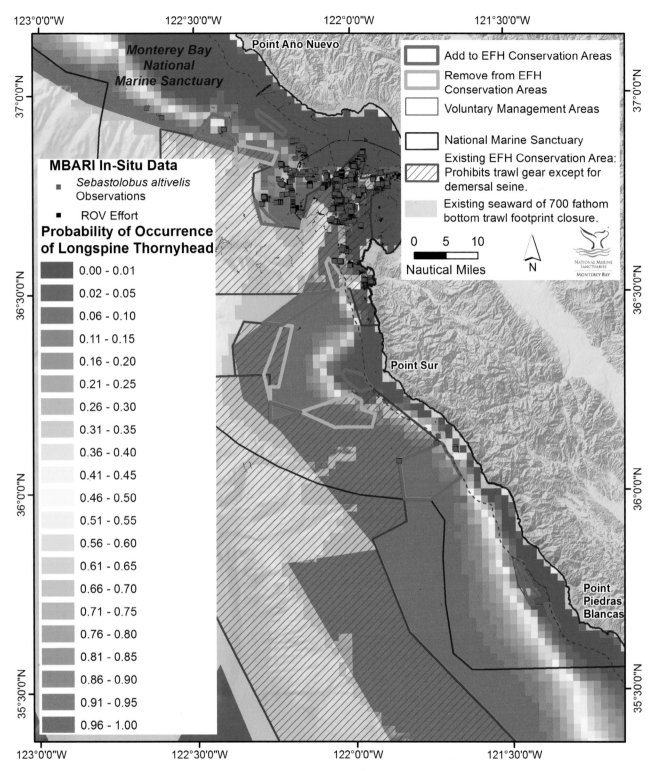

Figure 11.7. **National Marine Fisheries Service occurrence model for longspine thornyhead (*Sebastolobus altivelis*) and in situ observations by Monterey Bay Aquarium Research Institute within the Monterey Bay National Marine Sanctuary study area. The red color covers the continental slope and symbolizes high probability of occurrence.** Courtesy of NOAA Monterey Bay National Marine Sanctuary. Data from NOAA National Marine Fisheries Service, Monterey Bay Aquarium Research Institute, California Department of Fish and Wildlife, and Esri.

Conclusion

All participating stakeholders were dedicated to crafting a collaborative proposal that balanced protection of groundfish essential fish habitat with allowing for bottom-trawling opportunities. Engaging in this collaborative process was an important step in building a persistent cooperative relationship and trust between MBNMS, fishermen, and the conservation community. Using GIS as a main decision support tool allowed the collaborations to stay on track as it allowed stakeholders to have a shared understanding of the spatial distribution of resources and fishing efforts. By using live GIS maps, a wayward discussion could be refocused to a specific place to enable the group to move forward, or data layers could be added in real time to help the group see trade-offs between the proposed changes in the EFH conservation area boundaries. At the time of this writing, the proposal was still under consideration by PFMC for implementation.

The EFH data catalog provided streamlined access to data layers and associated information to all levels of GIS users. The GIS desktop user was able to download zipped map packages that contained thorough metadata. The online user could access the data via ArcGIS Explorer software, ArcGIS Online, Google Earth, or the Pacific Coast Ocean Observing System (PaCOOS) map viewer (Goldfinger and Romsos 2014). The non-GIS user was able to download prepared maps. The data catalog was a crucial support tool for developing the proposal by providing a range of data (e.g., biological, anthropogenic, and habitat) to a variety of users. However, augmenting the regional data with local data proved useful and insightful. MBNMS GIS staff also had copies of some of the regional data within the same GIS project so that the data could be symbolized based on local scale ranges, which were often not as extensive as the US West Coast ranges. This helped us better understand the local variability of the data. GIS staff were also well versed in the associated caveats of all the data so that it could be used appropriately. The collaborative proposal showcases the benefits of pairing stakeholders with different interests and knowledge together with GIS staff who were intimately familiar with the data caveats and limitations.

ArcGIS Online can be a valuable tool for stakeholder processes as it helped us reach consensus among stakeholders. It is user-friendly for nontechnical staff and stakeholders and is also a useful asset to those versed in GIS. MBNMS GIS support staff selected a finite number of layers for the ArcGIS Online project so that users would not be overwhelmed, as there was a tremendous amount of spatial data associated with this project (see the EFH Story Map link in the "Supplemental Resources" section at the end of this chapter). GIS staff also ensured that links functioned properly, attributes and attribute names were fully expounded, and images and descriptive graphs for each feature were included. It required time and dedication by the GIS staff but was an essential ingredient in the collaborative process.

MBNMS managers noted that the main reason the collaborative process was effective was because of the use of a full suite of GIS products. Figure 11.8 captures that truth as a conservation group and the trawling fleet have a discussion surrounded by various spatial products. First, they discuss an area projected on screen from a geoenabled PDF file. A large printed poster map of the study area is on the wall behind the trawl fleet representative. In addition, a computer running ArcGIS for Desktop is in the foreground. Having desktop GIS allowed MBNMS GIS staff to digitize and analyze areas in real time, which proved informative for the group because they had to evaluate all options. Running live ArcGIS during the meetings allowed stakeholders to ask and respond to questions efficiently, allowing them to make much greater progress than without real-time spatial data.

Figure 11.8. **Representatives from a conservation group (left) and trawling fleet (right) discuss their ideas while reviewing a geoenabled PDF as other stakeholders listen and study the projected map using ArcGIS for Desktop software.** Courtesy of NOAA Monterey Bay National Marine Sanctuary.

The proposed boundary modifications of Areas 1–15 (see figure 11.5) represent a package of areas that, if taken collectively, provide increased protection to rocky reef HAPCs. It is useful to include these additional hard-substrate areas in the package along with sensitive biogenic species, including corals and sponges, while also providing increased access to valuable fishing grounds. The package comprises 10 additions to EFH conservation areas that prohibit bottom-trawl gear and five modifications that reopen portions of current EFH conservation areas to bottom trawling. A total of 106 sq mi was added to EFH conservation areas within the MBNMS study area, and of that total increase, 39% is hard substrate. In the end, the stakeholders were able to agree on areas to reopen and close to trawling, and deliver a collaborative proposal to PFMC for Areas 1–13 (again, still under consideration). Two of the areas in MBNMS, Areas 14 and 15, were negotiated without the participation of Oceana, Ocean Conservancy, or the Natural Resources Defense Council. At the end of the stakeholder process, MBNMS finalized the design of these two areas with the California Risk Pool, as the full group did not have the time or opportunity to reach consensus on these areas.

In addition, the proposal identifies and proposes three "voluntary management areas" (see figure 11.5) that will not be affected by EFH regulation, yet will be recognized as areas to be avoided by trawlers because of sensitive habitat. The details on how long the voluntary measures will be in place, and the time scale for revisiting the agreement, will be worked out by the group in the year ahead. These areas are nonregulatory, will be managed and monitored by MBNMS, and should be considered as experimental or pilot areas to test the effectiveness of the concept of voluntary management areas for habitat protection.

In summary, the array of proposed modifications to EFH conservation areas would demonstrate a net benthic habitat benefit to groundfish EFH by

- further minimizing adverse impacts of bottom-trawl fishing to groundfish EFH to the extent practicable, based on new and newly available information.
- providing socioeconomic benefits at both the fishery and community scale.
- increasing access of bottom-trawl fishing activity to historically productive trawl fishing grounds in MBNMS that currently prohibit bottom-trawl activity while minimizing adverse impacts to EFH groundfish on a whole. The reopening comes in the context of an overall package of revisions that will yield a strong net increase in habitat protection.
- increasing the overall protection of hard substrate, including rocky reef HAPCs, biogenic habitats (e.g., corals and sponges), canyon habitat, and soft-substrate habitat, at various depths.

One measure of the success of this collaborative effort is that the proposal received eight support letters from a variety of interests. The support letters were not just from the stakeholders that had representatives engaged in the collaborative proposal, but also from other influential entities, including Julie Packard of the Monterey Bay Aquarium, Mark Stone of the California state Legislature, and US Congressman Sam Farr, representing California's central coast.

Letter of US Rep. Sam Farr (D-CA) to NMFS and PFMC, October 21, 2013

Habitat protection, based on sound science and active stakeholder and community participation, is a fundamental management tool that the Council is authorized to employ as it works to restore and sustainably manage our nation's fish populations. The Monterey Bay National Marine Sanctuary is one of the prized gems of our ocean. NOAA is authorized to protect and provide for multiple uses of the resources within the Sanctuary. To do so, they must continue to facilitate similar negotiations, bringing stakeholders to the table to work collaboratively from the beginning. This will not only foster trust and understanding between all parties, but also provide a forum to make the hard decisions about how to protect and use the most valued areas in our ocean.

In my view, habitat protection is a critical tool of fishery management, and we must encourage all stakeholders to review the best science, work collaboratively to build trust, and identify areas for protection that will help us sustainably protect and manage our fish populations. I call on the Council to endorse the outcome of the Sanctuary-facilitated negotiation, thereby empowering the Sanctuary staff, local officials, fishermen, and NGOs to continue the challenging work by returning to the negotiating table. There is much more work they must do together.

As Congressman Farr notes in his letter of support (see sidebar), there is much more that the stakeholder group must do together to achieve other resource protection and use goals. Also, in the event that the proposal does not go through PFMC's implementation process, the stakeholder group can still work together to identify other potential alternative resource management opportunities to achieve similar goals to those of the collaborative proposal. The group found the multitude of GIS tools employed during this proposal development process to be useful and reliable. As such, a similar stakeholder decision-making process could be used to consider alternate management applications, such as implementing other voluntary management areas. The spatial data provided a common language for stakeholders during the development of this proposal and will be a keystone to future collaborations. The development of this proposal, with the aid of various types of GIS tools, provides a strong foundation for fishermen, conservationists, and resource managers to work collaboratively toward other resource management goals in the future.

Acknowledgments

We thank our various collaborators at MBARI and CSUMB, as well as the various stakeholders from the commercial trawling, conservation, and resource management communities who worked with us on the EFH proposal. We are also grateful to PFMC and NOAA NMFS for their guidance and support. The editorial assistance of Dawn Wright, and the comments and suggestions of two anonymous reviewers, greatly improved this chapter.

References

Gass, J., M. D'Iorio, and H. Selbie. 2013. "The Pacific Regional Ocean Uses Atlas: Participatory GIS to Incorporate Traditional Knowledge into Ocean Management." In *Proceedings of the Esri Ocean GIS Forum*, Redlands, CA, November 5–7. http://bit.ly/1tOTTw9.

Gleason, M. G., S. McCreary, M. Miller-Henson, J. Ugoretz, E. Fox, M. Merrifield, W. McClintock, P. Serpa, and K. Hoffman. 2010. "Science-Based and Stakeholder-Driven Marine Protected Area Network Planning: A Successful Case Study from North Central California." *Ocean & Coastal Management* 53:52–68.

Goldfinger, C., and C. Romsos. 2014. PaCOOS: West Coast Habitat Server. http://pacoos.coas.oregonstate.edu. Last accessed July 26, 2014.

Lanier, A., C. Steinback, C. Burt, C. Macdonald, S. Fletcher, T. Haddad, T. Welch, and W. McClintock. 2011. "The Oregon MarineMap Project: An Online Tool to Assist in Oregon's Ongoing Marine Spatial Planning Processes." Paper C10 in *Proceedings of Coastal GeoTools*, Myrtle Beach, SC, March 21–24. http://1.usa.gov /UuUC9j.

NMFS (National Marine Fisheries Service). 2013a. *Appendix to Groundfish Essential Fish Habitat Synthesis: A Report to the Pacific Fishery Management Council*. Seattle, WA: NOAA NMFS Northwest Fisheries Science Center.

———. 2013b. *Groundfish Essential Fish Habitat Synthesis: A Report to the Pacific Fishery Management Council*. Seattle, WA: NOAA NMFS Northwest Fisheries Science Center.

Office of the Federal Register. 2006. "EFH Final Rule for Implementing the Regulatory Provisions of Amendment 19 to the Pacific Coast Groundfish Fishery Management Plan (71 FR 27408)." *Federal Register* 71(91). http:// www.gpo.gov/fdsys/granule/FR-2006-05-11/06-4357.

Office of National Marine Sanctuaries. 2008. *Monterey Bay National Marine Sanctuary Final Management Plan.* http:// montereybay.noaa.gov/intro/mp/mp.html. Last accessed July 26, 2014.

Pacific Coast Ocean Observing System. 2014. *5-Year Review of Pacific Coast Groundfish EFH, Consolidated Geographic Information Data Catalog and Online Registry.* http://efh-catalog.coas.oregonstate.edu/overview. Last accessed July 26, 2014.

Pacific Fishery Management Council. 2012. *Pacific Coast Groundfish 5-Year Review of Essential Fish Habitat Report to the Pacific Fishery Management Council Phase 1: New Information; Groundfish Essential Fish Habitat Review Committee.* http://www.pcouncil.org/groundfish/background/document-library/pacific-coast-groundfish-5 -year-review-of-efh. Last accessed July 26, 2014.

Paul, E., W. McClintock, and D. Wright. 2012. "SeaSketch for Oil Spill Response." *Journal of Ocean Technology* 7 (4): 130–31.

Shannon, C. E. 1948. "A Mathematical Theory of Communication." *The Bell System Technical Journal* 27:379–423, 623–56.

Stelzenmuller, V., J. Lee, A. South, J. Foden, and S. I. Rogers. 2013. "Practical Tools to Support Marine Spatial Planning: A Review and Some Prototype Tools." *Marine Policy* 38:214–27.

Supplemental Resources

Dawn J. Wright, ed.; 2015; *Ocean Solutions, Earth Solutions*; http://dx.doi.org/10.17128/9781589483651_d

A URL and QR code are provided for the digital content that goes with this chapter:
- Story Map app for essential fish habitat (EFH), at http://bit.ly/1a6yzqX

This content is also available on the Esri Press "Book Resources" webpage for chapter 11. Go to esripress.esri.com/bookresources. Then, in the list of Esri Press books, click *Ocean Solutions, Earth Solutions*. On the *Ocean Solutions, Earth Solutions* resource page, click the chapter 11 link to access that webpage and the hyperlink to view the EFH Story Map, based on the Esri Story Map app.

CHAPTER 12

More Than Maps: Connecting Aquarium Guests to Global Stories

Jennifer Lentz, Emily Yam, Alie LeBeau, and David Bader

Abstract

With a mission to instill a sense of wonder, respect, and stewardship for the Pacific Ocean, its inhabitants, and ecosystems, the Aquarium of the Pacific in Long Beach, California, aims to connect its diverse audience of 1.5 million annual visitors to beautiful animal collections, conservation messages, and ideas that inspire a better understanding of human relationships to environments and a changing global ecosystem. To promote exploration of complex issues such as climate change, ocean exploration, ocean acidification, extreme weather, and how global data is collected, the Aquarium makes use of various technologies and communication messages. Tools such as digital interactives, spherical projections of global visualizations, large-scale maps, geospatial applications, and ArcGIS software provide Aquarium guests context, visualizations, and time to explore the things necessary to construct an understanding of global systems and relationships between the pieces.

Although these innovative technologies are enticing and hold the attention of the audience, they are most effective when paired with skilled educators who ask questions and guide guests to make their own discoveries. As an Informal Science Education Institution, the Aquarium provides experience and exposure to its guests who may not have previously experienced these underwater worlds, or heard the stories from a global perspective. These experiences provide the base for understanding, appreciating, and taking action for effecting change in the future. By connecting people to science, and being the trusted conveyer of current information, the Aquarium is helping to create a more scientifically literate community.

As the Aquarium moves forward with its programs and initiatives, it will continue to strive to be "more than a fish tank" as it creates innovative, immersive experiences that make use of geospatial technologies that serve students, teachers, families, and adult audiences while inspiring an appreciation for the impact humans have on our planet and the future we can create together.

Dawn J. Wright, ed.; 2015; *Ocean Solutions, Earth Solutions*; http://dx.doi.org/10.17128/9781589483651

Introduction

The Aquarium of the Pacific

The Aquarium of the Pacific is a public, nonprofit 501(c)(3) organization, located along Rainbow Harbor, in downtown Long Beach, California. Since opening its doors in 1998, the mission-based facility has aimed to instill a sense of wonder, respect, and stewardship for the Pacific Ocean, its inhabitants, and its ecosystems. This mission is carried out by providing engaging, entertaining, educational, and empowering exhibits and programs about the Pacific Ocean and the people who interact with it, live near it, and depend on it for food, recreation, and inspiration. The Aquarium is an Association of Zoos and Aquariums (AZA) accredited institution that serves approximately 1.5 million annual visitors, making it the fourth most attended aquarium in the nation in 2014. According to a national marketing survey, the Aquarium serves the most ethnically diverse audience of any major aquarium (Morey Group 2014).

The Aquarium of the Pacific is home to more than 12,000 animals representing a range of habitats, including local, temperate Southern California waters; cool northern Pacific environments; tropical western Pacific islands; and the darkness of the deep sea. Guests visiting the Aquarium can touch sharks, sea jellies, and tide pool animals; stroll through an aviary of colorful lorikeet birds; watch a training enrichment session with sea otters; and walk through a tunnel into the Tropical Pacific habitat. Going beyond its live animal collection, the Aquarium of the Pacific prides itself on using experts in various fields of ocean science to develop programs that are then communicated through the Aquarium of the Pacific's educators, exhibits, and use of cutting-edge technologies. Ocean exploration, global climate change, ocean acidification, overfishing, and medicines from the sea are just a few of the topics explored through the Aquarium's exhibits, films, and programs.

Complementing the animals, a broad portfolio of technologies is used to deepen the learning experience at the Aquarium. Guests explore near real-time, historical, and modeled data projected on the Aquarium's Science on a Sphere (SOS) installation in the Aquarium's Ocean Science Center. Developed by the National Oceanic and Atmospheric Administration (NOAA), SOS is a 6 ft diameter sphere designed to display dynamic information on large spatial and temporal scales (McDougall et al. 2007; Randol et al. 2012). Here, guests can watch beautifully produced films and engage in discussions with educators about earth systems science, global events in weather, and climate change. A cross-site, summative evaluation of facilities with SOS technology (Goldman et al. 2010) indicates that facilitation through Science on a Sphere correlates with increased perception by learners of the value of the earth as well as a need to protect it. Results from a front-end research report on SOS interpreters at science museums show that there are challenges involved with SOS-related ongoing learning, especially with time and effort investments despite an interest from both managers and interpreters to stay up to date on information (Hart and Hayward 2011). Educator-facilitated experiences using this technology give learners the tools to interpret scientific visualizations and opportunities to connect science and data to their daily lives (Randol et al. 2012). The Ocean Exploration Hub features Hiperwall technology, a large array of flat screens that act as a portal into the world of exploration. Throughout the day, Aquarium guests can see a live feed of activities on ocean-going research vessels, including seafloor mapping, instrument deployment, and the robot's-eye view of the ocean depths as scientists search for specimens. At scheduled show times, the Aquarium broadcasts live interactions at the Ocean Exploration Hub, facilitated by educators, enabling Aquarium guests in Long Beach to interact in real time with scientists working around the world. The Aquarium has partnered with Cisco Systems to install 15 monitors throughout the Aquarium, adding a timely dimension to the timeless dimension for which aquariums are well known.

Technology is a central element in the Aquarium of the Pacific's redefinition of the modern aquarium. Pacific Visions, the Aquarium's expansion set to open in 2017, includes a two-story, dynamic digital theater with a 32 ft tall, 180° arc digital projection wall and a 30 ft diameter floor projection disc, immersing guests in a virtual ocean environment. An innovative partnership model in collaboration with leading scientists, educators, filmmakers, and artists has brought to life powerful stories to be featured in Pacific Visions, highlighting the great ocean animal migrations, the leading role of the ocean in the world's weather and climate, and the vast and undiscovered biodiversity of the ocean. Through interactive geospatial technology, guests guide the exploration of potential future scenarios. Once these scenarios are developed, guests are challenged to identify the actions citizens must take for a sustainable future. Importantly, these experiences will be accessible to educators, scientists, and the general public via digital platforms to allow for conversations, learning, and engagement to continue beyond the Aquarium visit. Technology is essential in enhancing a guest's experience at facilities such as the Aquarium because it allows for learning opportunities that are longer lasting, more powerful, and more accessible.

The Role of Informal Science Education Institutions

Informal Science Education Institutions (ISEIs) such as aquariums, museums, science centers, and zoos serve a critical role in today's society in improving public understanding of science (McDougall et al. 2007; Falk and Dierking 2010) and educating about conservation. ISEIs have the unique distinction of offering free-choice, learner-centered experiences that can change learning opportunities for people of all ages (Tran 2008; National Research Council 2009; Falk and Dierking 2010), promoting lifelong, life-wide, and life-deep learning. The Aquarium takes pride in its role as a supporter of traditional schools, or formal education, and strives to engage and inspire its guests and the next generation of scientists (Randol et al. 2012).

In free-choice learning environments, learners are more likely not only to absorb the information being presented, but also to continue to seek out information on their own time (National Research Council 2009). Learners who see science as fun are more likely to excel in informal science education settings (Jolly et al. 2004; Falk and Dierking 2010). Additionally, ISEIs are generally viewed by the public as a trusted and reliable source of scientific information (Falk et al. 2007; McDougall et al. 2007; Falk and Dierking 2010), with recent studies showing that the public looks to zoos, aquariums, and museums for information and guidance on environmental and conservation issues (The Ocean Project 2009a and 2009b; Centers for Ocean Sciences Education Excellence 2014). This is especially important when it comes to helping the public understand climate change (The Ocean Project 2009a).

Aquariums are particularly well suited to tell climate change and conservation stories because live animal collections establish interest and undeniable emotional connections between people, animals, and environments (Falk et al. 2007). An aquarium visitor who enjoys the penguin or sea otter exhibits is more likely to find out about the threats affecting these animals and their habitats. Additionally, the public's trust in ISEIs as a source of information gives aquariums the power to increase not only science literacy, but also the public's awareness, understanding, and acceptance of such issues as anthropogenic climate change, and ultimately inspire more responsible, conservation-minded ways of living (Coyle 2005; Falk et al. 2007; McDougall et al. 2007; Randol et al. 2012).

In an era dominated by information and technology, the use of such media is expected to support science learning, especially in spaces such as museums, zoos, and aquariums (Coyle 2005; McDougall et al. 2007). Technology, in particular, is an important mode of learning for people because it allows learners to engage in inquiry that is specific to the content (Linn et al. 2004). Ucko and Ellenbogen (2008) suggest that "technology-based tools can supplement and assist, but do not necessarily have to replace, direct experiences in a museum. Rather, they make objects and phenomenon a central component of an inquiry, and then can further support learners as they extend their investigations beyond simply reading exhibit labels or even beyond a one-time visit." Learners who interact with technology at the Aquarium are able to examine issues that are important to them, and explore these issues through conversations with educators (Randol et al. 2012). This is of particular importance when an understanding of large spatial and temporal scales is needed to learn about the complexities of global phenomena such as scales in climate, marine, and earth system sciences. Technology allows a facilitator to access prior learning and guide learning because content can be reviewed and rescaled as needed (Goldman et al. 2010). Geospatial technologies lend themselves well to this kind of knowledge building because they provide a specific context for learners and can be tied directly to displays such as live animal exhibits (Barstow and Hoffman 2007; Goldman et al. 2010).

Learning Theory and Next-Generation Science Standards

Using technology as a tool, the Aquarium of the Pacific employs best practices in learning theory and pedagogy to create a rich experience for learners. Inquiry-based methods in free-choice environments allow guests to build their own understanding and meaning-making. Also called *constructivism*, these learning experiences are an active process in which open-ended questions, manipulation of objects, and real-world experiences help guide learners on their individual journey to understand the world around them (Dewey 1938; Vygotsky 1986; Jonassen et al. 1999; Markham et al. 2003; Tran 2011). "Constructivist learning environments employing inquiry-based, student-led investigations; open-ended questions; and real-world experiences are increasingly seen as a key to educational reform (Jonassen et al. 1999; Markham et al. 2003)," according to Henry and Semple (2012). Although this foundational learning theory has been evolving for over half a century, it has been translated most recently in the development and nationwide implementation of the Next Generation Science Standards (NGSS) (Quinn et al. 2012; NGSS 2014).

ISEIs, including the Aquarium of the Pacific, are positioned to support NGSS adoption in formal or traditional classroom environments by designing experiences that are inquiry driven, data driven, and authentic (McDougall et al. 2007; Randol et al. 2012). The Aquarium has recently revitalized its inquiry approach with the ultimate goal of creating experiences that make its learners feel and behave like explorers. Aquarium educators create opportunities for all guests to ask questions; make observations; examine scale in time and space; and discuss, debate, and share ideas. Facilitation of all public presentations, school programs, and one-on-one interpretation is now focused on four main themes that mirror the learning cycle: wonder, explore, discover, and learn (WEDL). The WEDL focus enables staff to teach nonscientists how to think both critically and scientifically, rather than telling them what to think (Niepold et al. 2007; Tran 2011; Quinn et al. 2012). Thus, instead of being the conveyor of facts, Aquarium educators help learners process information relative to their personal experiences and contexts.

Interactive, Web-Based Technologies

Interactive, web-based technologies will continue to play an increasingly more important role in aspects of society by redefining how people learn, get information, engage with each other, and connect to the world around them (Linn et al. 2004; Ucko and Ellenbogen 2008). ISEIs must respond to these changes by using the strengths of the technology to stay relevant and hold the attention of today's youth (National Research Council 2009; The Ocean Project 2009b). To serve as a truly effective learning tool, the application, interactive tool, or website must provide more than just verbal information presented in a digital interface. Technology must have a visually enticing display coupled with an interactive interface that both captivates and maintains the user's interest through the use of inquiry-based tools (Linn et al. 2004). Though by no means a replacement for direct, hands-on experience and observation, the availability of such technologies has been shown to complement, supplement, and even assist in an individual's ability to "connect" with the given subject matter (Ucko and Ellenbogen 2008). Additionally, by making these technologies available remotely as either an Internet or smartphone application, ISEIs are able to extend their ability to connect, engage, and educate the public beyond the walls of the Aquarium (The Ocean Project 2009b).

Inquiry, constructivism, and technology complement one another in a way that creates unique opportunities for learning that also builds skills in the long term. The National Science Education Standards state, "Science requires different [human] abilities" (National Research Council 2006). Human cognitive abilities include literacy, numeracy, logic, and graphicacy. Success in each of these is dependent, to varying degrees, on a student's ability to think and imagine spatially, a concept known as *spatial ability* (National Research Council 2006; Martinez et al. 2009). In particular, learning about spatial systems provides opportunities for deep meaning-making and skill building (Bednarz 1995; Drennon 2005).

Geographic Information Systems

GIS provides ISEIs with the ability to use interactive, cutting-edge, spatial technologies to create an inherently constructivist learning environment that promotes scientific and spatial literacy in people of all ages by turning them into explorers. Geospatial visualizations help students use "critical thinking skills to examine real-world issues" and "provides geography, technology, and spatial analysis skills" in a science and/or humanities context (Demski 2011). Spatial thinking ability is a key component in students' ability to think critically, solve problems both inside and outside the classroom, and be an effective citizen in today's modern society (Libens et al. 2002; Bednarz et al. 2006; National Research Council 2006; Milson and Alibrandi 2008; Schultz et al. 2008; Martinez et al. 2009). Learners "need to be able to visualize the geospatial distribution of cultures, economies, and natural resources to fully understand the complexities of our global environment" (Guertin et al. 2012). Spatial content and geospatial technologies have been shown to improve scientific literacy, spatial reasoning, and higher-level thinking skills in people of all ages (Kidman and Palmer 2006; National Research Council 2006; Doering and Veletsianos 2007; Niepold et al. 2007; Schultz et al. 2008; Demski 2011; Hagevik 2011; Henry and Semple 2012; Kulo and Bodzin 2013). GIS is recognized as an interdisciplinary technology that supports high-level thinking and spatial reasoning, inquiry science, and open-ended investigations; directly correlates to real-world experience; allows students to visualize complex real-world problems; supports multiple modes of learning (Donaldson 2001; Bodzin and Anastasio 2006; National Research Council 2006); and is therefore a powerful tool for ISEIs.

Bednarz et al. (2006) contend that the ability to use images and geospatial technologies intelligently and critically is becoming a requirement to participate effectively as a citizen in modern society. Students and Aquarium guests can use GIS to solve real-life investigations and draw on skills crucial to developing higher-order thinking and problem solving (Ramirex and Althouse 1995; Sanders et al. 2002; Bodzin and Anastasio 2006). To provide contextual relevance and connect guests with a larger systemic understanding of their experience, the Aquarium has developed, implemented, and promoted different geospatial visualization programs.

Overview of the Aquarium's Geospatial Programs

Floor-to-Ceiling Exhibit Maps

The Aquarium of the Pacific uses various spatial platforms, ranging from simple static maps and images to user-driven, interactive geospatial visualizations and GIS, to connect guests to exhibits and ecosystems. Maps are used as geographical organizers throughout the Aquarium to connect the animals and exhibits to real places around the world. These stylized maps provide guests with context for the marine organisms and cultural aspects of the ocean region they are experiencing. Similar maps are found at the entrance to the Aquarium's three main galleries: Tropical Pacific (Western Pacific to Australia, figure 12.1a), North Pacific (Northern California to Alaska), and Southern California and Baja (temperate to subtropical regions of the Pacific Ocean, figure 12.1b). Adjacent to the June Keyes Penguin Habitat, a map of South America is marked with the typical distribution of Magellanic penguins (figure 12.1c). Many Aquarium guests are surprised to find that penguins live in environments other than Antarctica, especially the warmer climates of South America. The map also supports educators as they tell the story of four individual penguins in the collection, which were rescued when they became stranded off the coast of Brazil. Maps can also be found accompanying other major exhibits based on ecologically important but not widely known areas, such as the Gulf of California (figure 12.1d) and Coral Triangle (figure 12.1e). Both exhibits share a story for guests exploring the galleries: previously unknown locations with high endemism, high biodiversity, and tremendous threat from human impact.

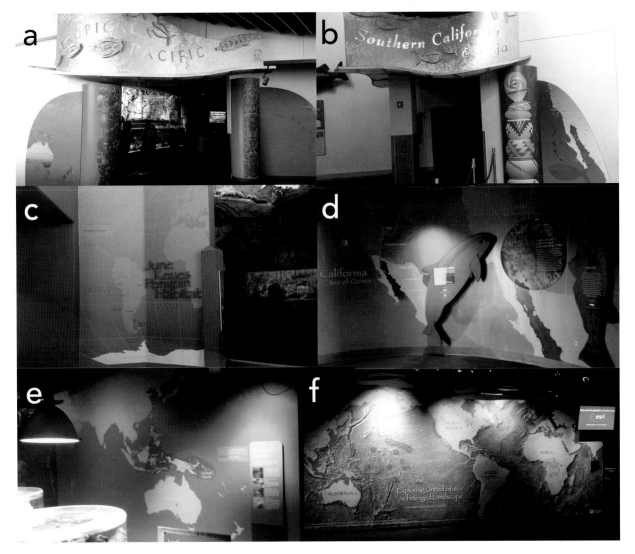

Figure 12.1. **Floor-to-ceiling maps located throughout the Aquarium of the Pacific. Stylized maps surround the entrances of each of the Aquarium's main galleries, including (a) the Tropical Pacific and (b) Southern California and Baja, to orient guests to the part of the world they are about to explore. A map is also used at (c) the entrance of the June Keys Penguin Habitat. Floor-to-ceiling maps are also showcased in the Aquarium's galleries, including (d) Gulf of California, (e) Coral Triangle, and (f) US Exclusive Economic Zones.** Photos courtesy of the Aquarium of the Pacific in Long Beach, California.

In addition to highlighting specific areas showcased in Aquarium exhibits, maps are used to generate discussion and broaden the public's awareness of the world ocean. In 2013, Esri partnered with the Aquarium in the creation of a large-scale (10 × 23 ft) map of the US Exclusive Economic Zones (EEZs) using its high-resolution Ocean basemap (figure 12.1f). The EEZ map, located toward the end of the Tropical Gallery, allows guests to be visually struck by the vast expanse of US territories worldwide, including those underwater. The image has become a source of conversation between parents and children, who can be heard discussing places around the world they have a connection to, through travel or other experiences. Using maps to set personal and place-based context has been a great success of the Aquarium.

Google Liquid Galaxy Exhibit

This blend of personal connection, maps, and technology has merged in the Liquid Galaxy installation (figure 12.2), provided to the Aquarium by Google. The exhibit was unveiled during the NOAA Office of Ocean Exploration and Research Forum, hosted by the Aquarium in July 2013. Google's Liquid Galaxy brings Google Earth to a large semicircle of screens, creating an immersive experience.

Figure 12.2. **Google's Liquid Galaxy exhibit at the Aquarium of the Pacific.** Photo courtesy of the Aquarium of the Pacific, Long Beach, California. Google Earth Map data ©2014 Google, Image ©2014 DigitalGlobe, Image ©2014 GeoForce Technologies; Data SIO, NOAA, US Navy, NGA, GEBCO.

The virtual globe interface of Google Earth allows people to view locations on Earth at any angle and from any distance (Schultz et al. 2008). The large surround screen displays of Liquid Galaxy provide guests the ability to explore, ask questions, and make discoveries as they fly around the world. Aquarium educators say guests have often expressed interest in continuing the experience once they leave, by installing the software on their home computers.

In February 2014, the Aquarium hosted a community event associated with a National Aeronautics and Space Administration (NASA)-funded teacher workshop. Interactive, educator-led stations around the Aquarium, including Liquid Galaxy, were designed to give teachers, parents, and students an opportunity to discover that data is everywhere. As expected, most guests instinctively used the interactive technology of Liquid Galaxy to find their own home, but more in-depth explorations were also observed. For example, a history teacher used it to introduce students to his home in Eastern Europe. He guided them on a three-dimensional journey through the history of his town; the layout of the present-day city, where castle walls once stood; and how the streets radiated from a central location, indicating where the castle had once stood. By pairing his personal experience with the engaging and illustrative technology, this teacher was able to use the Aquarium to provide a memorable learning experience for his students. These quality engagements with Liquid Galaxy are similar to those observed by studies of Google Earth in K-12 classrooms (National Research Council 2006; Doering and Veletsianos 2007; Schultz et al. 2008; Guertin et al. 2012; Henry and Semple 2012), suggesting that this exhibit is not only fostering spatial reasoning, but has practical applications for informal learning environments.

NOAA's Science on a Sphere Exhibit

In 2011, the Aquarium of the Pacific became the first major aquarium with an SOS installation (figure 12.3). The Aquarium has additionally become the first institution with an SOS to create content focused directly on the most pressing issues facing today's oceans and share this material with other members of the SOS Users Network. Today, the Aquarium continues to be one of the most prolific producers of content for SOS, having produced more than seven films, three of which were created entirely in-house. Content, including visualizations, datasets, films, and interpretive materials, is shared widely across the network of institutions with SOS, reaching 33 million people worldwide (NOAA 2014). To improve the effectiveness of the experiences offered, evaluation results are also shared among the institutions in the SOS network (Goldman et al. 2010).

Figure 12.3. **Science on a Sphere exhibit entrance and interactive sphere at the Aquarium of the Pacific.** Photos courtesy of the Aquarium of the Pacific, Long Beach, California.

The large-scale and high-definition quality images of SOS make it the centerpiece of the Aquarium's use of technology. Both attractive and adaptable, the sphere is a perfect platform for telling stories to Aquarium guests. Spherical films produced by the Aquarium of the Pacific are dynamic media pieces that combine datasets, imagery, and narratives about people to tell stories about a broad range of topics (Randol et al. 2012). These films, ranging from four to eight minutes, focus on the following topics: sea level rise; the role of seaports; biodiversity; ocean exploration; coastal issues; marine debris; and how global data is collected through satellites, buoys, and remote-sensing technologies.

In addition to films, the Aquarium's Education Department uses the SOS platform to engage the public in deeper conversations about the ocean, climate, and earth systems. Through classroom programs and interpretations with public audiences (including school groups, families, and small groups of adults), educators facilitate and navigate through a dialog matched with datasets that help people to connect to historical, and even recent, events, such as Hurricanes Katrina and Sandy, Typhoon Haiyan, and the Tohoku earthquake that led to the Fukushima nuclear power plant leak. Modeled data can show connectivity across the planet, such as the global connections of commercial airline traffic, Facebook Friend connections, and large-scale animal migrations. Animated datasets allow guests to see global processes over different temporal scales, such as water availability, warming from climate change, and changes in the spatial distribution of organisms. Conversations between educators and learners about these images build skills interpreting spatial data and false-color images, understanding changes over space and time, and how those changes affect ocean life and humans (Goldman et al. 2010). Ultimately, all the Aquarium's SOS programs are designed to maximize the sphere platform by telling compelling narratives that help guests feel connected to marine and terrestrial environments while also increasing science literacy and awareness of conservation issues. A 2012 evaluation designed to explore the impact of the Aquarium's SOS films on its guests found that "visitors identify and remember the main messages presented in the programs" and the "presentations encourage discourse with others outside the aquarium" (Randol et al. 2012).

Aquarium of the Pacific's ArcGIS Online Website

Using the user-friendly ArcGIS Online interface with hosting capabilities and templates, the Aquarium has quickly begun creating interactive mapping applications and Story Map apps, based on the Esri Story Map app, and making them available through an Esri-hosted site. The Aquarium's ArcGIS Online website (aop.maps.arcgis.com), shown in figure 12.4, currently has a number of interactive web maps and Story Map apps and provides links to other, non-ArcGIS Online geospatial online applications developed by and for the Aquarium.

AQUARIUM
OF THE PACIFIC®
A non-profit institution

OCEAN
EXPLORATION

Story Maps Website

Story Maps and other GIS-related projects

AOP's CELC Story Map

AOP's Magellanic Penguins Story Map

AOP's Visions '13 Cruise Story Map

CELCs & Extreme Weather Events by Coastal Region

The Aquarium of the Pacific's mission is to instill a sense of wonder, respect, and stewardship for the Pacific Ocean, its inhabitants, and ecosystems. Our vision is to create an aquarium dedicated to conserving and building Natural Capital (Nature and Nature's services) by building Social Capital (the interactions between and among peoples).

The Aquarium of the Pacific (AOP) is the fourth most-attended aquarium in the nation. It displays over 11,000 animals in more than 50 exhibits that represent the diversity of the Pacific Ocean. Each year more than 1.5 million people visit the Aquarium. Beyond its world-class animal exhibits, the Aquarium offers educational programs for people of all ages from hands-on activities to lectures by leading scientists. Through these programs and a variety of multimedia experiences, the Aquarium provides opportunities to delve deeper into ocean science and learn more about our planet. The Aquarium of the Pacific has redefined the modern aquarium. It is a community gathering place where diverse cultures and the arts are celebrated and a place where important topics facing our planet and our ocean are explored by scientists, policy-makers and stakeholders in the search for sustainable solutions.

Esri.com | Help | Terms of Use | Privacy | Contact Esri | Contact Us | Report Abuse

Figure 12.4. Screen capture of the Aquarium of the Pacific's ArcGIS Online home page, available online at aop.maps.arcgis.com. The interactive maps, Story Map apps, and applications referred to in figures 12.5 to 12.9 can be accessed using the "Story Maps and Other GIS-Related Projects" ribbon on the Aquarium's ArcGIS Online home page. Screen capture courtesy of Aquarium of the Pacific, Long Beach, California.

The Aquarium's ArcGIS Online home page showcases top geospatial applications, including the Coastal Ecosystem Learning Centers (CELC) Story Map (figure 12.5) created using the Esri Story Map Tour template and designed to introduce the Coastal America Partnership and its network of CELCs. The Coastal America Partnership is a partnership of federal, state, and local-agency governments and private organizations that work together to protect, preserve, and restore the US nation's coasts (The Coastal America Partnership 2014). The CELC Story Map begins with an explanation of the Coastal America Partnership, the mission of the partnership, and the location of CELCs within each coastal region (top panel). The user is able to travel between the different CELCs and see the name, address, website, image, and close-up map of each one (bottom panel).

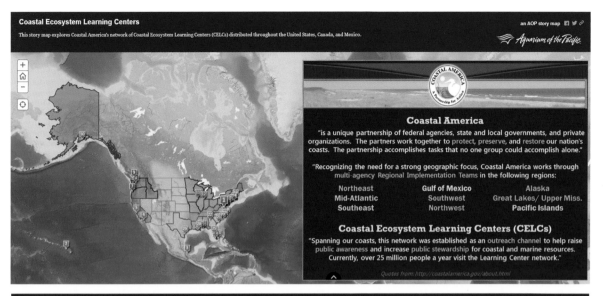

Figure 12.5. The Aquarium of the Pacific Coastal Ecosystem Learning Centers Story Map takes the user on an interactive, guided tour of aquariums of research centers throughout North America. The map begins by (top) showing a map of all the Coastal Ecosystem Learning Centers locations and explaining how they are all part of the Coastal America Partnership. The user can then (bottom) choose to go on an interactive tour to learn where specific learning centers are located and what they look like, with the images hyperlinked to the website for the given Coastal Ecosystem Learning Center. Screen captures courtesy of the Aquarium of the Pacific, Long Beach, California. Data sources: Coastal America Partnership and Coastal Ecosystem Learning Centers, GEBCO, NOAA NGDC, Natural Earth, and World Hydro Reference Overlay Map service.

Users who wish to learn more about CELCs may be interested in the CELC Extreme Weather by Coastal Region map app, which examines the frequency of extreme weather events for each of the nine coastal regions defined by Coastal America (figure 12.6). The goal of this Story Map app is to illustrate the partnership between these organizations and their efforts to increase public awareness and stewardship of marine and coastal ecosystems.

Figure 12.6. **The CELC (Coastal Ecosystem Learning Centers) Extreme Weather by Coastal Region Story Map is an interactive ArcGIS Online map app that enables the user to visualize the distribution of Coastal Ecosystem Learning Centers throughout North America and examine the frequency of extreme weather events within each coastal region.** Screen capture courtesy of the Aquarium of the Pacific, Long Beach, California. Data sources: Coastal America Partnership and Coastal Ecosystem Learning Centers, GEBCO, NOAA NGDC, Natural Earth, and World Hydro Reference Overlay Map service.

The Esri Story Map Tour template was also used to create a Story Map app about the Aquarium's Magellanic penguins (figure 12.7), which provides information about each of the penguins including nicknames, personality traits, their journey to the Aquarium, and a picture. Both the creation and release of this Story Map app were designed to coincide with the public debut of Heidi and Anderson, the first penguin chicks hatched at the Aquarium. Guests, who may already feel a connection with these charismatic animals, can use the Story Map app to learn more about the species, but also appreciate how all the penguins came to the Aquarium.

Figure 12.7. The Aquarium of the Pacific's Magellanic Penguins Story Map. Screen capture and images courtesy of the Aquarium of the Pacific, Long Beach, California, and Robin Riggs. Sources: Esri, DigitalGlobe, Earthstar Geographics, CNES/Airbus DS, GeoEye, i-cubed, USDA FSA, USGS, AEX, Getmapping, Aerogrid, IGN, IGP, swisstopo, and the GIS User Community.

Stand-Alone Geospatial Web Applications

In addition to the interactive maps and Story Map apps created using ArcGIS for Desktop software and ArcGIS Online service, the ribbon on the Aquarium's ArcGIS Online home page also includes links to the following stand-alone spatial applications created by the Aquarium. The Aquarium's Oceans in Motion application was designed to enable the Aquarium's guests to view some of the images, videos, and spatial datasets featured in the Science on a Sphere film "Oceans in Motion." The Sea Level Rise application provides an interactive, map-based interface in which the user can locate areas that are currently at risk of flooding and those that could be at risk in the future (figure 12.8), including the types and locations of infrastructures at risk from sea level rise.

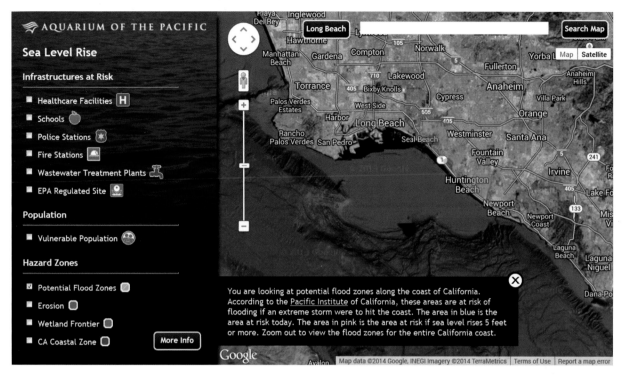

Figure 12.8. The Aquarium of the Pacific's interactive Sea Level Rise app. The interactive interface of this app allows the user to identify the locations of vulnerable populations, infrastructures, and hazard zones resulting from flooding at current sea levels (shown in blue) and with a 5 ft rise in sea level (shown in pink). This app can be accessed through a link on the Aquarium's ArcGIS Online home page, or directly at http://www.aquariumofpacific.org/webapps/slr. Screen capture and images courtesy of the Aquarium of the Pacific, Long Beach, California. Data sources: all data and research used in this app was based on a report prepared by the Pacific Institute, mapping support provided by Arc2Earth with a Google Earth basemap. Map data ©2014 Google, INEGI Imagery ©2014 TerraMetrics.

The whale research application Southern California Whale Research Project: Connecting People, Science, and Whales (figure 12.9) is a product of the Aquarium's partnership with the Cascadia Research Collective in Olympia, Washington. Since June 2010, the Aquarium has worked to collect data to help understand the movements of blue whales in Southern California. After photo identification of individual whales, the web-based app makes use of GIS and other software technologies to create a user-friendly geospatial database that provides access to whale sighting data for both scientists and the public. Although the focus of the project is on blue whales, the database also includes data collected on other cetacean species, including fin whales, gray whales, humpback whales, minke whales, orcas, Risso's dolphins, bottlenose dolphins, and common dolphins. The interactive interface allows the user to query by species, individual whale, time period, and observed behaviors such as fluking and lunge feeding. The application also has the option to overlay layers depicting the study area and shipping lanes of the large ships headed into the ports to see which whales are most often in danger. Although available to the public through the Aquarium's website, this application is also used as an interpretation tool to engage guests in discussions about the risks these endangered animals face and connections to conservation.

Figure 12.9. The Aquarium of the Pacific's Southern California Whale Research Project: Connecting People, Science, and Whales app allows users not only to view the locations of cetacean species common to the Southern California area, but also filter the data by date, species, behavior (fluking, lunge feeding, and so forth), or even a specific whale. This app can be accessed through a link on the Aquarium's ArcGIS Online home page or directly at http://whaleproject.aquariumofpacific.org.

Screen capture and images courtesy of the Aquarium of the Pacific, Long Beach, California. Data sources: all data used in this app was based on ongoing research at the Aquarium of the Pacific, in collaboration with Cascadia Research Collective and Harbor Breeze Cruises. The app was created for the Aquarium of the Pacific by JLOOP, using a Google Earth basemap. Map data ©2014 Google Imagery, ©2014 TerraMetrics.

Applications such as these help extend learning experiences beyond the walls of the Aquarium by enabling guests to learn more about the world's fisheries, marine protected areas, ocean currents, sea surface temperatures, dead zones, ocean acidification, and so forth. These applications also provide information on what we can do to preserve and restore marine ecosystems.

Geospatial Outreach through Teacher Workshops

The Aquarium hosts a number of annual workshops designed to offer classroom teachers professional development opportunities that provide new experiences, information, and resources. These Aquarium-facilitated workshops align with the needs of the teachers now transitioning to NGSS. The focus on experience, exposure, and access to scientists working in the field, rather than prewritten lesson plans, is in keeping with the goals of NGSS (National Research Council 2006 and 2009; Quinn et al. 2012). The Aquarium's annual teacher professional development opportunities include the five-day Boeing Teacher Institute (BTI), two-day NASA Our Instrumented Earth workshop, and a one-day NOAA Office of Exploration and Research workshop. These opportunities provide extensive resources to teachers at no cost and support the development of curricula and activities that include ocean sciences and data interpretation in the classroom.

Although teachers may be aware of GIS and the beneficial role it can play in the development of spatial reasoning skills in students, some are often overwhelmed by the idea of learning new software, intimidated by the idea of finding "real data," and unsure of how best to incorporate GIS into their current curricula (Doering and Veletsianos 2007). To help address these problems, access to data and simple GIS overviews have been incorporated into workshops. During the Boeing Teacher Institute in 2013, a step-by-step tutorial was provided (Lentz 2013b), and a 20-minute presentation, titled "Story Telling with Maps" (Lentz 2013d) introduced teachers to GIS and the power of the new Esri Story Map capabilities. A simple example showing the locations of schools represented by the workshop participants was provided for teachers. NASA's Our Instrumented Earth teacher workshop, in October 2013, had more data-focused content. During this two-day workshop, a 60-minute presentation, titled "An Introduction to GIS and Spatial Data," walked the participants through finding spatial data and bringing it into a GIS using the ArcGIS Online interface, including the location sightings and geographic range of California sea lions (Lentz 2013a and 2013c).

Preview of Future Geospatial Programs

The use of technology and geospatial programs has just begun and will continue to grow at the Aquarium of the Pacific. The planned expansion of Pacific Visions provides state-of-the-art, innovative resources, creating an immersive experience for Aquarium guests. In much the same way that IMAX and OmniMAX theaters gave people the virtual ability to fly, Pacific Visions turns people into deep-sea explorers, storm chasers, and even allows them to join pods of migrating whales. The Aquarium creates increasingly more advanced geospatial visualizations designed to both maximize the educational potential of existing resources, and prepare staff for the 2017 opening of Pacific Visions by providing training on how to most effectively interpret using interactive, GIS-based technologies.

Recognizing that narratives used in conjunction with visualization tools are an effective means of relaying scientific information to nonscientists (Ward 2006; Niepold et al. 2007), all upcoming geospatial materials are being designed to support specific stories.

New SOS Content
The Aquarium continues to develop its capacity to tell stories by building new interactive datasets designed to connect guests to important environmental issues, such as human population growth, ocean acidification, and marine animal migrations. Among the content currently under development are new SOS image sequences that show the Aquarium's sea turtle rehabilitation and tracking program, launched in 2006.

This story, in particular, allows the Aquarium to leverage the popularity of sea turtles to provide information about behaviors and migration. The story can include threats to sea turtles: habitat loss, turtle egg harvesting, turtle consumption, jewelry made from their shells, unintentional bycatch from commercial fisheries, and, perhaps most common, severe injuries and death caused by the consumption of plastic bags and other man-made materials. By showing guests geospatial visualizations produced using the authentic tracking data from sea turtles rescued, rehabilitated, and released, the Aquarium is able to provide positive examples of its mission and commitment to conservation. Additionally, educators can provide guests with specific actions to take to reduce threats currently facing these animals.

New Esri Story Map App

ArcGIS Online made it possible for organizations such as the Aquarium to quickly and easily produce attractive, interactive geospatial mapping applications and Story Map apps, using the Esri Story Map app. In 2013, links to specific Aquarium-produced Story Map apps were released through social networks (Facebook and Twitter) to gauge public interest in web GIS. Because of positive feedback from the test releases, this content is being made available directly through the Aquarium's website. With 180,000 unique monthly visits from around the world, Aquarium Story Map apps have the potential to be a global educational tool.

Recognizing the power of this new medium, the Aquarium plans to create templates better suited to the specific requirements of the upcoming Story Map apps. Drawing on the concepts presented in the Aquarium's current CELC Story Map (see figure 12.5) and CELC Extreme Weather by Coastal Region mapping application (see figure 12.6), the Aquarium has a working draft of a CELC Story Map designed to help people connect to climate change by addressing local impacts. The Story Map app presents the user with different extreme weather events by providing personal accounts from CELCs that experience or are at risk of these events. Users can select a specific extreme weather phenomenon (hurricanes, severe droughts, wildfires, blizzards, flooding, and so forth) to view a brief description, related images and diagrams, and an interactive map that allows the user to see past, present, and future at-risk areas. This map will also include the locations of CELCs, accompanied by photographic and written narratives of how that particular CELC was affected by the weather. For example, Audubon's Aquarium of the Americas, located in New Orleans, Louisiana, was impacted when Hurricane Katrina hit in 2005. Structurally, the facility was still intact, but subsequent flooding and electrical issues created unstable life support systems; most of the animal collection was lost. Incredibly, the facility was able to rebuild and open a year later (Los Angeles Times 2006). During Hurricane Sandy in 2012, another CELC, the New York Aquarium, flooded and sustained severe damage. Eighteen staff members cared for the collection of animals, and the facility reopened seven months later (Fears 2012). Combining interactive geospatial visualizations of extreme weather events with personal narratives, this Story Map app connects local places or places people have visited to the very real events and impacts of severe weather.

The Aquarium continues to create new Story Map apps to connect its guests with geospatial technologies; the possibilities are endless, and the stories are varied. Upcoming projects include a virtual tour of the Aquarium that pairs exhibit images with the actual geographic location to which the exhibit was modeled, a depiction of local watersheds and the story of where Southern California gets its water, and narratives of ambassador Aquarium animals that represent threatened or endangered counterparts. Still other visualizations could tell the great success stories of recovered populations, sustainable fisheries, and responsible management. Ultimately, the Aquarium plans to use Story Map apps to connect people to real places and issues by making topics such as climate change, habitat loss, and threatened and endangered populations easier to understand and relate to.

Conclusion

Our world is rapidly changing. The way people learn, communicate, and get information about our planet is evolving, and this evolution is happening more quickly than ever before. To stay involved, the public is in need of trusted sources of scientific data, information, and tools to critically evaluate this content. The Aquarium of the Pacific, and its fellow informal science education institutions, are the facilitators of this current science and must stay connected to new technologies and storytelling strategies to engage, connect, and inspire guests to take action. Through captivating animal exhibits and innovative technologies, the Aquarium of the Pacific is connecting people to the physical and biological world around them, increasing the capacity of its audience to think critically about conservation, and redefining what it means to be an aquarium.

Acknowledgments

The comments and suggestions of three anonymous reviewers greatly improved this chapter.

References

Barstow, D., and M. Hoffman. 2007. *Revolutionizing Earth System Science Education for the 21st Century: Report and Recommendations from a 50-State Analysis of Earth Science Education Standards.* Cambridge, MA: TERC Center for Earth and Space Science Education.

Bednarz, S. W. 1995. "Researching New Standards: GIS and K–12 Geography." In *GIS/LIS 1995 Conference Proceedings*, Nashville, TN, November 14–16.

Bednarz, S. W., G. Acheson, and R. S. Bednarz. 2006. "Maps and Map Learning in Social Studies." *Social Education* 70 (7): 398–404.

Bodzin, A., and D. Anastasio. 2006. "Using Web-Based GIS for Earth and Environmental Systems Education." *Journal of Geoscience Education* 54 (3): 297–300.

Centers for Ocean Sciences Education Excellence. 2014. *COSEE Ocean Inquiry Group Report: Opportunities for Creating Lifelong Ocean Science Literacy.* Boston, MA: University of Massachusetts, School for the Environment.

The Coastal America Partnership. 2014. Coastal America Partnership. http://www.coastalamerica.gov. Last accessed July 3, 2014.

Coyle, K. 2005. *Environmental Literacy in America.* Washington, DC: National Environmental Education and Training Foundation.

Demski, J. 2011. "Map Quests: GIS Technologies Allow Students to Tackle Real-World Issues while Developing Critical Thinking Skills." *T H E Journal 38 (8): 12–14.*

Dewey, J. 1938. *Experience and Education.* New York: Touchstone.

Doering, A., and G. Veletsianos. 2007. "An Investigation of the Use of Real-Time, Authentic Geospatial Data in the K–12 Classroom." *Journal of Geography* 106 (6): 217–25.

Donaldson, A. 2001. "With a Little Help from Our Friends: Implementing Geographic Information Systems (GIS) in K–12 Schools." *Social Education* 65 (3): 147–50.

Drennon, C. 2005. "Teaching Geographic Information Systems in a Problem-Based Learning Environment." *Journal of Geography in Higher Education* 29 (3): 385–402.

Falk, J. H., and L. D. Dierking. 2010. "The 95% Solution." *American Scientist* 98: 486–93.

Falk, J. H., E. M. Reinhard, C. L. Vernon, K. Bronnenkant, J. E. Heimlich, and N. L. Deans. 2007. *Why Zoos & Aquariums Matter: Assessing the Impact of a Visit to a Zoo or Aquarium.* Silver Spring, MD: Association of Zoos and Aquariums.

Fears, D. 2012. Water World: Photos of the Animals in the Flooded New York Aquarium. http://www.today.com/pets/water-world-photos-animals-flooded-new-york-aquarium-1C6841450. Last accessed July 3, 2014.

Goldman, H. K., C. Kessler, and E. Danter. 2010. *Science on a Sphere: Cross-Site Summative Evaluation*. Edgewater, MD: Institute for Learning Innovation. http://www.oesd.noaa.gov/network/SOS_evals/SOS_Final_Summative_Report.pdf.

Guertin, L., C. Stubbs, C. Millet, T.-K. Lee, and M. Bodek. 2012. "Enhancing Geographic and Digital Literacy with a Student-Generated Course Portfolio in Google Earth." *Journal of College Science Teaching* 42 (2): 32–37.

Hagevik, R. A. 2011. "Fostering 21st-Century Learning with Geospatial Technologies: Middle Grades Students Learn Content through GPS Adventures." *Middle School Journal* (J3), 43 (1): 16–23.

Hart, J., and J. Hayward. 2011. *EarthNow: Science Museum Presenters' Interest in Future Data Sets for Science on a Sphere (SOS): Front-End Research (Evaluation Series, Report #1)*. Northampton, MA: People, Places & Design Research.

Henry, P., and H. Semple. 2012. "Integrating Online GIS into the K–12 Curricula: Lessons from the Development of a Collaborative GIS in Michigan." *Journal of Geography* 111 (1): 3–14.

Jolly, E J., P. B. Campbell, and L. Perlman. 2004. *Engagement, Capacity, and Continuity: A Trilogy for Student Success*. Groton, MA: GE Foundation Report. http://www.campbell-kibler.com/trilogy.pdf.

Jonassen, D., K. L. Peck, and B. G. Wilson. 1999. *Learning with Technology: A Constructivist Perspective*. Toronto: Merrill/Prentice-Hall.

Kidman, G., and G. Palmer. 2006. "GIS: The Technology Is There But the Teaching Is Yet to Catch Up." *International Research in Geographical and Environmental Education* 15 (3): 289–96.

Kulo, V., and A. Bodzin. 2013. "The Impact of a Geospatial Technology-Supported Energy Curriculum on Middle School Students' Science Achievement." *Journal of Science Education Technology* 22:25–36.

Lentz, J. A. 2013a. *GIS Wildlife Data Tutorial: NASA Teacher Workshop Handout*. Long Beach, CA: Aquarium of the Pacific. http://jenniferalentz.info/Teaching/Tutorials/Tutorial_UsingGISwildlifeData_2013.pdf.

———. 2013b. *How to Make an Esri Storytelling Interactive Webmap: Boeing Teacher Institute (BTI) Teacher Workshop Handout*. Long Beach, CA: Aquarium of the Pacific. http://jenniferalentz.info/Teaching/Tutorials/CreatingStoryMapsOnline_2013.pdf.

———. 2013c. *An Introduction to GIS and Spatial Data: NASA Teacher Workshop Presentation*. Long Beach, CA: Aquarium of the Pacific. http://jenniferalentz.info/Teaching/Lectures/AOP-GuestLecture_NASA_2013.pdf.

———. 2013d. *Story Telling with Maps: Boeing Teacher Institute (BTI) Teacher Workshop Presentation*. Long Beach, CA: Aquarium of the Pacific. http://jenniferalentz.info/Teaching/Lectures/AOP-GuestLecture_BTI_2013.pdf.

Libens, L., K. Kastens, and L. Stevenson. 2002. "Real World Knowledge through Real World Maps: A Developmental Guide for Navigating the Educational Terrain." *Developmental Review* 22 (2): 267–322.

Linn, M. C., P. Bell, and E. A. Davis. 2004. Specific Design Principles: Elaborating the Scaffolded Knowledge Integration Framework." In *Internet Environments for Science Education*, edited by M. C. Linn, E. A. Davis, and P. Bell, 315–40. Mahwah, NJ: Lawrence Erlbaum Associates.

Los Angeles Times. 2006. "New Orleans Aquarium Reopens After Katrina." http://articles.latimes.com/print/2006/may/27/nation/na-briefs27.2. Last accessed July 3, 2014.

Markham, T., J. Larner, and J. Ravitz. 2003. *Project Based Learning Handbook: A Guide to Standards-Focused Project-Based Learning for Middle and High School Teachers*. Novato, CA: Buck Institute for Education.

Martinez, A. E., N. A. Williams, S. K. Metoyer, J. N. Morris, and S. A. Berhane. 2009. "A Geospatial Scavenger Hunt." *Science Scope* 32 (6): 18–23.

McDougall, C., J. McLaughlin, W. Bendel, and D. Himes. 2007. "NOAA's Science on a Sphere Education Program: Application of a Scientific Visualization System to Teach Earth System Science." In *Proceedings of the Fifth International Symposium on Digital Earth*, San Francisco, CA, June 5–9.

Milson, A. J., and M. Alibrandi, eds. 2008. *Digital Geography: Geospatial Technologies in the Social Studies Classroom*. Charlotte, NC: Information Age Publishing.

Morey Group. 2014. *Aquarium of the Pacific—2014 Los Angeles Market Study*. Charleston, SC: Morey Group Industry Report.

National Research Council. 2006. *Learning to Think Spatially: GIS as a Support System in the K–12 Curriculum.* Washington, DC: National Academies Press.

———. 2009. *Learning Science in Informal Environments: People, Places, and Pursuits.* Washington, DC: National Academies Press.

Next Generation Science Standards. 2014. The Next Generation Science Standards. http://www.nextgenscience.org /next-generation-science-standards. Last accessed July 3, 2014.

Niepold, F., D. Herring, and D. McConville. 2007. "The Case for Climate Literacy in the 21st Century." In *Proceedings of the Fifth International Symposium on Digital Earth*, San Francisco, CA, June 5–9.

NOAA (National Oceanic and Atmospheric Administration). 2014. What Is Science on a Sphere? http://sos.noaa.gov. Last accessed July 3, 2014.

The Ocean Project. 2009a. *America, the Ocean, and Climate Change: New Research Insights for Conservation, Awareness, and Action: Executive Summary.* http://theoceanproject.org/wp-content/uploads/2011/12/ExecutiveSummary _June2009.pdf. Last accessed July 3, 2014.

———. 2009b. *America, the Ocean, and Climate Change: New Research Insights for Conservation, Awareness, and Action: Key Findings.* http://theoceanproject.org/wp-content/uploads/2011/12/America_the_Ocean_and_Climate _Change_KeyFindings_1Jun09final.pdf. Last accessed July 3, 2014.

Quinn, H., H. Schweingruber, and T. Keller, eds. 2012. *A Framework for K–12 Science Education: Practices, Crosscutting Concepts, and Core Ideas.* Washington, DC: National Academies Press.

Ramirex, M., and P. Althouse. 1995. "Fresh Thinking: GIS in Environmental Education." *T H E Journal* 23 (2): 87–90.

Randol, S., M. Werner-Avidon, and C. Castillo. 2012. *Aquarium of the Pacific Ocean Science Center Final Evaluation Report.* Berkeley, CA: The Research Group, The Lawrence Hall of Science, University of California, Berkeley.

Sanders, R., L. Kajs, and C. Crawford. 2002. "Electronic Mapping in Education: The Use of Geographic Information Systems." *Journal of Research on Technology in Education* 34 (2): 121–29.

Schultz, R. B., J. J. Kerski, and T. C. Patterson. 2008. "The Use of Virtual Globes as a Spatial Teaching Tool with Suggestions for Metadata Standards." *Journal of Geography* 107 (1): 27–34.

Tran, L. U. 2008. "The Work of Science Museum Educators." *Museum Management and Curatorship* 23 (2): 135–53.

———. 2011. "Module 2. Session 1: Building Knowledge, Constructing Understanding." *Reflecting on Practice: The Lawrence Hall of Science.* http://reflectiveeducators.org.

Ucko, D. A., and K. M. Ellenbogen. 2008. "Impact of E-Technology on Informal Science Learning." In *The Impact of the Laboratory and Technology on Learning and Teaching Science K–16*, edited by D. Sunal, E. Wright, and C. Sundberg. Charlotte, NC: Information Age.

Vygotsky, L. 1986. *Thought and Language.* Cambridge, MA: Harvard University Press.

Ward, A. 2006. "The Glory of the Story: A Summary of Kendall Haven's Presentation at the May EPO Colloquium." *The Earth Observer* 18 (4): 17–20, 22.

Supplemental Resources

Dawn J. Wright, ed.; 2015; *Ocean Solutions, Earth Solutions*; http://dx.doi.org/10.17128/9781589483651_d

URLs and QR codes are provided for the digital content for this chapter, explained in the following sections.

Hyperlinks are also available on the Esri Press "Book Resources" webpage for chapter 12. Go to esripress.esri.com/bookresources. Then, in the list of Esri Press books, click *Ocean Solutions, Earth Solutions*. On the *Ocean Solutions, Earth Solutions* resource page, click the chapter 12 link to access that webpage and the hyperlinks to view the Story Map apps and visit the Aquarium website.

Coastal America Partnership and CELCs

- Story Map app of the Coastal America Partnership and its network of Coastal Ecosystem Learning Centers (CELCs), at http://bit.ly/1pQGvsR

The Coastal America Partnership is a partnership of federal, state, and local-agency governments and private organizations that work together to protect, preserve, and restore the US nation's coasts. The Story Map app, based on the Esri Story Map app, begins with an explanation of the Coastal America Partnership, the mission of the partnership, and location of CELCs within each coastal region. The user is able to travel between the different CELCs and see the name, address, website, image, and close-up map of each one.

Magellanic Penguins

- Magellanic Penguins Story Map, at http://bit.ly/1pQGO73

This Magellanic Penguins Story Map provides information about each of the penguins living at the Aquarium of the Pacific, including nicknames, personality traits, their journey to the Aquarium, and a picture. Both the creation and release of this Story Map app were designed to coincide with the public debut of Heidi and Anderson, the first penguin chicks hatched at the Aquarium. Guests, who may already feel a connection with these charismatic animals, can use the Story Map app to learn more about the species but also appreciate how each penguin came to the Aquarium.

Aquarium of the Pacific Website

- Aquarium of the Pacific website, at http://aop.maps.arcgis.com/home

The Aquarium has quickly begun creating interactive mapping applications and Story Map apps, based on the Esri Story Map app, and making them available through this Esri-hosted site, which currently has a number of interactive web maps and Story Map apps and provides links to other geospatial, online applications developed by and for the Aquarium.

CHAPTER 13

Uncovering the Oceans through Marinescape Geovisualization

Rosaline Canessa, Robert Newell, and Cathryn R. Brandon

Abstract

More often than not, geographic representation of the marine environment strips the ecosystem of its richness habitats, biodiversity, and verticality. Mapping the marine environment in GIS results in marine features represented as a two-dimensional plane on the sea surface or as bare bathymetry. These are geographically accurate but not visually pleasing nor comprehensive in representing marine ecosystems. In contrast, gaming and virtual reality programs produce attractive visuals but typically are not geographically referenced. This chapter explores marinescape geovisualization, the integration of visual simulation with map-based geography for marine environments. First, a review of progress from conventional two-dimensional maps to immersive and interactive four-dimensional models for marine environments is presented. Subsequently, a case study of a marinescape geovisualization model, integrating GIS and landscape visualization software of an offshore marine protected area, is described along with feedback provided by marine protected area planners and scientists. Finally, the chapter concludes with insights on key research and development issues to advance marinescape geovisualization.

Introduction

Marine environments are under stress from human-use activities (Hinrichsen 2011). These impacts are exacerbated, as marine environments are poorly understood in comparison with terrestrial environments. Other than scientists and those who scuba dive, the public and even decision-makers have relatively little knowledge of the marine environment. Maps have long been the predominant visual currency for land-use planning, but some maps are often too abstract or schematic, making it difficult for the layperson to relate mapped information to their real-world experience or cognition (Paar 2006), and, in fact, they can contribute to confusion and errors in orientation (Lewis and

Dawn J. Wright, ed.; 2015; *Ocean Solutions, Earth Solutions*; http://dx.doi.org/10.17128/9781589483651

Sheppard 2006). This suggestion is particularly relevant to the marine environment, for which there are no equivalents of satellite images or orthophotos to drape over bathymetry. One only has to look at the difference between Google Earth and Google Oceans to appreciate the limitations in mapping marine environments. Maps and virtual globes often depict the open ocean as a blank, blue space or with bare bathymetry that belies the dimensionality, fluidity, diversity, complexity, and significance that lie under the sea surface and could compromise informed decision-making and public understanding. There is a clear need for improved tools to communicate complicated and diverse ocean and coastal information to support marine planning and public outreach (Jude et al. 2007).

As an alternative to conventional maps and GIS, landscape visualization has been promoted as a new common communication currency integral to land-use planning (Lange 2001; Orland et al. 2001). Combining graphic technology and art to create varying degrees of realism, landscape visualization aims to better situate the user within a simulated landscape. Geovisualization is the integration of visual simulation with map-based geography. Geovisualization reflects rising spatial literacy among the public, which is increasingly exposed to simulated games and virtual reality. Thus, the public has high expectations for a more realistic, accurate, intuitive, engaging, and accessible representation of the world, rich with structure, texture, and movement (Paar 2006). Landscape visualizations have yielded positive results in development planning, forestry management, climate change visioning, changing public perceptions, and adaptive response (e.g., Lewis and Sheppard 2006; Burch et al. 2009). However, marinescape geovisualization lags far behind (Canessa 2008). Efforts such as Google's foray into a "street view under the sea" is an indicator that the marine environment is a new frontier of geovisualization.

One of the key challenges of depicting and visualizing marine environments is the inherent three-dimensional (3D), volumetric, and time-dependent aspects of marine environments, such as biological, chemical, and physical processes (Wright and Goodchild 1997; Gold et al. 2005; Wright et al. 2007; Goralski and Gold 2008). A 3D GIS or geovirtual environment allows not only the sea surface to be portrayed, but also pelagic and benthic environments altogether. Marine data structures have been designed to store depth and time values, but the visualization capabilities of displaying these dimensions have room for improvement (Wright et al. 2007), particularly in integrating spatial data with game engines and augmented reality.

This chapter explores the evolution of marinescape geovisualization to current innovation, presents a case study applying landscape visualization to a marine area, and charts a research and development agenda for marinescape geovisualization. The following sections provide an overview of the variety of coastal and marine geovisualizations. Each section refers to geovisualization that assumes a specific dimensionality and discusses the spatial aspects and potential applications associated with the dimensional representation.

Development of Marinescape Geovisualization

Exploring the Surface: Conventional Maps and 2D Dimensionality

Many conventional maps depict the ocean as the negative or contrasting space (graphically speaking) to land contours, and accordingly, they typically communicate marine information in reference to terrestrial geography (e.g., coastal sites of marine habitat, locations of coastal communities, oceanic transport routes, and so forth). Conventional maps are two-dimensional (2D) and assume an overhead perspective. Thus, they are limited in the amount of detailed information that can be visually conveyed regarding the water column and seafloor. However, because of their simplicity, maps are relatively easy to create and communicate information clearly; therefore, they have been (and continue to be) employed extensively to communicate a range of information regarding coastal and marine spaces (Kraak 2011; MacEachren and Kraak 2001).

Because of their 2D format, such maps can be particularly effective in clearly conveying surface characteristics of the ocean. For example, the work of the Geophysical Fluid Dynamics Laboratory (2008) in mapping dynamic ocean surface temperatures produced visualizations that provide a comprehensive and extensive view of dynamic heat flow across the oceans of the world. Maps are also useful for displaying vector data oriented along the ocean surface, and thus can clearly communicate essential nautical information, such as the directions of ocean currents along the surface (e.g., as seen through the interactive map system of Oceanweather [2014]). Furthermore, although 2D maps do not contain a depth axis, they are useful for identifying locational and area-based data, and thus can identify subsurface elements as they align with the ocean's surface, examples of this being research conducted through the US National Centers for Coastal Ocean Science on coral reef habitat ranges and reef fish distributions (e.g., Costa et al. 2014).

Maps have the ability to capture and convey biophysical data on extensive ranges of marine space (e.g., global temperatures, currents, and large areas of habitat). This positions maps as effective tools for synthesizing, analyzing, and communicating information on certain large-scale environmental issues. For example, the National Center for Ecological Analysis and Synthesis synthesized data on a comprehensive suite of environmental threats to marine environments (e.g., pollution, invasive species, and bycatch fishing) and mapped their cumulative impacts around the globe (Halpern et al. 2008) (figure 13.1). Although the map does not provide detailed information on subsurface impacts and which threats have the most impact at certain localities, the visualization does provide a broad, global-scale impression of marine environments at risk. In addition, because maps have the ability to clearly communicate surface information (as discussed), they are particularly effective in capturing and conveying potential environmental impacts that could result from surface pollutants and potential associated sources of oil such as marine vessels (Serra-Sogas et al. 2008). An example of this includes an interactive map depicting the British Petroleum (BP) oil spill of April 2010 posted online (accessible from http://nyti.ms/1o7IjNJ) through the *New York Times* (Aigner et al. 2010). Because of its simple 2D format and clear graphical contrast between oil and water, the visualization clearly illustrates the extents of oil spills, and thus marine spaces at risk.

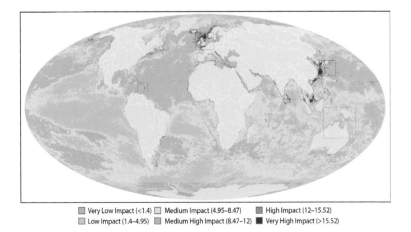

□ Very Low Impact (<1.4)   □ Medium Impact (4.95–8.47)   ■ High Impact (12–15.52)
□ Low Impact (1.4–4.95)   □ Medium High Impact (8.47–12)   ■ Very High Impact (>15.52)

Figure 13.1. A global map of human impacts to marine ecosystems. Image courtesy of US National Center for Ecological Analysis and Synthesis.

Another common ocean-based environmental issue depicted through map visualizations is sea level rise induced through climate change. However, because 2D maps do not include a depth axis, they cannot geometrically display rise in water; therefore, sea level rise is depicted through the adjacent topological relationship between the terrestrial and marine contours. The effect of sea level rise in a 2D plane is the ocean essentially "eating" away the land, making these maps still meaningful (if not entirely geometrically accurate) tools for communicating climate change impacts. Examples of this include the interactive maps of varying sea level rises developed through the University of Arizona's Department of Geosciences, made publicly available through the ArcGIS Viewer for Silverlight web application (methodology described in Weiss et al. 2011).

Advancements in, and increasing availability of, GIS technology over the last couple of decades have provided new opportunities for creating innovative geovisualization projects that integrate biophysical data with social, economic, and cultural information associated with marine and coastal environments (Rodríguez et al. 2009; Green 2010; Souto et al. 2012). An example of this includes the work of Longdill et al. (2008), who used ArcGIS software to map a variety of information associated with the Bay of Plenty, New Zealand, including biophysical characteristics, economic activities, and local cultural values, and used this integrated mapping to help identify suitable and sustainable sites for aquaculture development. Another example includes the work of Ruiz-Frau et al. (2011), who, also using ArcGIS, developed spatial representations of ecological, economic, and cultural values associated with the coast of Wales and held by diverse stakeholders for the purpose of informing site-specific marine management practices.

Exploring the Seafloor: Moving to 2.5D and 3D Dimensionality
2.5D Dimensionality
Advancements in multibeam bathymetry techniques and software throughout the late 1970s and early 1980s have made it possible to capture and characterize large areas of the seafloor, allowing humans to virtually explore the terrain below the ocean's surface (Tyee 1986). Many map-type visualizations have incorporated bathymetric information and expressed it through the two-and-a-half dimensional (2.5D) perspective, a method of conveying seafloor terrain through contours or relief surfaces. Geovisualizations in 2.5D are limited in that they do not effectively capture bathymetric perspectives. However, they can be useful for providing seafloor information to people that readily use and understand map formats, such as those in the nautical professions (Scheepens et al. 2014). As can be seen with the Esri Story Map app featured in figure 13.2, the 2.5D perspective holds the advantage of being able to convey planar information related to ocean surface (in this case, the route of the *HMS Titanic*), while still capturing and displaying some subsurface features.

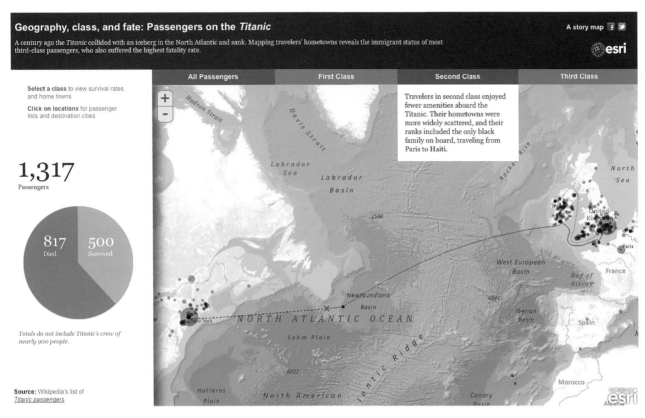

Figure 13.2. **Esri Story Map app displaying the travel route of the Titanic.** By Esri; data sources: Esri, GEBCO, NOAA, National Geographic, DeLorme, Geonames.org, and other contributors.

Public dissemination of and exposure to information on the shape and character of the seafloor has increased through the development of open-source ocean visualization applications that assume a 2.5D perspective. Virtual Oceans, developed through the Lamont-Doherty Earth Observatory at Columbia University, is an example of a GeoMapApp-based application that conveys multibeam bathymetry data in 2.5D format, using an interactive globe browser supported by the National Aeronautics and Space Administration (NASA) World Wind (Marine Geoscience Data System 2011). Of more significance in terms of public outreach is the 2.5D representation of the ocean floor incorporated into Google Earth. Google Earth allows anyone with an Internet connection access to detailed information of the planet's geography (Shufeldt et al. 2012). Google Earth's inclusion of seafloor details holds important implications for public understanding of the ocean because it represents a shift from strictly 2D, surface-focused perspectives of marine spaces (that historically have dominated conventional maps) to perspectives that provide glimpses of what lies below the ocean's surface.

3D Dimensionality

Relief-type visualizations can convey coarse information regarding the seafloor; however, they are limited in that they do not employ a depth axis, and thus cannot portray the full perspectives of bathymetry. To explicitly capture the geomorphological features of the seafloor, a geovisualization must be completely 3D. Many 3D marine geovisualizations assume a "sandbox" perspective (figure 13.3), allowing users to examine and compare different terrain features of the seafloor (e.g., displays of the seafloor shelves in the same frame as the slopes). Such a perspective is particularly useful for gaining insights on subsurface geomorphology, local ecology, and sensitive habitats (Lundblad et al. 2006).

Several different applications have been used to construct 3D bathymetric models, including Geological Surveying and Investigation in 3 Dimensions (GSI3D) (Kessler et al. 2009), 3D GeoModeller (Calcagno et al. 2008), and Advanced Visual Systems (AVS) (2014). However, Fledermaus (a product of Quality Positioning Services, or QPS) is arguably among the most popular software used specifically for creating 3D bathymetric visualizations, and its application in marine resource management has been the subject of much research. Fledermaus applications include the work of Cooper et al. (2011) on modeling the outcomes of potential strategies for gravel habitat restoration and the work of Weaver et al. (2006) on modeling and characterizing the marine habitat in the Tortugas South Reserve (Florida Keys National Marine Sanctuary). An important feature of Fledermaus is that it can be used to create digital elevation models compatible with ArcGIS, allowing for in-depth spatial analysis using bathymetric models. An example of this workflow includes the study of Rengstorf et al. (2012) on the use of different bathymetric models for marine habitat suitability analysis, which used Fledermaus to create the bathymetric models and ArcGIS to analyze terrain attributes.

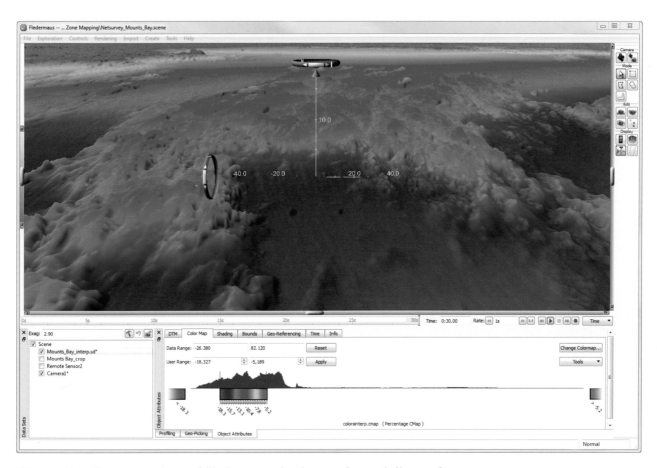

Figure 13.3. Screen capture of Fledermaus bathymetric modeling software. Courtesy of Quality Positioning Services BV (QPS).

The modeling software noted here is highly specialized and requires particular skill sets to use effectively. However, effective coastal management requires multidisciplinary approaches and input from a range of different actors and stakeholders (Cicin-Sain and Belfiore 2005). Recognizing this, Collaborative Ocean Visualization Environment (COVE) was developed as a user-friendly, 3D marine geovisualization system that can facilitate collaborative work among the scientific community (COVE 2009). COVE was specifically designed through the feedback of different scientists and engineers to ensure its user-friendliness and utility in multidisciplinary projects (Grochow et al. 2008). Systems like COVE allow for the use of sophisticated visualizations in management projects, while simultaneously maintaining inclusivity and collaboration.

Exploring Place: Immersive Geovisualizations

Geovisualizations that aim to capture and convey bathymetry using the 3D, sandbox perspective often focus on the seafloor, excluding elements found in the water column and above the surface. However, a comprehensive and accurate portrayal of the marine environment must incorporate all dimensions of marine space, and thus include elements and interactions of the water column and transsurface. The challenge in constructing such a visualization is that marine environments consist of elements spanning multidimensional space; thus, adding certain elements to a marine/coastal geovisualization can potentially obscure others. Immersive geovisualizations provide a possible solution to this issue by allowing users to navigate through and around elements, permitting visual access to all aspects of the visualization and thus visually accommodating for multidimensionality. Accordingly, immersive geovisualizations can be constructed with a high degree of realism, detail, and accuracy (in terms of capturing more real-life elements).

The application of immersive geovisualizations in natural resource planning and land-use management has been explored in the terrestrial context (Lewis and Sheppard 2006; Schroth et al. 2006; Schroth et al. 2009; Sheppard et al. 2011). However, research and applications of immersive coastal and marine geovisualizations are still in their infancy. Among the little work done are projects such as Ecosystem Science and Visualization Group (EcoViz), the Nereus Program, and the North West Shelf Joint Environmental Management Study (NWSJEMS) (2010). EcoViz is a research project based at California State University that builds visualizations of marine protected areas (MPAs) (figure 13.4) for the purposes of public education, informing decision-makers, and supporting marine management practices of MPAs (EcoViz 2011). The Nereus Program is a collaboration between the University of British Columbia and Nippon Foundation that focuses on fisheries impacts research. Nereus employs immersive visualizations for the purpose of communicating complex scientific information to the broader public (Nereus Program 2013). NWSJEMS is a research project based in Western Australia, aimed at developing and conducting methods that support sustainable management practices of marine environments. Similar to Nereus, NWSJEMS uses immersive geovisualizations (namely, an animation referred to as "shelf life") to communicate outcomes of their work and educate the public on multisector management in marine ecosystems (NWSJEMS 2010).

Figure 13.4. **Immersive visualization of Point Lobos Marine Protected Area.** By Fred Watson, Alberto Guzman, Tom Thein; footage by Kip Evans; courtesy of University Corporation at Monterey Bay.

An important similarity among the applications of immersive geovisualizations described here is that they all were used to communicate scientific information to broader groups (i.e., the public). Immersive geovisualizations are powerful tools for communicating place-based information, meaning they have the ability to situate someone in a "place" and provide them with a better sense of represented location (Turner et al. 2013). Such a feature is important for increasing understanding of the environment and making sound management decisions (Lewis and Sheppard 2006; Phadke 2010). This feature of immersive visualizations will become increasing more significant for public understanding of the marine environment as Google's Ocean Street View project progresses. Launched in 2012, this project involves the development of publicly available interactive visualizations (accessible from https://www.google.com/maps/views/streetview/oceans?gl=us) that immerse the user in the marine environment. As Google continues to build this virtual underwater world, the public's access to marine "places" increases, which potentially could contribute to public familiarity and understanding of marine environments and thus make the world below the ocean's surface seem less "alien."

Exploring the Living Sea: Incorporating Time and Interactivity

Marine and coastal environments are dynamic places with complex interactions; therefore, accurately capturing the "reality" of these environments in geovisualizations requires incorporation of fourth-dimensional properties (Beegle-Krause et al. 2010). Such a task can be data intensive and challenging; thus, visualizations that involve temporal dimensions often focus on patterns or trends of specific aspects of the marine environment while excluding or simplifying other temporal dynamics. For example, the Nereus Project constructed a visualization specifically designed to convey the impacts of overfishing from 1950 to the present, and it does this by displaying a marine scene with fish populations fluctuating in density as a counter representing year increases (Nereus Program 2010). In another example, the BP oil spill visualization posted in the *New York Times* is interactive, allowing users to select different dates to examine the oil spill pattern and expansion throughout the Gulf of Mexico over time (Aigner et al. 2010).

Over the last couple of decades, the computer game industry has propelled technological advancements in the development of dynamic virtual spaces. Accordingly, game engines have been suggested as potentially powerful tools for constructing geovisualizations with 3D and four-dimensional (4D) properties (Gold et al. 2004). *Infinite Scuba* (a diving simulation game) serves as an illustrative example of how a game platform could be used to develop a dynamic and sophisticated marine geovisualization. *Infinite Scuba* was developed through partnership between the Professional Association of Diving Instructors (PADI) and Cascade Game Foundry for the purpose of emulating a realistic and enjoyable virtual diving experience. The virtual environment in *Infinite Scuba* is dynamic and contains a multitude of interacting elements (such as marine life; figure 13.5), demonstrating the potential that computer game technology has for developing realistic, immersive geovisualizations.

Figure 13.5. **Screen capture from the *Infinite Scuba* video game.** Courtesy of Cascade Game Foundry Corp. Copyright © Cascade Game Foundry Corp., 2014.

Marinescape Geovisualization Case Study: SGaan Kinghlas–Bowie Seamount

Background

SGaan Kinghlas–Bowie Seamount rises from 3,000 m depth to within 25 m of the sea surface, 180 km west of Haida Gwaii on Canada's Pacific coast (figure 13.6). Seamounts are productive habitats as deep, cold waters upwell when encountering the seamount structure, bringing in nutrients. Seamounts can also serve as stepping stones for migratory species. SGaan Kinghlas–Bowie Seamount is home to a diversity of species from deep water to coastal species, brought to the seamount through coastal gyres (Canessa et al. 2003). In recognition of these characteristics, SGaan Kinghlas–Bowie Seamount was declared a federal MPA in 2008. However, given its remoteness and expense in getting there, SGaan Kinghlas–Bowie Seamount has remained largely unexplored and inaccessible, except to a handful of scientists and fishermen. SGaan Kinghlas–Bowie Seamount is not only physically remote, but also cognitively remote with a lack of information on which to make management and planning decisions and for the public to engage. For both these reasons, the objectives of the SGaan Kinghlas–Bowie Seamount MPA include scientific study and public engagement.

Figure 13.6. SGaan Kinghlas–Bowie Seamount (after Canessa et al. [2003]). By the authors, based on Canessa et al. (2003). Data from National Geophysical Data Center, National Environmental Satellite Data and Information Service.

Initially, information on SGaan Kinghlas–Bowie Seamount was gleaned from videos and photographs taken by scientists. These comprised snapshots of small portions of the vast seamount complex. This changed significantly with the acquisition of multibeam data, which enabled the development of a 2.5D model in ArcGIS of the top 200 m of the seamount (figure 13.7). For the first time, it was possible to see the "landscape" of the seamount, at least the top 200 m. The images clearly show features such as the ridge, plateaus, and hills. Furthermore, it was possible to navigate through different perspectives of the seamount as in a "dive-through." Although the 2.5D perspective showed more detail in the physical form of the summit, it is devoid of geophysical texture and biological richness. Thus, it can relay the impression that SGaan Kinghlas–Bowie Seamount is barren and mislead the impression of its ecological value. Whether users are politicians, planners, policy makers or the public, this can lead to uninformed priorities and decisions.

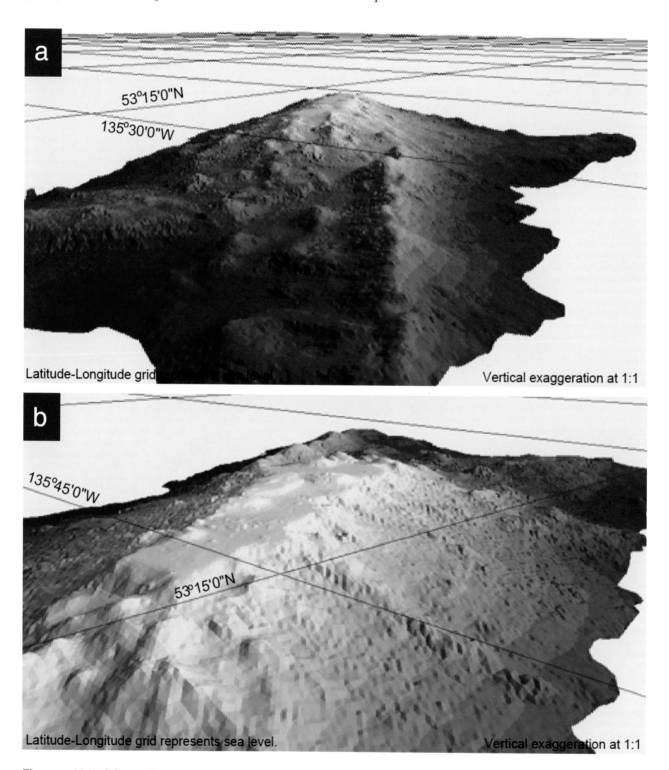

Figures 13.7. SGaan Kinghlas-Bowie Seamount perspective views (a) and (b) (after Canessa et al. [2003]). By the authors, based on Canessa et al. (2003). Bathymetry data from the Canadian Hydrographic Service and US National Oceanic and Atmospheric Administration; bathymetry model from AXYS Environmental Consulting.

This section describes the development of a "proof of concept" marinescape geovisualization model of SGaan Kinghlas–Bowie Seamount to bridge the experiential and knowledge gaps of this remote MPA. The project aimed to enhance the 2.5D bathymetric model with photorealistic and geographically accurate marinescape geovisualizations. The model was viewed by individuals involved with SGaan Kinghlas–Bowie Seamount in a variety of capacities to acquire their feedback.

Data and Methods

Data and information were acquired from a variety of sources. Fundamental was the multibeam data that captured detailed depth. Multibeam data with 5 m resolution was available for the top 200 m of Bowie Seamount. Available georeferenced video and still photography enabled geographic understanding and mapping of habitats and species. Field data, species lists, and other observations and reports were used to additionally characterize oceanographic conditions, geological composition, and species assemblages.

The first step was to build a bathymetric model with the multibeam data in ArcGIS. Because of file size and processing time, 10 m resolution was used for a small area near the pinnacle. The bathymetric model is a good representation of the overall topography. However, use of that resolution results in a relatively smooth surface. Video and photographs, in fact, show a rugged and uneven surface composed of weakly cemented volcanic tephra and volcanic sand. The bathymetric model was imported into 3D Nature's Visual Nature Studio (VNS), in which a surface that more realistically portrays ruggedness was created by modifying the terrain effect in the software using parameters such as fractal depth and vertical displacement. Color and pattern were also applied to simulate vegetation ground cover (figure 13.8a). Images of ground cover were extracted from video and photos and mosaicked to further enhance visual representation of actual benthic habitats (figure 13.8b).

Figure 13.8. **SGaan Kinghlas–Bowie Seamount substrate geovisualization in (a) simulation with pattern, size and color; and (b) a mosaic of images extracted from photos.**

Once the bathymetry and substrate were modeled, attention turned to the water column and animal biota. Field reports described the water as clear and deep blue with a horizontal visibility of approximately 15 m. These conditions were simulated using atmospheric effects such as fog and haze (figure 13.9). The last step was to add species assemblages and individual organisms. The seamount ecosystem was divided into four ecotypes based on depth ranges of 0–30 m, 30–60 m, 60–200 m, and below 200 m. These ecotypes were defined based on the distribution of algal cover as described in field reports. Sufficient information was available to model all but the ecotype below 200 m. Each ecotype was divided into overstory and understory to inform the vertical distribution of organisms. The understory contains small ground cover plants and benthic species. The overstory contains tall plants and pelagic organisms. In this proof of concept, only understory was modeled. Finally, species groups and individual species were identified from field reports and species lists. These were assigned to the appropriate ecotype to build a species library. Each species identified requires a graphic image of an individual organism so that it can be placed on the ground cover. Gradually, a species image library is created. The result was an interactive 3D model of the seamount pinnacle that can be rotated and viewed from different angles. A short dive-through video has been created of one path over the seamount. As the dive-through proceeds, the changing longitude, latitude, and depth (noted as camera elevation) is displayed, confirming that the seascape model is georeferenced. The marinescape visualization video is available on the Esri Press "Book Resources" webpage for chapter 13, at esripress.esri.com/bookresources. For more information, see the "Supplemental Resources" section at the end of this chapter.

Figure 13.9. SGaan Kinghlas–Bowie Seamount simulation of visibility (a) without atmospheric effect and (b) with atmospheric effect (e.g., fog and haze).

Evaluation

The SGaan Kinghlas–Bowie Seamount proof of concept marine geovisualization was distributed to individuals engaged with SGaan Kinghlas–Bowie Seamount in varying capacities, such as biologists, planners, and members of the Advisory Board. Users were presented with an audiovisual description of the development of the model and a link to the dive-through video, which is also available on the book resource webpage for chapter 13 (see "Supplemental Resources" at the end of the chapter). Users then completed an evaluation survey.

Geovisualizations often must balance being informative, visually realistic, and spatially accurate, often requiring trade-offs among the three. Respondents were asked to indicate the importance they placed on the need for maps and visualizations to be informative, realistic, and georeferenced. With respect to using maps and visualizations for planning and management of SGaan Kinghlas–Bowie Seamount, respondents said that being informative, realistic, and georeferenced were all important, although being informative was rated most important. With respect to using maps and visualizations for public education and outreach, respondents said that being informative and realistic were important but were relatively neutral on the importance of visualizations being georeferenced. On a scale of 1 to 5, from very unrealistic to very realistic, respondents rated the SGaan Kinghlas–Bowie Seamount proof of concept marine geovisualization a 4, somewhat realistic. On a scale of 1 to 4, from misinformative to very informative, respondents rated the SGaan Kinghlas–Bowie Seamount proof of concept marine geovisualization a 4, very informative.

Although this was not a comprehensive evaluation of the SGaan Kinghlas–Bowie Seamount proof of concept marine geovisualization, it provided some insight and direction, particularly with respect to the need for a more complete representation of the ecosystem, including the water column (e.g., near-surface currents, areas of upwelling, and high-productivity areas). On the other hand, all respondents acknowledged that the available data for SGaan Kinghlas–Bowie Seamount is critically limited. One respondent noted that field data needs to be collected with marinescape geovisualization in mind, as opposed to retroactively trying to generate marinescape geovisualization with whatever data is available. Several respondents noted the importance of validating the geographic accuracy and overall validity of the marinescape geovisualization model, without which it may serve to misinform rather than inform. One respondent also commented that the haze effect made the visualization blurry and difficult to see. This comment reflects the challenge of balancing realism and abstraction to create an effective and informative geovisualization. Despite some of these limitations, respondents noted

environment, acquired through films, photographs, and books. Interactivity and immersion take the perspective of the user into account in the visualization scene.

Interactivity reflects the way the user directs the navigation and exploration of the scene. Interactivity might allow users to manipulate the perspective view, interrogate features, or modify elements of the scene. The way in which interaction occurs can be thought of as interaction metaphors (MacEachren et al. 1999; Germanchis et al. 2004). Current interaction with marine spatial data and marine geovisualizations is based on the same metaphors we use for terrestrial and urban environments. New interaction metaphors may be required for marine environments, especially when increasing a sense of immersion (Germanchis et al. 2004). For example, when using a marinescape geovisualization for coastal planning, one could interact with the data through a stakeholder metaphor, decision-maker metaphor, or scientist metaphor such that the environment and information would get structured around the type of metaphor chosen, which in turn dictates one's interaction.

Immersion takes into account the perspective of the user in the visualization scene by embedding the user directly in the scene as a participant or avatar or providing the view through the eyes of the user. In terrestrial visualization, this may be a person who is walking, running, driving, or flying through a scene. The marine analogues are floating, swimming, diving, or driving a remotely operated vehicle (ROV). That immersive experience could be further enhanced by accentuating the entry into the marine environment (e.g., walking across the coast from land to sea or jumping in from a boat).

User Needs Assessment

A critical step in the advancement of marinescape geovisualization is ensuring that the development of such models and tools is tightly tuned to the needs of intended users. Scientists, marine planners, and the general public will likely have different needs on a range of factors such as level of realism, amount and type of information, spatial accuracy, immersion, interactivity, and collaboration, to name a few. A user-needs assessment early in the process should address these and other key factors. Given that marinescape geovisualization is novel, a user-needs assessment may require a prototype being developed for users to begin to understand and imagine how marinescape geovisualization might be applied to achieve their goals. Questions in a user-needs assessment might include:

- Who are the intended users?
- What is their level of knowledge of the marine area and key issues?
- How much marine experience do users have?
- What types of visualizations have they used previously for the study area?
- How important is realism?
- How important is information?
- How important is spatial accuracy?
- Is the focus on the sea surface, water column, seabed, or below the seabed?
- How important is object movement and fluidity of the surrounding environment?
- How important is it for the users to feel that they are immersed in the scene?
- What extent of interactivity and self-direction in the scene is required?
- Will the marinescape geovisualization be primarily used for viewing or analysis?
- Will the visualization be used by an individual or collaboratively in a group?

With respect to the last question, a growing body of research exists on geovisualization for collaborative use of map-based visualizations in groups (Brewer et al. 2000; MacEachren 2004).

A number of case studies in terrestrial and urban environments discuss collaborative uses of 3D GIS and the geovirtual environment (Stock et al 2005); but few, such as COVE discussed earlier, exist for marine contexts. Collaborative uses of GIS for marine spatial planning are still confined to 2D GIS case studies (Alexander et al. 2012; Merrifield et al. 2013). 3D visualizations for marine spatial planning contexts are dominated by static 3D landscape visualization software (Shaw et al. 2009; Phadke 2010), rather than navigable 3D GIS or game engines.

Usability, Evaluation, and Effectiveness

Finally, as marine geovisualization advances, we need to empirically evaluate both its usability and effectiveness in achieving its case-specific goal (Jude 2008). Ultimately, we are interested in answering questions such as:

- Was available data adequate for an effective marinescape geovisualization?
- Is it important to communicate the difference between simulation and reality, or the level of uncertainty in the data, to the user?
- Was information intensity a barrier?
- Did some elements create more of a distraction than a focus?
- To what extent did the marinescape geovisualization
  - provide insight that was otherwise missing from other available data, information, or visualizations?
  - facilitate planning and management decisions?
  - create a connection to place?
  - engage the public?
- What limitations did users face?

Conclusion

Marinescape geovisualization offers an exciting medium to enhance our connection to the marine environment and ability to make more informed decisions that reflect its spatiotemporal dimensions. Advances in landscape visualization offer a starting point (e.g., simulating turbidity with fog and haze and creating 3D object libraries). However, the comparatively limited amount of georeferenced marine data and the integral 3D dimensionality, along with the fluidity of the water column, require more sophisticated techniques such as those used in game engines. Thus, we are leading toward an integration of standard practices in landscape visualization and innovation of game engines for virtual reality with GIS for geographically accurate place-based visualizations. However, we do not want to lose the geographic connection to marine visualization that may result with an emphasis on game engines. Geography in general can have a unique impact on the use of computer-game research by bringing the real world into these simulations. In addition, GIS could greatly benefit from enhanced interactivity and immersion of game engines. In addition, we must not lose sight that the aim is not to develop marinescape geovisualization for the "wow" factor. Instead, the primary goals are, arguably, to make more informed planning and management decisions and enhance marine literacy among the public. Therefore, technological advancement should be complemented with research to understand how users engage with marinescape geovisualization and the impact the marinescape geovisualization has on meeting the users' goals.

Acknowledgments

We would like to thank Carolin Kadner and Manuel Burckhardt for their work on the development of the proof of concept marinescape geovisualization model of SGaan Kinghlas–Bowie. We are also grateful to two anonymous reviewers for their comments.

References

Aigner, E., J. Burgess, S. Carter, J. Nurse, H. Park, A. Schoenfeld, and A. Tse. 2010. "Tracking the Oil Spill in the Gulf." *New York Times*. August 2.

Alexander, K., R. Janssen, G. Arciniegas, T. O'Higgins, T. Eikelboom, and T. Wilding. 2012. "Interactive Marine Spatial Planning: Siting Tidal Energy Arrays around the Mull of Kintyre." *PLoS One* 71:e30031. doi:10.1371/journal.pone.0030031.

AVS (Advanced Visual Systems). 2014. Geospatial Case Study. http://www.avs.com/solutions/express/geospatial -mapping-using-data-visualization. Last accessed February 2, 2014.

Beegle-Krause, C. J., T. Vance, D. Reusser, D. Stuebe, and E. Howlett. 2010. "Pelagic Habitat Visualization: The Need for a Third and Fourth Dimension: HabitatSpace." In *Estuarine and Coastal Modeling: Proceedings of the 11th International Conference*, edited by M. L. Spaulding, 187–200. Reston, VA: American Society of Civil Engineers.

Brewer, I., A. M. MacEachren, H. Abdo, J. Gundrum, and G. Otto. 2000. "Collaborative Geographic Visualization: Enabling Shared Understanding of Environmental Processes." In *Proceedings of Information Visualization (InfoVis)*, 137–41, Salt Lake City, UT, October 9–10. doi:10.1109/INFVIS.2000.885102.

Burch, S., A. Shaw, S. Sheppard, and D. Flanders. 2009. *Climate Change Visualization: Using 3D Imagery of Local Places to Build Capacity and Inform Policy*. State of Climate Visualization, CSPR Report-2009:4.

Calcagno, P., J. P. Chilès, G. Courrioux, and A. Guillen. 2008. "Geological Modelling from Field Data and Geological Knowledge: Part I. Modelling Method Coupling 3D Potential-Field Interpolation and Geological Rules." *Physics of the Earth and Planetary Interiors* 1711:147–57.

Canessa, R. 2008. "Seascape Geovisualization for Marine Planning." *Geomatica* 62:23–32.

Canessa, R., K. Conley, and B. Smiley. 2003. "Bowie Seamount Marine Protected Area: An Ecosystem Overview." *Canadian Technical Report Fisheries and Aquatic Sciences* 2461.

Casas, S., S. Rueda, J. V. Riera, and M. Fernandez. 2012. "On the Real-Time Physics Simulation of a Speed-Boat Motion." Paper presented at GRAPP '12: International Conference on Computer Vision, Imaging, and Computer Graphics Theory and Applications, 121–28, Rome, Italy, February 24–26.

Cicin-Sain, B., and S. Belfiore. 2005. "Linking Marine Protected Areas to Integrated Coastal and Ocean Management: A Review of Theory and Practice." *Ocean & Coastal Management* 48 (11–12): 847–68.

Cooper, K., S. Ware, K. Vanstaen, and J. Barry. 2011. "Gravel Seeding: A Suitable Technique for Restoring the Seabed Following Marine Aggregate Dredging?" *Estuarine, Coastal and Shelf Science* 911:121–32.

Costa, B., J. C. Taylor, L. Kracker, T. Battista, and S. Pittman. 2014. "Mapping Reef Fish and the Seascape: Using Acoustics and Spatial Modeling to Guide Coastal Management." *PLoS ONE* 91:1–17.

COVE (Collaborative Ocean Visualization Environment). 2009. A Visual Environment for the Oceans. http://cove.ocean.washington.edu/story/About. Last accessed March 7, 2014.

EcoViz (Ecosystem Science and Visualization). 2011. About EcoViz. http://ecoviz.csumb.edu/home/about.htm. Last accessed March 6, 2014.

Fritsch, D., and M. Kada. 2004. "Visualisation Using Game Engines." In *Proceedings of the International Society for Photogrammetry and Remote Sensing (ISPRS) Technical Commission V Symposium*, 621–25, Riva del Garda, Italy, June 23–25.

Geophysical Fluid Dynamics Laboratory. 2008. Visualizations—Oceans. Ocean Surface Temperature. http://www.oceanweather.com/data/index.html. Last accessed February 2, 2014.

Germanchis, T., C. Pettit, and W. Cartwright. 2004. "Building a Three-Dimensional Geospatial Virtual Environment on Computer Gaming Technology." *Journal of Spatial Science* 491:89–95. doi:10.1080/14498596.2004.9635008.

Gladstone, W. 2011. *Unity 3.xGame Development Essentials*. Birmingham, UK: Packt Publishing.

Gold, C. M., M. Chau, M. Dzieszko, and R. Goralski. 2004. "The Marine GIS: Dynamic GIS in Action." *International Archives of Photogrammetry, Remote Sensing and Spatial Information Sciences* 35:688–93.

———. 2005. "3D Geographic Visualization: The Marine GIS." *Developments in Spatial Data Handling*, 17–28.

Goralski, R., and C. Gold. 2008. "Marine GIS: Progress in 3D Visualization for Dynamic GIS." In Headway in Spatial Data Handling: *13th International Symposium on Spatial Data Handling*, edited by A. Ruas and C. Gold, 401–16. Berlin: Springer.

Green, D. R. 2010. "The Role of Public Participatory Geographical Information Systems in Coastal Decision-Making Processes: An Example from Scotland, UK." *Ocean & Coastal Management* 5312:816–21.

Grochow, K., M. Stoermer, D. Kelley, J. Delaney, and E. Lazowska. 2008. "COVE: A Visual Environment for Ocean Observatory Design." *Journal of Physics: Conference Series* 125:12092–98.

Halpern, B. S., S. Walbridge, K. A. Selkoe, C. V. Kappel, F. Micheli, C. D'Agrosa, J. F. Bruno, K. S. Casey, C. Ebert, E. Fox, R. Fujita, D. Heinemann, H. S. Lenihan, E. M. P. Madin, M. T. Perry, E. R. Selig, M. Spalding, R. Steneck, and R. Watson. 2008. "A Global Map of Human Impact on Marine Ecosystems." *Science* 319 (5865): 948–52.

Hinrichsen, D. 2011. *The Atlas of Coasts and Oceans: Ecosystems, Threatened Resources, Marine Conservation*. Chicago: University of Chicago Press.

Jude, S. 2008. "Investigating the Potential Role of Visualization Techniques in Participatory Coastal Management." *Coastal Management* 36:331–49.

Jude, S. R., A. P. Jones, A. R. Watkinson, I. Brown, and J. A. Gill. 2007. "The Development of a Visualization Methodology for Integrated Coastal Management." *Coastal Management* 35 (5): 525–44.

Kessler, H., S. Mathers, and H. G. Sobisch. 2009. "The Capture and Dissemination of Integrated 3D Geospatial Knowledge at the British Geological Survey Using GIS 3D Software and Methodology." *Computers and Geosciences* 356:1311–21.

Kraak, M. J. 2011. "Is There a Need for Neo-cartography?" *Cartography and Geographic Information Science* 382:73–78.

Lange, E. 2001. "The Limits of Realism: Perceptions of Virtual Landscapes." *Landscape and Urban Planning* 54:163–82.

Lewis, J. L., and S. Sheppard. 2006. "Culture and Communication: Can Landscape Visualization Improve Forest Management Consultation with Indigenous Communities?" *Landscape and Urban Planning* 77:291–313.

Lloret, J. R., N. Omtzigt, E. Koomen, and F. S. de Blois. 2008. "3D Visualizations in Simulations of Future Land Use: Exploring the Possibilities of New, Standard Visualization Tools." *International Journal of Digital Earth* 11:148–54.

Longdill, P. C., T. R. Healy, and K. P. Black. 2008. "An Integrated GIS Approach for Sustainable Aquaculture Management Area Site Selection." *Ocean & Coastal Management* 518 (9): 612–24.

Lundblad, E. R., D. J. Wright, J. Miller, E. M. Larkin, R. Rinehart, D. F. Naar, B. T. Donahue, S. M. Anderson, and T. Battista. 2006. "A Benthic Terrain Classification Scheme for American Samoa." *Marine Geodesy* 292:89–111.

MacEachren, A. M. 2004. "Moving Geovisualization toward Support for Group Work." In *Exploring Geovisualization*, edited by J. Dykes, A. M. MacEachren, and M. J. Kraak, 445–61. Amsterdam: Elsevier.

MacEachren, A.M., R. Edsall, D. Haug, R. Baxter, G. Otto, R. Masters, S. Fuhrmann, and L. Qian. 1999. "Virtual Environments for Geographic Visualization: Potential and Challenges." In *Proceedings of the ACM Workshop on Paradigms in Information Visualization and Manipulation*, Kansas City, MO, November 6. http://www.geovista .psu.edu/publications/NPIVM99/ammNPIVM.pdf.

MacEachren, A. M., and M. J. Kraak. 2001. "Research Challenges in Geovisualization." *Cartography and Geographic Information Science* 281:3–12.

Marine Geoscience Data System. 2011. Explore Our Planet with Virtual Ocean. http://www.virtualocean.org. Last accessed February 2, 2014.

Merrifield, M. S., W. McClintock, C. Burt, E. Fox, P. Serpa, C. Steinback, and M. Gleason. 2013. "MarineMap: A Web-Based Platform for Collaborative Marine Protected Area Planning." *Ocean & Coastal Management* 74:67–76.

Nakamura, K., I. Suzuki, M. Yamamoto, and M. Furukawa. 2011. *Advances in Artificial Life: Darwin Meets von Neumann, Part I: Lecture Notes in Computer Science 5778*, edited by G. Kampis, I. Karsai, and E. Szathmary, 99–106. Berlin: Springer-Verlag.

Nereus Program. 2010. *Nereus Ocean Visualization* [video]. http://vimeo.com/17379108. Last accessed March 7, 2014.

———. 2013. Scope of Impact. http://www.nereusprogram.org/#!project/c1mhs. Last accessed March 7, 2014.

NWSJEMS (North West Shelf Joint Environmental Management Study). 2010. Background. http://www.cmar.csiro .au/nwsjems/about/background.htm. Last accessed March 3, 2014.

Oceanweather. 2014. Current Marine Data. http://www.oceanweather.com/data/index.html. Last accessed February 2, 2014.

Orland, B., K. Budthimedhee, and J. Uusitalo. 2001. "Considering Virtual Worlds as Representations of Landscape Realities and as Tools for Landscape Planning." *Landscape and Urban Planning* 54:139–48.

Paar, P. 2006. "Landscape Visualizations: Applications and Requirements of 3D Visualization Software for Environmental Planning." *Computers, Environment and Urban Systems* 30:815–39.

Phadke, R. 2010. "Steel Forests or Smoke Stacks: The Politics of Visualisation in the Cape Wind Controversy." *Environmental Politics* 191:1–20.

Rengstorf, A. M., A. Grehan, C. Yesson, and C. Brown. 2012. "Towards High-Resolution Habitat Suitability Modeling of Vulnerable Marine Ecosystems in the Deep-Sea: Resolving Terrain Attribute Dependencies." *Marine Geodesy* 354:343–61.

Rodríguez, I., I. Montoya, M. J. Sánchez, and F. Carreño. 2009. "Geographic Information Systems Applied to Integrated Coastal Zone Management." Geomorphology 1071 (2): 100–105.

Ruiz-Frau, A., G. Edwards-Jones, and M. Kaiser. 2011. "Mapping Stakeholder Values for Coastal Zone Management." *Marine Ecology Progress Series* 434:239–49.

Scheepens, R., H. van de Wetering, and J. J. van Wijk. 2014. "Contour-Based Visualization of Vessel Movement Predictions." *International Journal of Geographical Information Science* 28 (5): 891–909.

Schroth, O., U. W. Hayek, E. Lange, S. R. J. Sheppard, and W. A. Schmid. 2006. "Multiple-Case Study of Landscape Visualizations as a Tool in Transdisciplinary Planning Workshops." *Landscape Journal* 30 (1): 53–71.

Schroth, O., E. Pond, S. Muir-Owen, C. Campbell, and S. R. J. Sheppard. 2009. *Tools for the Understanding of Spatio-temporal Climate Scenarios in Local Planning: Kimberley BC Case Study*. Swiss National Science Foundation Report PBEZP1-122976.

Serra-Sogas, N., P. O'Hara, R. Canessa, C. P. Keller, and R. Pelot. 2008. "Visualization of Spatial Patterns and Temporal Trends for Aerial Surveillance of Illegal Oil Discharges in Western Canadian Marine Waters." *Marine Pollution Bulletin* 56:825–33.

Shaw, A., S. Sheppard, S. Burtch, D. Flanders, A. Wiek, J. Carmichael, J. Robinson, and S. Cohen. 2009. "Making Local Futures Tangible: Synthesizing, Downscaling, and Visualizing Climate Change Scenarios for Participatory Capacity Building." *Global Environmental Change* 194:447–63.

Sheppard, S. R. J., A. Shaw, D. Flanders, S. Burch, A. Wiek, J. Carmichael, J. Robinson, and S. Cohen. 2011. "Future Visioning of Local Climate Change: A Framework for Community Engagement and Planning with Scenarios and Visualisation." *Futures* 434:400–412.

Shufeldt, O. P., S. J. Whitmeyer, and C. M. Bailey. 2012. "The New Frontier of Interactive, Digital Geologic Maps: Google Earth-Based Multi-level Maps of Virginia Geology." *Geological Society of America Special Papers* 492:147–63.

Souto, H., N. Gomes, and R. Carvalho. 2012. "Development of a GIS for the 'Celebration of Coastal Culture.' " *Journal of Coastal Conservation* 164:431–37.

Stock, C., I. Bishop, and A. O'Connor. 2005. "Generating Virtual Environments by Linking Spatial Data Processing with a Gaming Engine." In *Proceedings of the Sixth International Conference for Information Technologies in Landscape Architecture*, Dessau, Germany, May 26–28.

Turner, P., S. Turner, and L. Burrows. 2013. "Creating a Sense of Place with a Deliberately Constrained Virtual Environment." *International Journal of Cognitive Performance Support* 1 (1): 54–68.

Tyee, R. C. 1986. "Deep Seafloor Mapping Systems: A Review." *Marine Technology Society Journal* 20 (4): 4–16.

Wang, X. 2005. "Integrating GIS, Simulation Models, and Visualization in Traffic Impact Analysis." *Computers, Environment and Urban Systems* 29:471–96.

Weaver, D. C., D. F. Naar, and B. T. Donahue. 2006. "Deepwater Reef Fishes and Multibeam Bathymetry of the Tortugas South Ecological Reserve, Florida Keys National Marine Sanctuary, Florida." In *Emerging Technologies for Reef Fisheries Research and Management*. NOAA Professional Paper NMFS 5, edited by J. C. Taylor, 48–68. Seattle, WA: National Oceanic and Atmospheric Administration.

Weiss, J. L., J. T. Overpeck, and B. Strauss. 2011. "Implications of Recent Sea Level Rise Science for Low-Elevation Areas in Coastal Cities of the Conterminous USA." *Climatic Change* 1053 (4): 635–45.

Wright, D. J., M. J. Blongewicz, P. N. Halpin, and J. Breman. 2007. *Arc Marine: GIS for a Blue Planet*. Redlands, CA: Esri Press.

Wright, D. J., and M. F. Goodchild. 1997. "Data from the Deep: Implications for the GIS Community." *International Journal of Geographic Information Science* 11 (6): 523–28.

Yano, K., T. Nakaya, Y. Isoda, and T. Kawasumi. 2009. "Virtual Kyoto as 4D-GIS." In *Virtual Geographic Environments*, edited by H. Lin and M. Batty, 69–86. Redlands, CA: Esri Press.

Ware, C., R. Arsenault, M. Plumlee, and D. Wiley. 2006. "Visualizing the Underwater Behavior of Humpback Whales." *IEEE Computer Graphics and Applications* 26:14–18.

Supplemental Resources

Dawn J. Wright, ed.; 2015; *Ocean Solutions, Earth Solutions*; http://dx.doi.org/10.17128/9781589483651_d

For the digital content for this chapter, listed below in this section, go to the Esri Press "Book Resources" webpage at esripress.esri.com/bookresources. Then, in the list of Esri Press books, click *Ocean Solutions, Earth Solutions*. On the *Ocean Solutions, Earth Solutions* resource page, click the chapter 13 link to access that webpage and view the videos.

Geovisualization Videos

Two geovisualization movies can be observed on the chapter 13 webpage:
- *Explorations in Marinescape GeoVisualization*
- *Dive-Through of a Geovisualization Model of the Pinnacle of SGaan Kinghlas–Bowie Seamount*

CHAPTER 14

Approaches to Visualizing Complex Ocean Data Using Worldwide Telescope

Rob Fatland

Abstract

Trends in oceanography toward burgeoning model complexity and resolution, profusion of sensor platforms and instruments, expanding remote-sensing products, and new data-intensive analytical laboratory methods are outstripping scientists' ability to keep pace with their own data. Advances in visualization methods based on the Worldwide Telescope virtual globe can help narrow the gap between data acquisition and data insight. Three data visualization scenarios from oceanographic research are described, all motivated by current research on marine microbial ecology in relation to dissolved organic matter and physical transport processes. The role of visualization is placed in the broader context of data analysis and data management with the accompanying suggestion that both data services and resource confederation are important supporting philosophies, and heterogeneous data resources are amenable to a complementary-tools approach that includes GIS. The data visualization scenarios described here use the Worldwide Telescope capacity for large (on the order of 10 million data values) time-series datasets, together with three-dimensional spatial rendering and elapsing time, to create data exploration environments. Results of this work include implementation of a curtain plot as a data selection tool and visualization controller for Worldwide Telescope, potential for experimental design based on estuary model output, and identification of new ideas for the deconstruction of dissolved organic matter spectra.

Introduction

A challenge of oceanographic research data is that research programs are increasingly able to collect many different types of data with the idea of arriving at new insights, but the tools for organizing,

Dawn J. Wright, ed.; 2015; *Ocean Solutions, Earth Solutions*; http://dx.doi.org/10.17128/9781589483651

analyzing, comparing, and understanding these datasets lag behind and hinder that path to insight. This chapter describes oceanographic research visualization methods that incorporate time and accommodate trends in complex data. This also extends to considerations of data analysis and data management in relation to visualization and, more broadly, in relating research questions to data.

Although oceanographic datasets occupy a vast conceptual space, this chapter keeps to a modest length by briefly considering data visualization in three illustrative scenarios. To address these scenarios, I describe methods developed using the Worldwide Telescope (WWT) visualization engine, a freely available virtual globe that supports large (~10 million-element) datasets and time-series playback. I also present three broader ideas suggested by the relationship between data visualization, analysis, and data management: (1) heterogeneous data resources are amenable to a multiple-pillar or complementary-tools approach that includes GIS, virtual globes, and graphing tools; (2) a data-as-a-service technical philosophy can be a viable complement to the more traditional method of downloading data to a local computer system; and (3) research as a process is hindered by delays, by various forms of latency at many time scales, from seconds to years, and this state of affairs can be significantly improved by learning, developing, and applying emerging technology. These remarks are inspired by the book *The Fourth Paradigm* (Hey et al. 2009), which presents a deeper look at data-intensive science and the gaps in converting the capacity to gather increasing volumes of data into scientific insight.

The three example visualization scenarios described in this chapter are motivated by research on marine microbial ecology in relation to dissolved organic matter (DOM) and physical transport processes. Microbial ecology in the physical ocean is studied by the Monterey Bay Aquarium Research Institute (MBARI) and collaborators, particularly through synthesis of observational time-series datasets such as those produced by conductivity-temperature-depth (CTD) in situ observations, metagenomic/metabolomic methods (in situ and laboratory), remote sensing of sea surface state, acoustic Doppler current profiler (ADCP) measurements of local currents, in situ fluorescence, backscatter and nutrient observations, and so forth (e.g., MBARI 2010; Das et al. 2011). The water transport example described here is based on the Regional Ocean Modeling System (ROMS) (Arongo and Shchepetkin 2014) computational fluid dynamics model applied to estuaries, particularly in work done at the University of Washington (e.g., Geyer and MacCready 2014). DOM research includes particularly spectral analysis of DOM as carried out at the Old Dominion University and collaborating institutions, herein with particular emphasis on the technique of Fourier Transform Ion Cyclotron Resonance Mass Spectrometry (commonly abbreviated FTMS) (Sleighter and Hatcher 2007). In the resulting visualization scenarios, I suggest a research question addressable in some measure through visualization and then present approaches to visualization developed using WWT. WWT is a free graphics processing unit (GPU)-powered virtual globe (Microsoft Research 2014b) that runs on the Windows operating system. WWT and associated geoscience tools, collectively called Layerscape (Microsoft Research 2014a), are provided by Microsoft Research as research and education tools for general use.

This chapter seeks to describe a research-to-data relationship framework and then to describe three notional experiments A, B, and C, and the visualization methods developed to address them.

- Experiment A: The Environmental Sample Processor (ESP) built and operated by MBARI is allowed to drift for five days along the continental shelf off Monterey Bay. This drift experiment is further supported by a robotic autonomous underwater vehicle (AUV), an ADCP, and other sensors and platforms. Research question: Does ADCP data provide a usable constraint on autochthony/allochthony estimation of water measured by drifting sensors?

- Experiment B: A computational fluid dynamics model is driven by observational data to produce a 72-hour forecast of water conditions across a three-dimensional (3D) grid. Derived parameters include temperature, pH, dissolved oxygen, and phytoplankton and zooplankton concentration estimates. Research question: Can this fluid dynamics model suggest an in situ validation-observing campaign?
- Experiment C: analysis of DOM using FTMS. Research question: Can animated visualization, specifically extension of traditional two-dimensional (2D) graphs into three and four dimensions, suggest relationships in FTMS molecular formula sets?

A Research-to-Data Relationship Framework

A model of dynamic oceans can be assembled beginning with a static structure and adding dynamism (for example, currents as a set of smoothly moving conveyor belts, seasonal variability, turbulence, Coriolis effects, tides, atmospheric and bathymetric coupling, estuarine input, and so forth). This assembly is often presented by means of flat diagrams, maps, graphs, and equations on a static page, but we now have the means to bring the oceans to life through computer graphics and animation. Computer-enabled dynamism is an expanding discipline in oceanography research as a complementary enhancement to traditional approaches. To expand briefly beyond the topic of strict visualization, I present a three-component framework (figure 14.1). The first component is a hierarchy of models, questions, and problems in research that can be said to represent the state of scientific understanding of the oceans in earth system science. Placed opposite the research is the topic of data, particularly as a multiplicity of sources that are both complex and time varying. Research and data are then linked by processes: analysis, visualization, and data management. This framework is intended to indicate that challenges in data visualization exist not in isolation, but in the context of analysis and data management.

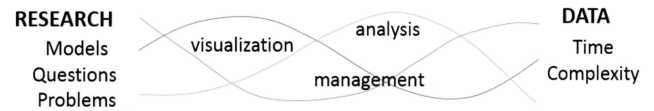

Figure 14.1. A research framework, in which research questions are connected to data complexity in terms of interrelated visualization, analysis, and data management processes.

Latency hinders progress in research on different time scales, from a user interface that introduces a few seconds of delay (perhaps discouraging data exploration) to an arduous validation process that slows data publication and reuse for years. If a data resource can provide several hundred datasets of interest but each requires 30 actions (cursor drags, clicks, text entries, and so forth) to make the datasets readable in a graphing application, two questions typically arise: "Is there a way to automate this?" and "How many of these files are really essential to analyze?" In other words, latency can divert time and energy, and it can alter the course of inquiry. Opportunities exist to develop technical solutions for various latencies across analysis, visualization, and data management. I describe one of these briefly, to wit, data services, and the related idea of confederating visualization and analysis applications.

Data Services and Monolithic versus Confederated Visualization Environments

Data as a Service (DaaS) is a developing alternative to working with local ("on this computer") copies of data. DaaS means datasets provided as an on-demand resource rather than files saved on a local storage device such as a hard drive, an early example being the Open-source Project for a Network Data Access Protocol (OPeNDAP) (2014). More recent technologies such as the Open Data Protocol (OData) (OPeNDAP 2014); web map services (WMS) (Open Geospatial Consortium 2014); and ArcGIS Online service, which combines both DaaS and Software as a Service (SaaS), contribute to a growing pool of solution templates. Many of these developments are reflected in the charter of the National Science Foundation–funded Data Observation Network for Earth (2014) to support "new innovative environmental science through a distributed framework and sustainable cyberinfrastructure."

The historical practice of manipulating local discrete files is typically quick and easy to understand, provided the datasets are neither too numerous nor too large. A negative aspect of local data management arises when data volume and/or complexity increases. Files must be identified, opened, read, modified, saved, closed, copied, and renamed, often in a rigid (possibly brittle) manner. This suggests a complexity horizon where datasets become too complex to be reliably managed as local files, leading to remote access (DaaS), database implementations, machine-independent formats such as network Common Data Format (netCDF) (Rew and Davis 1990), migration of computational tools to the cloud, and other SaaS approaches (e.g., ArcGIS Online at arcgis.com). DaaS, of course, is not a panacea, introducing as it does its own form of potential latency via dependency on consistent fast responses from remote computers and on the Internet in general. At this time, DaaS can be seen, therefore, as a complementary, rather than substitutive, means of obtaining data, with the following positive attributes:

- Datasets can be fetched as desired subsets rather than en masse.
- Datasets can be locally abandoned, rather than saved and curated, and refetched as needed.
- Datasets can be queried and fetched programmatically.
- Datasets can be updated by the DaaS provider, together with version metadata.

DaaS suggests an expanded approach to data visualization and analysis, to wit, confederation. Once a heterogeneous ensemble of data becomes available, a corresponding ensemble of analysis and visualization tools can be connected to it and to one another. Traditional visualization environments follow the rule of "begin by loading a dataset." A single application reads data in a predetermined format, often provides for some configuration steps, and then presents a graphical representation of that data. Built-in configuration controls such as a color map editor, scale exaggeration control, marker symbol menus, and filters then support the iterative exploration of the data in that application. Such monolithic software anticipates and implements as many useful data views as possible. Examples include the Collaborative Ocean Visualization Environment (COVE) (Center for Environmental Visualization 2014), older versions of Microsoft Excel, and the venerated Generic Mapping Tools (GMT) (Wessel and Smith 1991). In contrast, a confederated visualization/analysis environment (figure 14.2) provides DaaS pathways for data import and distributes visualization and analysis tasks across a number of applications.

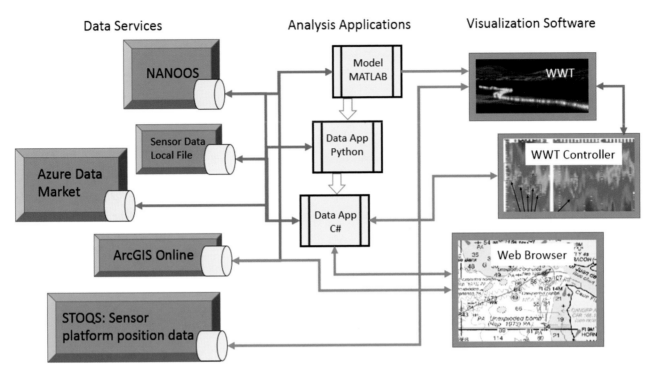

Figure 14.2. A confederated approach to data analysis and visualization: Data as a Service is represented as several source blocks (left, in blue) with data access ports, as well as ArcGIS Online as both Data as a Service and Software as a Service. NANOOS is the Northwest Association of Networked Ocean Observing Systems. STOQS is Spatial Temporal Oceanographic Query System. Analytical and visualization software treat these datasets (including local files) as on-demand resources. Domain expertise is built into the analysis applications. The WWT Controller (right, center) is described further in figure 14.5. Author illustration; map image courtesy of NOAA.

Software development in a data service environment can become very modular in nature. Data to be provided as a service can be systematically prepared for use, in which "data preparation" is an umbrella term for quality assurance, gap filling, coordinate and time transformation, provision of descriptive ancillary information, format standardization, and so forth. Data preparation by the service provider is abstracted away from data use (analysis and visualization) by the researcher. The researcher, in turn, may develop software that fetches prepared data from multiple services and builds synthetic new services. As an example, suppose a computational fluid dynamics model for an estuary needs source data from stream gauges as well as tidal, wind, and precipitation forecast data. An intermediary program could tap appropriate DaaS resources and synthesize a data service to drive the estuary model without having to modify the model code itself. This modularization would be a step toward making it easier to implement the model for many estuaries.

Visualization engines that are a component of a confederated environment need not do "everything," as in all possible visualizations, plus analytics. In the research examples to follow, visualizations developed in WWT use its native primitives: markers, lines, raster images, geographic coordinates, and time. WWT supports basic geographic, geometric, and temporal information (latitude; longitude; elevation; depth; x, y, z; date; time; time-elapse rate), in addition to a free viewing perspective. It provides graphical attributes for raster images, lines, polygons, and points; specifically, color, opacity, brightness, and time persistence. Hence, in this work, WWT renders, but does not manipulate, data. It does not calculate running averages or produce new color maps or create filtered views. Instead, a separate confederated application performs these tasks and sends WWT preconfigured representations of data. A separate application of this type is unconstrained in terms of how it is built. It could, for example, be Matrix Library (MATLAB) code; Microsoft Excel pulling data from a SQL Server database; a Python script; a C# application built on the WWT developer library *Narwhal;* or ArcGIS software run on the desktop, in the new ArcGIS Pro app that accompanies ArcGIS for Desktop software (Glennon 2014), or in the cloud in ArcGIS Online. Decoupling domain knowledge, data preparation, programming métier, and data visualization acts to separate tool building from the research process.

Two desiderata follow from latency minimization: First, a useful visualization can suggest a hypothesis amenable to testing via analysis. It is consequently desirable to build analytical tools that can easily identify or latch onto data subsets of interest. Second, it is desirable to build interface controls to quickly edit visual presentation: for example, turn graphical elements on and off, increase transparency, exaggerate spatial dimension, and so forth. This narrative now turns to the three visualization examples of interest with some inherited context from these prefatory remarks.

Experiment A: A Five-Day Marine Drift Experiment

Experiment A is a drift experiment carried out by MBARI and collaborators in the fall of 2012 off the coast of Monterey Bay. This is one in a series of experiments under the rubric of the Controlled Agile and Novel Observing Network (CANON) (MBARI 2010; Das et al. 2011). A central objective of the experiment is to drift with prevailing currents, remaining within a coherent water mass and observing microbial populations and metabolic processes over the course of several days. In practice, it is readily acknowledged that the central drifting platform, ESP (MBARI 2013), is subject to both depth-varying ocean currents and winds that act on the ESP surface float, and hence may move from one water mass to another, or from one microbial ecosystem to another, over the course of the experiment. In consequence, a view of the experiment centered on ESP as it drifts is called quasi-Lagrangian (Price 2006).

An ideal time-series picture of the water volume surrounding ESP would be synoptic (i.e., representing a comprehensive picture of the state of the system over time). MBARI researchers place many sensors in the water toward producing such a synoptic view, particularly by tasking a long-range AUV to orbit ESP once every three hours at a radius of approximately 1 km while continuously diving to 100 m and returning to the surface. Figure 14.3 shows the resulting orbiting/porpoising AUV trajectory surrounding the stately drift trajectory of ESP.

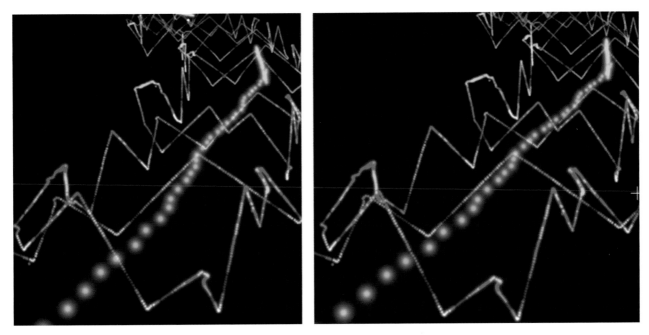

Figure 14.3. Experiment A, excerpted from the Worldwide Telescope visualization engine. Uncrossed stereo view of the Environmental Sample Processor track (orange) in relation to the path followed by an autonomous underwater vehicle in a Monterey Bay Aquarium Research Institute Controlled Agile and Novel Observing Network drift experiment. The underwater vehicle track shows nitrate concentration encoded by color.

Using WWT to view the time evolution of these different trajectories, it becomes clear that the datasets are difficult to compare visually. Although both the AUV and ESP carry CTD sensors, the porpoising path of the AUV contrasts with the relatively fixed depth of ESP. Taking an assumption of a stratified water column, I created a visualization (figure 14.4) in which ESP and AUV measurements are emphasized and made comparable using radiating horizontal ripples. AUV ripples can be triggered (as in figure 14.4) by a particular depth, or they can be triggered by a range of sensor measurements X. The latter case produces a visual approximation of surface (or volume) of constant X, whereas the former case allows the researcher to visually compare sensor measurements between platforms, AUV to ESP.

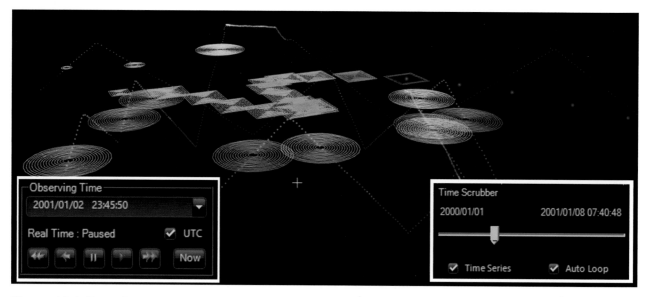

Figure 14.4. Experiment A, excerpted views of Worldwide Telescope. This is a volume-filling representation of a stratified water column, showing an autonomous underwater vehicle path (zigzag) in relation to the Environmental Sample Processor drift path (red dots, center right). Concentric rings show measurements made by corresponding instruments (in this case, salinity). Insets are the Worldwide Telescope internal clock (lower left) and time slider (lower right). The drag handle on the time slider allows the researcher to move the Worldwide Telescope clock and corresponding data view backward and forward in time.

ESP consists of a surface float and an in situ biological laboratory suspended at a depth of 10 m, together with a CTD as noted. The surface float mounts a downward-looking ADCP that continuously measures relative lateral currents at 2 m intervals from the surface to a depth of 90 m. If this lateral flow is fairly uniform and time-invariant for a period, it might be a suitable proxy for current flow nearby, particularly at the location of the AUV. This would, in turn, suggest that the orbiting AUV might sample the same water twice, once on an initial pass and again later as its orbit actively moves it to where the water is passively advected by relative lateral currents. This idea suggests a Lagrangian evaluation through visualization and analysis. To what extent is this drift experiment truly Lagrangian? That is, is the drifting ESP staying within a self-consistent water mass; or if not, how gradually does it move from one water mass to another with a distinctly different recent history? The question can be addressed by using ADCP measurements to advect AUV positions with time. The AUV itself is not advected, but the water through which it passes is. Figure 14.5 shows a confederated approach to visualizing this idea. A graph called a *curtain plot* is created by software that is independent of WWT. This graph shows time and depth (horizontal and vertical axes, respectively) for the AUV time-series data. Chlorophyll concentration measurements are represented using a heat map, with red representing greatest concentration. Note that the time range spans five days and that the porpoising trajectory of the AUV is compressed in this time period to give a visual impression of continuous measurement at all depths. The graph shows isolated areas of high concentration in the first two days, which then consolidate and migrate upward into a chlorophyll lens in the upper 20 m of the water column in the latter part of the drift interval.

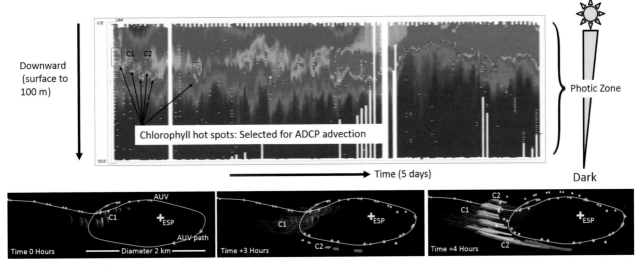

Figure 14.5. Experiment A, a graph (aka *chart*) control for Worldwide Telescope, from the Monterey Bay Aquarium Research Institute Controlled Agile and Novel Observing Network 2012 drift experiment. The time-depth curtain plot represents five days of Environmental Sample Processor drift, in which color indicates chlorophyll concentration with depth (red is maximum) as measured by an autonomous underwater vehicle orbiting the sample processor. Hence, this dataset folds back on itself in the Lagrangian frame as the underwater vehicle orbits the processor. The chart is a Worldwide Telescope control in two ways. First, a left-click cursor touch on the chart sets the Worldwide Telescope clock, so that a cursor drag across the chart is equivalent to dragging the telescope time slider. Second, a right-click cursor drag selects a region of the curtain plot as shown in the black boxes at the upper left. An acoustic Doppler current profiler-mounted downward facing on the sample processor float measures lateral current relative to the sample processor from the surface down to 100 m. In consequence, a selected data block from the chart control can be advected in time, presuming it follows the same current as measured at the sample processor (i.e., at the center of the autonomous underwater vehicle orbit). In the first Worldwide Telescope frame, the underwater vehicle has approached from the left (white dots), and some of the data (faint color chevrons, selected as C1 in the chart control) has begun to advect. Upon completing the first orbit (upper middle frame), a second block of data, C2, has been selected (brighter color patches) and also begins to advect. The fact that C1 and C2 are spatially coincident under this advection scheme, combined with the underwater vehicle location with time, suggests that the underwater vehicle is moving through a coherent water mass with a locally high concentration of chlorophyll.

The curtain plot connects to WWT via an application programming interface (API) described in more detail below. This allows the plot to be used as a data selector. Clicking and dragging over an area of interest in the curtain plot extracts that data subset from the complete AUV dataset. This subset is advected using the ADCP current measurements as a proxy. The results are installed in WWT as a visualization layer. The lower part of figure 14.5 shows two such selections from near the start of the drift experiment. Both their location and advection suggest that two high-concentration chlorophyll signals in the AUV curtain plot may be part of the same water mass.

A confederated (broad) approach to the Lagrangian evaluation problem extends beyond the in situ sensor work described here. Availability of remote-sensing data for near-surface chlorophyll concentration, sea surface temperature, wind, and currents means that a GIS would provide an independent approach to characterizing the Lagrangian fidelity of the ESP trajectory. Note also that the number of sensors (across platforms and sensor types) exceeds one hundred in the 2012 CANON experiment, creating a considerable data management challenge. MBARI researchers have implemented the Spatial Temporal Oceanographic Query System (STOQS) (MBARI 2014) to carry forward their ongoing effort to provide data in the DaaS paradigm.

Data-to-Visualization Mapping

Before proceeding with experiments B and C, I describe mapping time-series ocean data to visualization in a many-to-one sense; that is, from many types of oceanographic data to the virtual globe WWT engine. Relevant capabilities of the WWT engine include:

- Métier: three spatial dimensions and supporting Cartesian and rectilinear coordinate systems
- Time support: date, time, time elapse rate, a time slider control, time looping
- User-controlled viewing perspective
- Support for 3D models and mesh surface rendering
- Blank canvas space free of geospatial context markers
- Raster/vector/point rendering support
    - Raster draping
    - Segmented lines, polygons
    - Point rendering with multiple marker options
    - Control of color, opacity, and (for markers) brightness
- GPU-dependent data capacity (one million to 30 million data points)
    - Comparatively large data capacity, enabling particle trajectory interpolation (i.e., continuous visual paths for individual entities)
- Reproducible perspectives and perspective sequence capture for storytelling
- Text, image, and audio annotation embedding
- A peer-to-peer API for automated configuration and data import
- Web service connectivity to data sources (WMS, OData)
- Cut-and-paste data table input
    - Columns enumerating data attributes; rows corresponding to data values

In experiment A, the datasets include many types of measurement made along the AUV platform trajectory. Here, the basic CTD has been enhanced with additional sensors to measure fluorescence, dissolved oxygen, pH, and other parameters. A single marker can color-encode one type of measurement—say, salinity—but what of the other measurement types? Each may be mapped to a structure such as a cube (12 edges and eight vertices) that translates with time tracking the AUV location. This, however, presents the difficulty of remembering what each component of the structure corresponds to, in turn suggesting toggle switches for simplifying the view. Such a construction would require practice on the part of the researcher to use and interpret, an open topic. This example of using representational visual structures has an additional implicit degree of freedom, in marker decay time. As the AUV platform moves in space and time, the corresponding measurements are rendered in place; and a global time decay determines how long (on the WWT clock) that representation persists before fading out. Long persistence times tend to appear as long tails, suggesting a more spatially extended picture of the water volume. A short decay time also has an advantage, reducing visual clutter when many data streams are playing out in proximity. Similarly, opacity can be reduced to cut down on visual clutter. In some instances, a very faint, static monochrome view of a trajectory can be placed underneath a brighter dynamic view to provide a reference without obscuring the data/trajectory playback.

The WWT feature list cited here includes a time slider, as shown in figure 14.4. This reflects a current time within an active (looping) time range and can be dragged to a moment of interest. This, in combination with duration settings, gives the scientist a "browse through time" capability. Although WWT does not have an active data filter, the time control may be manipulated into providing that function as follows. A particular measurement type is mapped into a numerical time range, which is then declared to be time in the WWT table configuration for this data layer. By mapping data values to pseudotime in this way, the WWT time slider behaves as a data-windowing filter. For example, an AUV CTD time series for salinity in the range of 33.2 to 33.8 practical salinity units (psu) can be mapped to a time range from March 1 to March 31. The time slider setting will now select and illuminate locations where the salinity measurement fell within a narrow range, showing all such points simultaneously, irrespective of when those measurements were made.

Experiment B: Computational Fluid Dynamics Model

ROMS produces a time series of ocean state cells, in which state includes the flow field, water temperature, salinity, pH, dissolved oxygen, chlorophyll concentration, and other parameters (Arongo and Shchepetkin 2014). Figure 14.6 shows an oblique view of the continental shelf west of the Strait of Juan de Fuca, where the ROMS model was used to produce a flow field (Sutherland et al. 2011). A visualization in WWT was produced to explore the idea that coastal upwelling provides nutrient-rich water to Puget Sound via Juan de Fuca Canyon. Point markers are placed at selected locations within the model volume and allowed to circulate according to the derived flow field with an additive random walk component mediated by local estimated turbulence. The resulting particle trajectories are interpolated and imported as tracer data in WWT.

Figure 14.6. Experiment B, excerpted from the Worldwide Telescope visualization engine. The Regional Ocean Modeling System computational fluid dynamics model produces a time-varying vector field. Tracers placed and advected within the solution field show upwelling water entering the Strait of Juan de Fuca along Juan de Fuca Canyon (yellow arrow) and possibly via three other nearby canyons (white arrows). Author illustration; map image courtesy of NOAA.

As a candidate research insight question: How might the ROMS model results suggest a validating experiment in the field? The WWT visualization shows not only flow up Juan de Fuca Canyon, but additionally tracers flowing up two of three nearby canyons (figure 14.7).

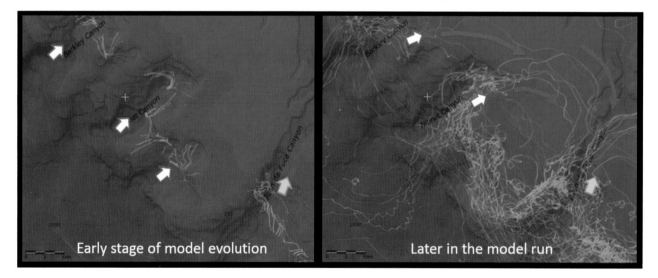

Figure 14.7. Experiment B, excerpted from the Worldwide Telescope visualization engine. Tracers show that in addition to Juan de Fuca Canyon, the Regional Ocean Modeling System model predicts that two of three smaller canyons to the north contribute upwelling water along the indicated (green) pathways.

One could proceed from what amounts to initial suggestions from this simple visualization to explore a more extended series of model runs, and from there to derive an expected flux along various paths. This expectation result might, in turn, motivate a dye trace experiment or other in situ observations to validate the model results.

Experiment C: Dissolved Organic Matter Analysis Using Fourier Transform Ion Cyclotron Resonance Mass Spectrometry

Finally, in experiment C, I consider DOM measurements using FTMS. This method of analyzing complex carbon-based molecules dissolved in water—with typical masses from 100 to 800 daltons (Da)—produces a sequence of extremely narrow peaks corresponding to ionized molecules. Each peak can be assigned a probable mass and, in many cases, a molecular formula. Molecular formula assignment is a Diophantine problem that operates on a small set of candidate atoms: notably, carbon, hydrogen, oxygen, nitrogen, phosphorus, and sulfur (Sleighter and Hatcher 2007). Formula assignment does not resolve molecular structure, so a single FTMS peak from a naturally occurring water sample is assumed to correspond to the presence of multiple isomers with the same formula. Peak amplitudes in an FTMS spectrum are retained for intersample comparison without any presumption that they correlate with relative concentrations of a specific compound. A question naturally arises: Can new visualization strategies suggest relationships not apparent in existing views and analysis of FTMS peaks?

Operating only on the subset of peaks within an FTMS spectrum that are assigned molecular formulas, I refine the question: What do molecular mass and mean peak heights suggest about relationships between different resolved molecular formulas? With no presumption about the large number of *possible* isomers present, a number that can exceed one million for a single molecular formula (Hertkorn et al. 2007; Hertkorn et al. 2008; Ruddigkeit et al. 2012), the general task of resolving molecular compounds in a DOM assembly remains open. Turning to visualization, a commonly used 2D graphing method places a marker for each FTMS molecular formula on a scatter chart with axes for hydrogen-carbon and oxygen-carbon ratios, also known as a *Van Krevelen diagram* (figure 14.8). In WWT, the Van Krevelen scatter chart can be constructed as a distribution of markers with the addition of a z-axis (for example, showing molecular mass or peak intensity as height for each molecular formula marker as shown in figures 14.9 and 14.10, respectively). Furthermore, the markers can be color coded depending on atomic constituents (for example, yellow if sulfur is present, green for phosphorus, and so forth). The result is a cloud of several thousand colored dots arrayed in three dimensions, which can be viewed from any perspective and flattened to z = 0 to produce the traditional Van Krevelen diagram view.

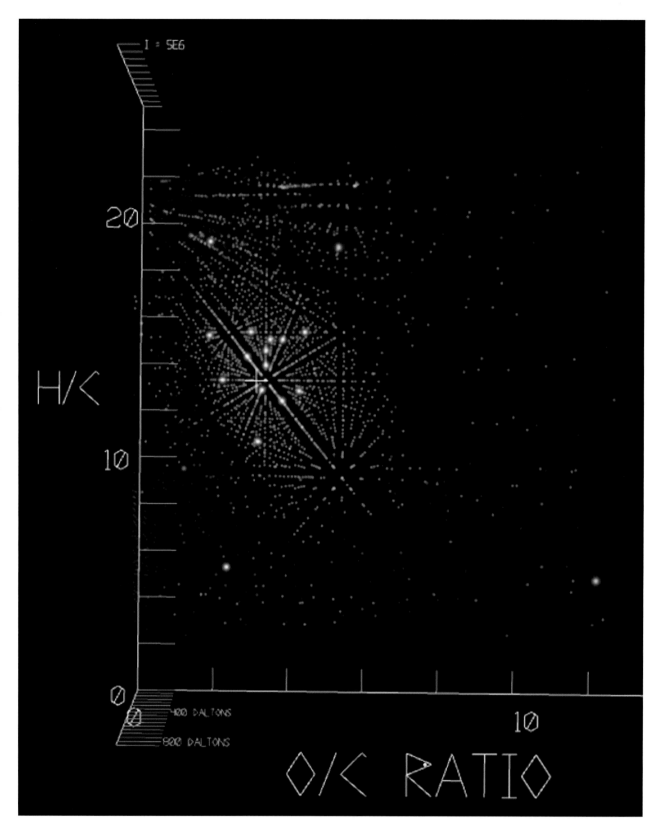

Figure 14.8. Experiment C, excerpted from the Worldwide Telescope visualization engine. Flat representation of a Van Krevelen diagram. Each dot is a molecular formula graphed according to its ratios, oxygen to carbon and hydrogen to carbon. Yellow indicates the formula contains sulfur; green, phosphorus; and blue, nitrogen.

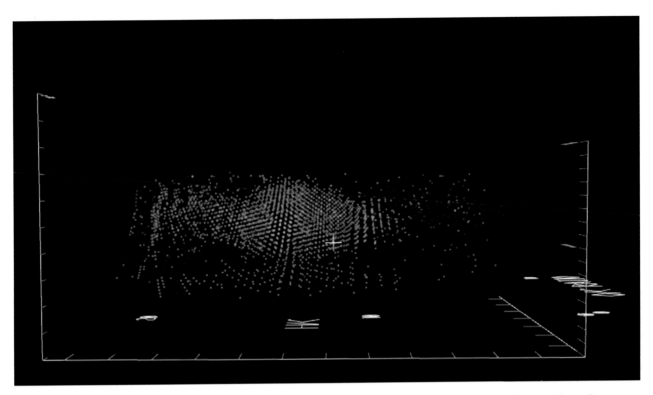

Figure 14.9. Experiment C, excerpted from the Worldwide Telescope visualization engine. A second Van Krevelen diagram view with z-axis scaled to molecular formula mass. Note the five aligned green molecules at upper left, part of a group of six related in formula by differences of CH_2. Prevalence of sulfur-bearing compounds (yellow) can be seen biased to large hydrogen-carbon ratios.

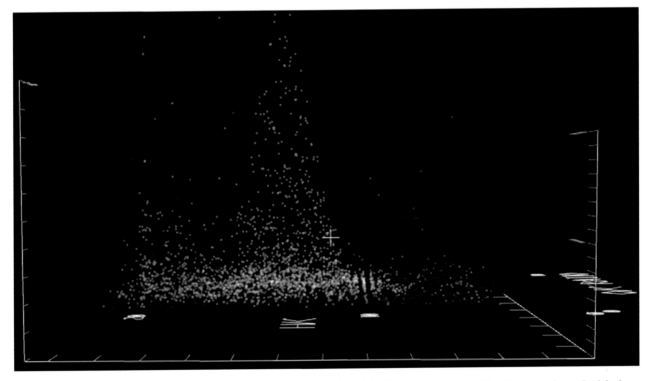

Figure 14.10. Experiment C, excerpted from the Worldwide Telescope visualization engine. A third Van Krevelen diagram view with z-axis scaled to Fourier Transform Ion Cyclotron Resonance Mass Spectrometry peak intensity.

Consider a smooth sequence of transitions (views) of the data, from traditional/flat (z = 0) to mass oriented (z-coordinate scaled by molecular mass) to FTMS peak intensity oriented (z-coordinate scaled to mean FTMS peak intensity) to finally, for the sake of looping, a transition back to the flat view (z = 0). The sequence is accomplished by creating keyframes, a sequence of four corresponding view structures with associated sequential time tags. Transition frames are added by interpolating z-axis coordinates and time. That is, a particular marker (molecular formula) will have a z-coordinate in flat, mass-oriented, and peak-oriented views, and interpolated frames will transition each marker between its respective z-coordinates in short time steps. The time slider in WWT then becomes a view control, used by a researcher to move between views while showing the transitions of each individual molecular formula from one view to the next. At this configuration of FTMS peaks in WWT, a further difficulty ensues. The inspection of the marker cloud works well when the researcher can shift the viewing perspective, but it is cumbersome to change perspective while also manipulating the time slider. Therefore, the Auto Loop control is toggled on, and the time elapse rate is increased until the full visualization cycle repeats in perhaps 10 seconds. Upon reaching the end point, the view returns to the start (as time is looping), which looks the same as the end point; so the researcher is presented with a smoothly and continuously cycling sequence of three views of the data, and can therefore pan, zoom, yaw, and pitch the viewing perspective at will to focus on regions and molecular formulas of interest. This viewing strategy uses Cartesian coordinates in a sandbox reference frame.

In this visualization, a sequence of six similar phosphorus-containing molecules separated by 14 Da intervals (CH_2 methylene groups) "fly apart" in an unexpected manner during the transition from mass to peak intensity representation (figure 14.11). Hence, this visualization suggests a pattern that could be investigated further using analysis code.

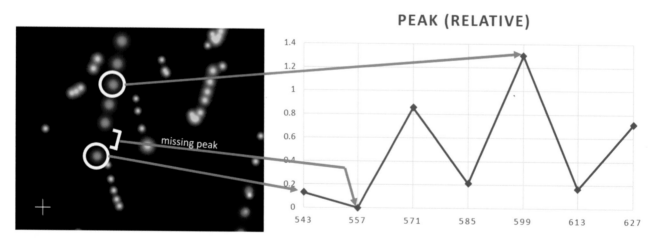

Figure 14.11. Experiment C, excerpted from the Worldwide Telescope visualization engine. Left: A Van Krevelen space detail. Six molecular formulas related by mass differences in increments of CH_2, with one missing. Right: Peak graph shows relative Fourier Transform Ion Cyclotron Resonance Mass Spectrometry peak intensity with increasing molecular mass. The sawtooth shape suggests that, in this source water sample collection, there is a variability that overrides a simple correlation between peak intensity and monotonic isomer count.

Conclusion

This chapter touches on a small set of data visualization examples from oceanography. It places visualization alongside analysis and data management as interrelated processes that connect research problems to data. There are premium incentives to shift data access methods, specifically building data services with a corresponding move toward resource confederation that can lower the barrier to data access, and consequently help meet a growing imperative for data reuse. Toward the goal of visualization, the methods and examples provided here, based on the WWT visualization engine, support the contention that the oceanographic research data challenge—how to effectively analyze and understand new harvests of large complex datasets—is amenable to a technical solution. That is, increasingly complex datasets can be explored via advanced visualization software such as WWT. Although WWT has capacity for large datasets existing in a four-dimensional space, its primary strengths are its program-accessible interface and its generic approach to manipulating data (Microsoft Research 2014a and 2014b). WWT and heterogeneous data resources in oceanography will become increasingly amenable to a complementary-tools approach that includes GIS, which will extend 3D visualization and manipulation to the realm of analysis in virtualized environments such as Citrix XenApp and XenDesktop, Microsoft Hyper-V VDI, and VMWare Horizon View (Meza 2014).

Acknowledgments

The cooperation and collaboration of researchers at MBARI, as well as Parker MacCready of the University of Washington School of Oceanography, are gratefully acknowledged. The comments and suggestions of editor Dawn Wright and two anonymous reviewers greatly improved this chapter.

References

Arongo, H. G., and A. F. Shchepetkin. 2014. Regional Ocean Modeling System. http://www.myroms.org. Last accessed July 5, 2014.

Center for Environmental Visualization. 2014. Collaborative Ocean Visualization Environment—University of Washington. http://cove.ocean.washington.edu/. Last accessed July 4, 2014.

Das, J., T. Maughan, M. McCann, M. Godin, T. O'Reilly, M. Messié, R. Bahr, K. Gomes, F. Py, J. Bellingham, G. S. Sukhatme, and K. Rajan. 2011. "Towards Mixed-Initiative, Multi-robot Field Experiments: Design, Deployment, and Lessons Learned," 3132–39. IEEE/RSJ International Conference on Intelligent Robots and Systems (IROS 2011), San Francisco, CA, September 20–25.

Data Observation Network for Earth. 2014. DataONE. http://www.dataone.org. Last accessed June 21, 2014.

Geyer, W. R., and P. MacCready. 2014. "The Estuarine Circulation." *Annual Review of Fluid Mechanics* 46:175–97.

Glennon, R. 2014. *Announcing the ArcGIS Pro Beta Program.* http://blogs.esri.com/esri/arcgis/2014/05/13/arcgis-pro-beta-program. Last accessed July 5, 2014.

Hertkorn, N., M. Frommberger, M. Witt, B. P. Koch, P. Schmitt-Kopplin, and E. M. Perdue. 2008. "Natural Organic Matter and the Event Horizon of Mass Spectrometry." *Analytical Chemistry* 80 (23): 8908–19.

Hertkorn, N., N. Ruecker, M. Meringer, R. Gugisch, M. Frommberger, E. M. Perdue, M. Witt, and P. Schmitt-Kopplin. 2007. "High-Precision Frequency Measurements: Indispensable Tools at the Core of the Molecular-Level Analysis of Complex Systems." *Analytical and Bioanalytical Chemistry* 389 (5): 1311–27.

Hey, T., S. Tansley, and K.M. Tolle, eds. 2009. *The Fourth Paradigm: Data-Intensive Scientific Discovery*. Redmond, WA: Microsoft Research.

MBARI (Monterey Bay Aquarium Research Institute). 2010. Controlled Agile and Novel Observing Network (CANON). http://www.mbari.org/canon. Last accessed June 21, 2014.

———. 2013. Environmental Sample Processor. http://www.mbari.org/esp. Last accessed June 21, 2014.

———. 2014. Spatial Temporal Oceanographic Query System. https://odss.mbari.org/odss. Last accessed June 21, 2014.

Meza, J. 2014. *ArcGIS Pro in Virtualized Environments*. http://blogs.esri.com/esri/arcgis/2014/06/30/38908. Last accessed July 5, 2014.

Microsoft Research. 2014a. Layerscape: Esri Ocean GIS Data Visualization Supporting Material. http://www.layerscape.org/Content/Index/31279. Last accessed July 3, 2014.

———. 2014b. Worldwide Telescope. http://worldwidetelescope.org. Last accessed July 3, 2014.

OData (Open Data Protocol). 2014. http://www.odata.org/. Last accessed June 21, 2014.

OPeNDAP (Open-source Project for Network Data Access Protocol). 2014. http://www.opendap.org. Last accessed June 21, 2014.

Open Geospatial Consortium. 2014. Web Map Service. http://www.opengeospatial.org/standards/wms. Last accessed June 21, 2014.

Price, J. F. 2006. *Lagrangian and Eulerian Representations of Fluid Flow: Kinematics and the Equations of Motion*. Woods Hole, MA: Woods Hole Oceanographic Institution Technical Report. http://www.whoi.edu/science/PO/people/jprice/class/ELreps.pdf.

Rew, R., and G. Davis. 1990. "NetCDF: An Interface for Scientific Data Access." *Computer Graphics and Applications* 10 (4): 76–82.

Ruddigkeit, L., R. van Duersen, L. C. Blum, and J.-L. Reymond. 2012. "Enumeration of 166 Billion Organic Small Molecules in the Chemical Universe Database GDB-17." *Journal of Chemical Information and Modeling* 52 (11): 2864–75.

Sleighter, R. L., and P. G. Hatcher. 2007. "The Application of Electrospray Ionization Coupled to Ultrahigh Resolution Mass Spectrometry for the Molecular Characterization of Natural Organic Matter." *Journal of Mass Spectrometry* 42 (5): 559–74.

Sutherland, D. A., P. MacCready, N. S. Banas, and L. F. Smedstad. 2011. "A Model Study of the Salish Sea Estuarine Circulation." *Journal of Physical Oceanography* 41:1125–43.

Wessel, P., and W. H. F. Smith. 1991. "Free Software Helps Map and Display Data." *Eos, Transactions of the American Geophysical Union* 72:441.

CHAPTER 15

Managing the Visual Landscape of Oregon's Territorial Sea

Paul Manson and Andy Lanier

Abstract

New and emerging proposals for ocean renewable-energy development have raised questions about how to responsibly and sustainably develop these resources. These uses have motivated states and nations to undertake coastal and marine spatial planning. Planning efforts often require collecting new data on various human uses of the ocean. This chapter describes the development and application of a participatory GIS method for assessing visual resources. This is based on participatory planning and democratic theory, arguing that engaging the politics and values of environmental controversies must be done in "hybrid forums" that use participatory codevelopment of assessment methods. This unique codevelopment approach allows for new ideas, concerns, and values to be captured in the final assessment method, and ultimately to allow for more durable agreements. The method uses a group-based field assessment to score views based on local knowledge, elements of visual quality, and computer modeling. This chapter details the development of this method with a stakeholder committee, the methods application the Oregon Territorial Sea Plan process, and the GIS analysis steps that support this method.

Introduction

Coastal and marine spatial planning (CMSP) efforts have emerged as a key policy priority for coastal nations and regions. This initiative is tied to an international focus on balancing biodiversity and resource conservation with community resilience and economic development. Common to many CMSP efforts is a dedication to participatory processes for planning (Ehler and Douvere 2009). These often take the form of participatory deliberation on the coproduction of data for decision-making. This chapter details the development and application of a data coproduction system to collect spatial data on the importance of visual resources in a CMSP context. Called

Dawn J. Wright, ed.; 2015; *Ocean Solutions, Earth Solutions*; http://dx.doi.org/10.17128/9781589483651

the Community-based Marine Assessment of Aesthetic Resources (C-MAAR), this method is a community-driven model for visual resource assessment that includes a field survey component tied to GIS modeling for assessing visual resources. C-MAAR is unique in its combination of a coastal and marine application, community deliberation tools, and ease of application in rapidly developing policy settings. C-MAAR was developed as part of the Oregon Territorial Sea Plan (TSP) update process. The development history is presented here as it informs ways to transfer this method to other settings.

A critique of quantification systems for values such as visual resources is that they overessentialize a human experience into abstract numbers (Hacking 2000). Further, these numbers are highly dependent on the social settings and various meanings in public decisions. They are not simple market-based values that can be aggregated (Sagoff 2011). Key to these critiques is that the control of the quantification is concentrated in a group of experts or interests that prevent broadening input and that the standard of proof is different in public debates over how society should use natural systems. The C-MAAR experience in Oregon has been a response by the planning agency, the Oregon Department of Land Conservation and Development (DLCD), to increase participation and bring community values into the spatial framework for decision-making to avoid some of these pitfalls. These values are critical in resource management because they provide the basis for understanding how differing communities define and conceptualize successful management or sustainability (Norton 2005; Gross 2010) and how policy agreements can form (Sabatier and Jenkins-Smith 1999). The approach detailed in this chapter is an effort to engage visual resource assessment with an acknowledgement that assessments are laden with politics. These politics reach back to the original questions of how and where to locate new ocean renewable-energy developments. Although assessment processes often try to separate the science from the politics, this approach creates a space for deliberation over visual resources that allows for the experts, community representatives, and the public to meet in a "hybrid forum" (Callon et al. 2009). These forums are defined by a mixing of types of knowledge from laymen to experts and across many issues. These forums allow for controversies to reveal what is important, the possible future that might be planned, and how to build agreement.

This chapter introduces the Oregon TSP update, its policy history, and the context for the visual inventory. It also discusses the challenges involved with obtaining, and options for soliciting, visual resource values from communities. The tension between expert-driven methods and experiential methods for assessing visual values is discussed. Next, the chapter explores development of the method, including the participatory process used. Oregon's ocean-planning experience is defined by a dedication to distributed data development with engaged participation by community and user groups. This dedication to public involvement is captured in Oregon's statewide planning goals (codified in state law), and it influenced the design of C-MAAR and its participatory components. The chapter discusses how the C-MAAR method was codeveloped with community members in joint method development workshops in a live policy-process setting. The C-MAAR approach, with thorough stakeholder committee involvement, appears to offer a way to successfully bridge these two interests: scientific rigor and responsiveness to public and policy needs. Finally, the chapter presents the technical details of C-MAAR, how it was applied, and discusses the outputs included in the planning process. The discussion and conclusion explore the challenges and strengths of the C-MAAR method and how to resolve key tensions in participatory GIS applications.

Background

The Oregon coast is an internationally recognized tourist destination. Over 20 million visits are made to Oregon's coastal parks each year (OPRD 2005). Scenic enjoyment is the third-most commonly stated primary recreational activity that visitors say they engage in on Oregon's coastal beaches (Shelby and Tokarczyk 2002). In addition, the Oregon coast highway (Pacific Coast Scenic Byway) has been federally recognized by the National Scenic Byways program, established by Congress and administered by the US Department of Transportation Federal Highway Administration. In addition to being one of the first scenic byways in the country, it has also been designated as an "All American Road," which recognizes US 101 as possessing "multiple intrinsic qualities that are nationally significant and have one-of-a-kind features that do not exist elsewhere" (CH2MHill 1997). Oregon's coastline is also unique in that it has over 70 state parks running along the highway, providing "public access and resource protection in a way that is unrivaled by any other US coastline park system" (CH2MHill 1997).

TSP Context

Ocean renewable-energy developers came to Oregon in 2007 seeking to deploy wave energy devices. These developers were taking advantage of a series of policy goals to expand renewable energy by the states of the West Coast, as well as the massive energy potential of the Pacific Ocean off Oregon. This interest resulted in a series of claims on areas of the ocean for developers to use, but outside an organized planning approach. The initial lack of a coordinated framework to understand the impacts of these claims led to a multiyear ocean planning effort by the State of Oregon through its statewide planning agency, DLCD. DLCD was charged with understanding the current human uses and natural resources that might need protection or consideration before siting of energy devices. DLCD's planning jurisdiction is the territorial sea, a 3 nm ribbon of ocean along the coast. The new planning policy was developed through an update to the state's TSP. This plan identifies the state priorities for management of the territorial sea and guides partner agencies in implementing state policy for the ocean.

The TSP update provided the process to consider new energy uses in a comprehensive framework. This process was required to assess resources and uses that might be in conflict and consider how best to locate new renewable-energy projects to minimize negative impacts. At the beginning, data for the planning process was not readily available, and coverage often did not include the full extent of the territorial sea. DLCD's first planning task was to develop new spatial data products with partners, ocean users, and community groups. A hallmark of the Oregon planning law is extensive public involvement. The TSP experience expanded this concept through participatory data development. These data products included public participation GIS (PPGIS) datasets, building on previous regional efforts to develop stakeholder and community-driven data (Bonzon et al. 2005). The commercial and sport fishing data collection was based on a participatory model of sitting down with boat captains to map out areas of important fishing effort (Steinback et al. 2010). Likewise, an effort to map the nonconsumptive recreation ocean-user community was also based on a PPGIS model (LaFranchi and Daugherty 2011). In addition to the PPGIS projects that occurred, the planning process was guided by a series of citizen-involvement committees that reviewed and contributed to the research and data collection. These elements formed the basis of the C-MAAR process during planning. Assembling these data products allowed for analyses of interactions among different user groups and evaluation of potential conflicts with new energy developments. As the planning process moved forward, it became clear that local community residents were concerned that future energy development on the ocean would potentially impact the quality of their views. Public testimony and feedback during meetings underscored the importance of considering aesthetic (e.g., viewshed) impacts for marine renewable-energy siting. In response to this concern, which supported concerns previously presented by the Oregon Parks and Recreation Department (OPRD), the decision was made to develop and implement a visual resource inventory for the Oregon coast.

This decision was supported by language in Oregon's Statewide Planning Goal 19, which is reiterated in TSP, that states agencies, through programs, approvals, and other actions, shall "protect and encourage the beneficial uses of ocean resources such as … aesthetic enjoyment." Additionally, Oregon's Ocean Shore Management Plan, a Federal Energy Regulatory Commission–approved "comprehensive plan," notes that OPRD (2005) "may identify important 'scenic features' that should be protected from development or other impacts for their scenic value." To support the development of a visual resource inventory, community involvement by an advisory committee of local community representatives, user group representatives, and management agencies was convened. Through this committee, the C-MAAR methodology was developed.

A final feature of the Oregon TSP update is that the process had to manage for unknown future ocean energy technologies. Ideas existed about the potential size and shape of some technologies, but most of these were unproven, and uncertainties existed about the likelihood of actual industrial-scale development. Other visual-resource efforts in the waters off the United Kingdom and Europe, as well on the US East Coast, were based on modeling possible visual impacts with known technologies. For example, offshore wind development uses mature turbine and pylon designs that can be modeled. At the time, wind was not being considered within Oregon's territorial sea because of the steep seafloor and depths. Similarly, there were no clear high-priority areas for energy developers, given the wide range of possible technologies being considered at the time. Instead, the Oregon TSP process assessed inventoried and assessed coastal visual resources to identify possible areas where conflicts would be low. This reflects an ongoing tension in marine spatial planning and ocean renewable energy: a project-based approach versus a programmatic zoning approach. Project approaches can be more certain in their assessments but also require assumptions on technology and placement. This approach can also be riskier for developers, who make the first move in identifying their priorities and then defending that option. The programmatic approach, on the other hand, allows for a comprehensive examination but with less specificity on the actual impacts.

Visual Resource Assessment Techniques and Theory

Visual resource assessments have a long tradition in various intellectual traditions, each with different assumptions about the value systems and rationale for including visual values (Dakin 2003). From an economic or utilitarian standpoint, visual resources are part of a bundle of benefits that include other economic values such as recreation or housing. Another perspective is to understand visual resources as part of a sense of place. In the sense of place framework, visual resources are part of defining identity, values, and community. These two perspectives can occupy opposing positions, the first focused more on individual and market-type values while the second is focused on communal and shared values. In between are other variations on value, such as behavioral or environmental psychological theory. The economic model is more commonly found in policy settings as it matches the logic of efficiency and effectiveness that dominate cost-benefit rationale in these decision-making settings (Frank 2000).

This tension has methodological implications. Economic or utilitarian models are focused on tools to discover hidden values from individuals and reveal them for analysis. Examples of this type of assessment include developing policy priorities driven by housing values versus community values. This perspective assumes the individual as the unit of analysis and that, in aggregate, social value can be assessed through tools to replicate the values that would be found in market exchanges (Stone 2001). This leads to an expert-driven assessment method that models value. Increasingly, the economic or utilitarian models for visual resources are expanding to include more complex understandings of value. From the economic model, natural systems provide aesthetic and visual amenities for human communities through sights, sounds, smells, and touch (Millennium Ecosystem Assessment 2003; De Groot et al. 2010). Visual resources are found in a wide diversity of settings that are not necessarily "natural" or "pristine" settings, but can be anthropogenic green spaces in urban areas; restoration sites; or managed landscapes such as forestry, agriculture, or open space and protected areas. These benefits can stand alone as a service or contribute to other benefits such as recreation, sense of place, or cultural heritage. Visual resources also provide important contributions as amenities in market values such as real estate values for homes or vacation properties (Lutzenhiser and Netusil 2001).

The communitarian perspective is more focused on interactions among community members. Here, the visual resources are not consumed as commodities, but likely are constitutive values that define a place and its people (Tuan 1977). In this assessment type, the methods tend to be experiential, in which participants are asked to share experiences with the goal of understanding the meaning behind views (as opposed to the utilitarian value). This expert-experiential tension has an impact on how tools for visual resources can be developed. Sense of place has been proposed as a critical tool for understanding the values and meanings communities hold in natural resource decisions. Sense of place is particularly powerful as it connects physical space, human values, and individual identities with rich layers of meaning (Relph 1976; Tuan 1977; Williams and Stewart 1998). It commonly suggests personal intimate relationships between "self" and "place." Thus, a "sense of place" is often understood largely in terms of positive affective qualities of place-attachment. Traditional and familial knowledge are often strongly linked to places. Stories and events are tied through locations, and locations can become symbols for the values that flow through the family history (Basso 1996). Place has also come to stand as a measure of civic community or community attachment (Kemmis 1992; Clark and Stein 2003). This is a primary concern of any visual or aesthetic assessment that is being included in local deliberations on the use of land or sea space.

Visual resource management has had a more expert-driven focus over time, in part because the biophysical nature of views is more readily amenable to spatial modeling. Visual assessments can use expert opinion, using objective categories and measures, or they can rely on individual perceptions and experiences (Dakin 2003). Critiques of expert-driven methods suggest a need to capture the experiential perspective of the viewers (Daniel 2001). Expert-based models try to fit the environment into categories and scales that simplify the environment. In general, GIS has a tendency toward this expert model versus the experiential one (Duncan 2006; Lejano 2008). The challenge for expert-based models is that the simplification of views or sense of place can over essentialize elements in the biophysical world because of data availability, the choice of categories, ease of analysis and visualization, or response to immediate political concerns. Thus, experiential and participatory systems can help overcome simplifications and generalization that might mis- or underrepresent local preferences and goals (Duncan 2006). Far less attention has been given to the local community and residents, so working in participatory contexts can open up new possibilities to reconfigure conventional understandings of knowledge that traditional GIS may produce (Elwood 2006).

At the same time, sense of place has been noted as richly theorized, but not fully developed empirically (Stedman 2003). Efforts to add empirical tools have allowed for quantitative investigations of place (Shamai 1991). However, limits have been identified in the progress of these examinations, especially in converting qualitative accounts of place to the quantitative indices. A chief concern is that sense of place can overestimate a common meaning versus contested meaning, and that scale of sense of place is often not investigated (Cheng and Daniels 2003). This recognition suggests current efforts to understand sense of place need to better understand levels of agreement over sense of place within communities, the contingency of this meaning on larger relationships in a region, and the role of scale in the individual perception of sense of place. This is particularly true of coastal and ocean settings, in which local community preferences can be complex. In the Oregon case, debate occurred about types and levels of appropriate industrial use of the ocean and potential impact on views. For example, commercial fishing fleets introduce cultural modifications on otherwise "natural" views, including bright floodlights and vessels. However, this view has become iconic of economic resilience for some community members, but not all. This value is tested when new ocean-based users, such as ocean renewable energy, emerge.

This review of some key issues in visual and aesthetic theory and practice helps describe a unique challenge for the application of visual resource assessments, namely, the tension between utilitarian-based and experience-based models. Utilitarian-based models suggest that comparisons can be made across different value sets based on the level of use. Experiences are not as easy to compare. Efforts to distill metrics such as visitor counts or visitor days show how the experience is lost with simple use metrics. However, experiences also more accurately capture what many feel about a place, more than a measurement of use. Each applies to logic in public policy that can be difficult to resolve.

Once the basis of the valuation is identified, another methodological challenge returns. The expert versus participatory models provides a similar common challenge in planning processes. The experts have access to unique tools and knowledge and are expected to act for the common good; however, for such experiential values as views, this expertise is easy to challenge because it evades usual conceptions of rationality and standardization (Daniel 2001). Although the C-MAAR approach proposed here does not solve these challenges, it attempts to respond to these concerns and balances experts and participants, and also endeavors to meet the time and resource constraints that exist in planning processes. By engaging in evaluating views through local participation, the experiential component of place can be captured and described in more defined methods using the constituent parts of the view experience.

Methods

Federal land management agencies use several accepted methodologies for managing scenic resources (BLM 1980a and 1980b; Smardon et al. 1988; US Department of Agriculture Forest Service 1995; Gobster 1996). These methods, which involve conducting inventories of scenic resources and evaluating potential changes based on established criteria and objectives, reflect the expert and use-based model of assessing views. The degree to which a renewable-energy facility (or other development) in Oregon's territorial sea has an impact on aesthetic resources depends on a variety of factors, some of which are comparable to those used in the land-based scenic impact assessments. Modeling and adapting these visual subordination standards for projects proposed in the territorial sea will allow the state to "provide time-tested qualitative benchmarks that can be measured using objective methods" (Apostol 2009). The existing models rely on identifying key physical attributes of a place and its views. These attributes are scored based on how they are affected or interacting with other human or naturalized features. The scoring is then combined across groupings of the attributes to provide composite scoring. Additionally, the perspective of the view and its distance are incorporated to ultimately derive a value for the view. These methods were developed for terrestrial applications and also with a focus on Western US landscapes.

Several factors had to be incorporated to adapt the federal landscape model to state waters for use in the Oregon TSP update. One issue was the different attributes and features of a coastal and marine view that are not typically the focus of other federal visual assessment processes. The other main factor was the desire to incorporate public involvement into the assessments. The methodology settled on a participatory field team, followed by expert-driven spatial analysis and scrutiny of the results in a public setting during multiple public meetings. The following section explores each of these factors and how the tool was developed and deployed on the Oregon coast.

Adoption and Adaptation of Established Methods

The C-MAAR application for the TSP update occurs with two discrete processes, the planning phase and the project phase. During the planning phase, work is done to collect baseline visual resource information and adopt the standards that will be applied in any review of a project during the project phase. These processes are described below. The project phase involves applying the planning tool to assess proposals for renewable-energy development. The project phase tiers from the planning level analysis but may also include focused reapplication of the tool to respond to particular proposal details. In 2013, the planning process for the Oregon TSP ended, leading to the project phase.

The TSP update process relied on a heavily involved advisory committee of local citizens, regional interests, and agency representatives. The committee provided support in the development of the C-MAAR methodology. The design of the field protocol, inventory of the sites, and application of the method were all developed with and reviewed by the committee. This approach allowed for the joining of the expert and experiential components of visual resources. Expert representatives provided insight on the technical options and latest research methods. The experiential perspective of the committee contributed the substantive concerns and values on viewsheds. This contribution was essential in identifying the new coastal and marine viewshed features and attributes included in the final field datasheet.

Planning the Inventory

The C-MAAR tool starts with creating an initial inventory of viewpoints in the study area. For Oregon, this included almost 150 viewpoints on public land. The choice to include only public viewpoints was an important one. The citizens and agencies involved made an early decision that views included in the plan must be of a public nature: they must be views from places that all Oregonians and coastal visitors could use and are likely to visit. This choice also assured that access to conduct field surveys would be easy and that others could evaluate the views for themselves as part of the public process. An initial inventory of viewpoints was developed from an inventory from OPRD. This inventory included approximately one hundred potential viewpoints. In addition to the OPRD inventory, an additional 50 potential sites were identified from a review of county and city planning documents, as well as through outreach to local staff and elected officials. These other sites were located in county parks, on federal lands, and on public right-of-way or beach access points.

Conducting the Assessments

C-MAAR employs a field survey instrument to evaluate publicly accessible viewpoints and collect data on their visual quality. The survey examines seven key features to evaluate on a numeric scale from 0 to 5, with one exception. The features are as follows:

- Landform: the steepness of vertical relief and diversity of landform features such as rocks, cliffs, and headlands.
- Vegetation: the mix and diversity of species and communities of vegetation and how they contribute to the scene (e.g., interesting forms, patterns, and so forth).
- Water: additional water features aside from the ocean, meant to capture rivers, lakes, waterfalls, coves, tide pools, unique wave breaks, waves breaking on offshore rocks, and so forth.
- Color: the diversity of colors in multiple natural media, sand, soil, water, and rock.
- Adjacent Scenery: the surroundings that contribute to the view being evaluated. This is terrestrially focused and tends to evaluate the natural or urban setting.
- Scarcity: a relative judgment of how common or scarce the view is as it relates to other views along Oregon's coastline.
- Cultural Modifications: the beneficial and detrimental human activities and developments within the view. The scoring for these features is the only one that allows for a negative score, in cases of anthropogenic impacts to the view.

In the field, these categories are reviewed by the survey team one at a time and then scored. The teams may be composed of state planning staff, local planning staff, and elected officials. In the Oregon case, the recruitment of teams from these groups was intended to balance a need for public participation and assessing views that are representative of the larger community. Teams could have been developed with local residents or visitors, but the sampling or recruitment strategy would have been problematic. This is especially true in communities with high rates of absentee owners, such as communities with many summer homes. By using elected officials and senior planning staff, representation and accountability were built into the review teams. Planners, as staff in these small communities, were accountable to their elected officials, something demonstrated by cases in which elected officials checked in on planners through the review process. Team size also varied; in some communities where there was greater interest in participating, teams grew to eight people. In areas with lower levels of interest and availability, the team size was at least three evaluators.

The first task for the team was to select the place to assess the view from, and delineate the extent of the view. Clues for identifying the point to assess from include developed facilities, such as interpretive signs or constructed viewpoint platforms. Observing where visitors tend to stand is another option. On beaches, standing on the sand below the vegetation line is recommended to capture the full view, although in some cases the tops of dunes or berms may be appropriate. Once the point is selected, the team discusses how far to the left and right to include in the view. Usually, the view is limited to the top of the beach where the terrestrial vegetation begins but sometimes further inland if unique features attract the eye. This creates an arc from the top of the sandy portion of the beach on one side of the viewer to the end of the sandy portion on the other side. Only features inside this view arc are considered in the scoring, with the exception of the features captured in the evaluation category "Adjacent Scenery," which includes the view inland from the viewpoint, the setting in which the viewpoint exists.

The features are discussed and documented in a narrative developed with the review team. A group facilitator or recorder notes the descriptions, features, place-names, and comments from the review team. This provides important documentation for any future questions or reviews. The team then openly discusses how they should score each category. The team works toward a consensus on the values. In the Oregon case, this was almost always successful. Where agreement cannot be reached, two assessments can be recorded and passed on for review in the decision-making process. Each site is totaled to produce a composite scenic quality value across the seven attributes. These values are then used to discuss how the method is capturing the view based on local ideas about the aesthetics of the site. In the application for Oregon, the 150 sites on the coast generated scores ranging from 7 to 30.5. High-scoring sites had a tendency to be higher-elevation locations, in naturalized settings, with steep relief, often on headlands or capes. Because water features were a separate element in the scoring, sites without freshwater streams, waterfalls, or other interesting water features (such as waves crashing on close-to-shore rocks or tide pools) often scored slightly lower, even if the view seemed otherwise iconic. Presence of seasonal or temporal features such as tide pools may vary over the year, so previous experience with, and local knowledge about, the sites by evaluators was important so that those features could be incorporated into the assessment. The field protocol requires that similar season and weather conditions are present throughout the study period. For the application in Oregon, the summer months were used and assessments occurred on sunny days with good visibility. Wind, sea state, and weather conditions are recorded in the assessment documentation in case these conditions need to be reviewed later. Time of day was also noted, and reviewers were asked to consider that as well.

During the fieldwork, participants easily agreed on most of the scores and how to describe and classify what they saw. The primary area of contention was around features that fall under the Cultural Modifications category. These features include all anthropogenic items in the view, such as historic lighthouses, jetties, signage, riprap, and significant automobile use on the beach. The last item raised contention as some participants in the surveys objected strongly to large numbers of vehicles on the beach, while in some cases it is tied to traditional beach uses such as surfing and boat launching for a historic dory fleet. Previous experience with the sites and local knowledge was important because vehicular use varies over time, and only sites that are known to consistently have large numbers of vehicles, such as boat launch areas, typically got points docked for this particular cultural modification.

With all the attributes collected, the results were then summed to assess the view. The assessment used the scenic quality rating scoring categories used in the Bureau of Land Management (BLM) methodology. Views scoring 19 or greater were given an A rating; 12–18, a B rating; and below 12, a C rating. This allows for a comparison in the field, often in terms of how the view compares to others surveyed with the same rating. Additionally, the committee asked for an emotional measure to be included in the field survey instrument. This experiential question (as the first question in the survey) asked field surveyors to evaluate how they felt upon seeing the view. The group used a 10-point scale to evaluate the general "feel of the place." Although this was not included in the data analysis (for scoring purposes), it was meant to be a check on the quantitative model and provides a balance with previous experience that people may have had at the sites. This emotional scale is also compared to the score rating to see if there is an approximate match. The scoring of the individual attributes represented constituents of the experiential view, and the emotional ranking was a second measure to test the validity of these measures. With the site level data collected, the next requirement was to convert these to spatial datasets that could interact with the rest of the planning data.

Field data was brought back to the office and processed for analysis. The datasheet includes the quantitative data on scoring for each attribute. It also includes narrative data on the location and attributes. The field teams document the particular features and elements of the view that justify the scoring for each attribute. Additionally, the general site conditions and special observations are noted on the datasheet. For example, the water feature attribute section often included notes naming features in the view, and the type of wave break or action present. All the data was entered into a database to allow for reporting. The results of the fieldwork are then processed into a summary report for public review and comment. Figure 15.1 presents an example of a report on one of the sites. This document includes a panoramic photo of the view, the scoring data, and narratives on the site. This easy-to-review document facilitated the review of the fieldwork and allowed for public involvement, even after site reviews. The document also facilitated easy comparisons of sites, which built support for the findings.

Ecola Point Viewpoint

SITE NAME		SITE ID:	COUNTY
Ecola Point Viewpoint		OPRD091	Clatsop

DESCRIPTION
Accessed via a short trail from Crescent Beach parking lot at Ecola State Park leading out to Ecola Point. A viewing deck provides expansive (iconic) ocean, shoreline, seastack and Tillamook Rock Lighthouse views.

DATE: 5/7/12	TIME: 15:20	WEATHER: Partly cloudy	WIND: N/A

Panorama Photos

ATTRIBUTE	SCORE	DESCRIPTION
Landform	5	Coastal headlands to the south (Neahkanie Mountain and Cape Falcon), coast range mountains, numerous large seastacks along the beach (many of which are vegetated), including Haystack Rock, various small to mid-sized offshore rocks (e.g., Sea Lion Rock) at Ecola Point, spires, caves, sandy and cobble beaches, coastal bluffs, steep vertical cliffs, coastal lowlands, sand spit (Ecola Cr.). The expansive ocean view to the north includes Tillamook Rock offshore, various stages of landslides, Indian Beach, Tillamook Head, coastal bluffs with a mix of sandstone and basaltic materials.
Vegetation	5	Coastal scrub shrub in the foreground with salal, wild cucumber, native roses, and elderberry along coastal bluff grasslands and sitka spruce forest, distant forests prevalent on headlands. Bull kelp and other intertidal and subtidal seaweeds visible at Ecola Point. Trees clinging to rocks add visual variety in the foreground.
Water	5	Waves crashing on offshore rocks, especially at Ecola Point in the foreground, coves in both directions (Crescent and Indian Beach), mouth of two creeks.
Color	5	Blue ocean with lots of variety due to reef and rocky shore/kelp beds punctuated by dark colored offshore rocks and white surf. Bright green of coastal shrubs and grasslands, dark green forests all contrast with tan sandy beach. Rust colored bluffs (some from landslides) provide interest, pink flowers in vetch and rose provide seasonal interest.
Adjacent	3	Park setting including trail, lookout, boardwalk, sitka spruce forest, grassy bluffs, day use parking lot.
Scarcity	5	Iconic views to the north and south, including Haystack Rock and Nehakanie Mountain/Cape Falcon, Ecola Point, Tillamook Head and Tillamook Rock/Lighthouse.
Cultural	1	Cannon Beach tucked into shoreline in distance, parking lot at Indian Beach, signs and rusting fence in immediate foreground (below the edge of the view), all provide for small detractions from the view. However, Tillamook Head Lighthouse in the distance provides interest due to its uniqueness.
Total	**29**	

Figure 15.1. **Example of a data report from a field survey.** Photo by Laurel Hillmann, Oregon Parks and Recreation Department.

The quantitative data provided values for the points, and the policy guidance required that this information be transformed into spatially explicit data to inform decisions made about particular spaces on the ocean surface. After conducting the fieldwork, the values from the data collection are then used to produce spatial outputs that can interact with other planning data. The C-MAAR tool used ArcGIS 10 software to develop viewshed maps based on the points surveyed and assign the various scoring rules to the polygons. There are three basic steps to this analysis: first is the generation of visible areas, then the application of distance zones, and finally adding the values from the field data.

The viewshed analysis tool uses a model of the earth's terrain to measure the distance from a point that is visible based on the curvature of the earth, physical obstacles, and light refraction. The elevation model had a large role in determining the areas that might be visible. The topography of the coast and shoreline were modeled in a digital elevation model (DEM) acquired from federal data providers. For much of the Oregon coast, the National Elevation Dataset provides 10 m horizontal resolution models. The viewpoints, as measured with GPS in the field, were captured with a different level of accuracy. Therefore, a very detailed model could add uncertainty to the model, creating visual artifacts in viewshed analyses. The points could fall in between the DEM's modeled world, creating artificial shadows or obstacles to views. Using too coarse a model would oversimplify the world, making views overly expansive and inclusive of areas. This was managed in the GIS by using a random distribution of "viewers" (computer-generated points) to simulate vantage points.

Viewpoints were buffered by 50 m, and then 40 new points were randomly distributed within this area. This method was developed for C-MAAR, and the specifications were based on iteratively testing different buffers and point numbers with the stakeholder committee. The viewsheds calculated by the software created 40 overlapping view areas, shown in the map area with a varying intensity of blue (figure 15.2). The more intense the color, the greater the number of overlapping views. Darker colors in the figure show areas in which more views overlap, and thus with a higher certainty of the actual viewshed being covered.

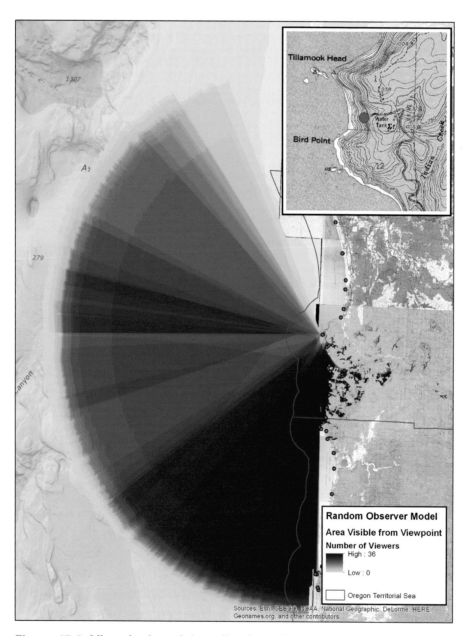

Figure 15.2. **Viewshed model application with random viewpoints and density of views measured as seen from Ecola State Park on the northern Oregon coast. The inset map shows the location of the viewpoint within the state park.** Data sources: Esri, GEBCO, NOAA, National Geographic, DeLorme, HERE, Geonames.org, and other contributors.

For each of the 150 viewpoints, this process was applied, creating more complex and overlapping representations of the possible views on the Oregon coast. These overlapping arcs were then attached to the field-based score and ranked in scenic quality ratings, of *A*, *B*, and *C*, based on the BLM rating system. Multiple views overlap, each with a different score. The scoring was developed so that it links to a design review standard for any future energy proposals. The higher the score, the more visually subordinate the development must be. The highest scores may exclude development unless it can be shown not to attract attention. In future energy design reviews, proposals will be evaluated from each of the overlapping ratings and viewpoints. The varying levels of quality are also broken down by distance from the viewer to capture intensity of potential impacts from any development. The BLM method identifies the fore-, middle (mid), and backgrounds and an additional seldom seen area (BLM 1980a and 1980b). These distances were reviewed with the committee and adapted based on stakeholder input on visible objects at various distances. The committee decided that the fore- and midground distances were visually difficult to separate in the ocean context. These two distances were collapsed into one. The scenic quality ratings from the field are combined with these distance ranges to create several classes of views based on their importance and sensitivity.

The field data is compiled in a point feature that contains the attributes from the surveys. Figure 15.3 shows the location of the inventoried viewpoints, as well as the initial values collected in the field for each view before the viewshed analysis. The model iteratively moves from row to row and runs the model for each viewpoint. Each raster viewshed output is stored and named using the point feature class data. The final output of the model run is a viewshed for each viewpoint and a raster with the viewshed value and distance zone. Processing each of the viewshed areas for the 150 survey points requires processing 40 individual viewshed models for each of the "random" observers placed at each site visited. These 40 random viewshed results can be combined as a weighted measure of density based on the number of overlapping views, or the maximum extent of views can be used. In the Oregon application, the maximum extent was used as a more conservative measure of possible visual impact. This was a decision by the stakeholder committee and is a policy preference to take a more precautionary approach. With this new composite viewshed area, the model applies rules for scores based on distance zones. Initial testing with the review committee revealed that the many steps involved required the GIS analyst to be heavily involved in processing the views. To resolve this issue, C-MAAR was developed in the ArcGIS ModelBuilder application to automate the processing (see "Supplemental Resources" at the end of this chapter).

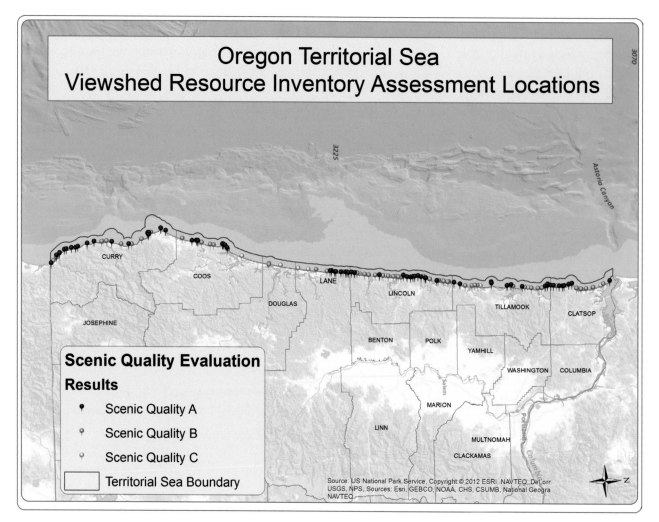

Figure 15.3. Viewpoint inventory and initial scenic quality rating values from field surveys. Source: US National Park Service, Copyright: © 2012 Esri, NAVTEQ, DeLorme, USGS, NPS; data sources: Esri, GEBCO, NOAA, CHS, CSUMB, National Geographic, and NAVTEQ.

As the method was developed with the advisory committee, implications of choices were tested against potential future scenarios. For example, if future potential development areas looked like they were going to be too easily visible from high scoring sites, future development could be limited in those places. The committee discussed the implications of the standards and strove to balance development and view conservation priorities. As the viewshed maps were generated and the field data values assigned to them, a final step the committee dealt with was assembling multiple maps into a single composite coastwide dataset. This was necessary to have the data fit with the other data types but required flattening the complexity of the multiple sites into a single representation, disconnecting the coastwide data from the reality it was meant to capture. Some participants in the planning process raised concerns that this representation seemed to restrict development almost entirely. Developing online access to the visual resource data has helped allow for interactive queries of the data, but the representation of the results is almost as important as the results themselves.

Another policy choice made in the development of the tool was the designation of a special class of views that have exceptionally high visual values. These views were predominantly from iconic vantage points. For most views, the methodology developed provided for decreasing visual resource classes as the distance from the viewpoint increases. For these "special areas," it was determined that the distance from the view should not affect the resource class, creating a large area of a high-value visual resource class on the ocean. In part, this was because the committee came to recognize that Oregon's territorial sea did not already have the types of management situations described in the BLM policy manual to qualify as special areas. Also, this was a compromise that resulted from the committee's modification of the BLM Visual Resource Class table. A cutoff score rating of 24 was chosen by the committee, and vetted in multiple public meetings, as an appropriate break for sites that should fall into this viewpoint category. As discussed in the next section, these areas represented the most protected class of visual resources.

Translating Models to Support Policy

Once the Visual Resource Inventory Assessments were conducted and the spatial data generated to extend the policy out to sea, the results had to be translated into proscriptions on uses and impacts. The committee used the C-MAAR tool to craft a series of visual resource classes that connected the scenic quality ratings to the level of development that would likely be allowed once proposals and visualizations occurred down the line. This follows a similar logic to the terrestrial visual resource tools developed by BLM, but the feature classes were adapted to recognize how development on the ocean is more likely to attract attention because of the lack of physical variation at sea (and inability to modify design to minimize impact because of safety-at-sea considerations).

The scores from the field are translated into four visual resource classes: I–IV as shown in figure 15.4. Class I areas are those with the highest value and are either in the foreground/ midground of high scenic quality viewpoints or within the viewshed of a special area viewpoint. The objective of class I areas is to preserve the existing seascape character, and although development is not precluded, it is not allowed to attract attention. Class II areas are expected to allow limited development as long as the existing character of the view is maintained and the new development attracts only minimal attention. Class III allows for more development, and it may attract the attention of viewers. However, the development may not dominate the view. The final class, class IV, allows development, and it may be the dominant feature, although mitigation is expected to manage the impacts.

Visual Resource Viewshed Class Derivation (I-IV)

Special Areas		I	I	I
Scenic Quality	A	I	II	II
	B	II	III	III
				IV
	C	III	IV	IV
		Fore/Midground (0 – 5 miles)	background (5-15 mi)	Seldom Seen (15 mi and beyond)
		Distance zones		

Figure 15.4. **Visual resource scenic class structure and scoring.**

A final step in the C-MAAR process is presenting the results. Cartographically, the outputs can become difficult to represent. There are three values per viewpoint, and thus in the Oregon case, over 450 unique view and value combinations are possible. The results then must be carefully presented, based on the needs of the audience for the product. Because the highest-value view data is the most important, the data was simplified into the maximum value and converted to polygons for cartographic symbolization. In Oregon, it was important to share the data products with the review committee and discuss sensitivity to concerns, such as the inherent meaning of color values (e.g., green meaning "ready for development"). The decision was made to present the visual resource overlaid in shades of gray, with hatched polygons overlaid for special areas. This best represented the visual resource class values while helping to keep interpretation of their meaning neutral from a policy standpoint.

Results

High visual resource classes tended to correspond with views from headlands and capes along the Oregon coast. These viewpoints are higher in elevation, increasing the distance the viewsheds extend out to sea. This resulted in the highest-value views also having the largest area on the ocean categorized in high-priority areas for viewshed protection. The distribution of these high-value areas was focused on the north and south coasts. Larger headlands and capes with miles of sandy beach separating them dominate the north coast. The headlands tend to be protected parklands and have relatively well-developed access to the high points. The sandy beach areas tend to be locations of coastal communities. In the transition from the north to central coast, larger bays and estuaries emerge as a new feature. These bays and rivers contribute to the visual value, although increased human impacts along these features are also found inland along the waterways with lower-intensity urbanization. The geological pattern changes on the south central coast from the prominent headlands at Cape Perpetua, followed by a large area dominated by dune fields. Here, the visual scores also decreased with the more simplified natural landscape, and the elevations decreased reducing the extent of views out to sea. This dunal area then transitions again close to Cape Arago, where the coastal mountains and coastline become closer and the larger rocky headlands reemerge as the dominant feature. In this southern portion of the coast, the coastline is primarily rocky with steep headlands and offshore rocks. This again increases the scenic quality ratings, although developed access to high points is not as common so the extent of views out to sea is not as great as on the north coast.

Figure 15.5 presents the composite map showing the total extent of views cataloged in the C-MAAR application. These are the viewshed values from the field before applying the different distance zones. With the scenic quality ratings applied, this map provides a weighted density of the views, as the higher-scoring views increase the value of the area for mapping display purposes. Multiple overlapping, highly rated views create the darkest hues in the figure. Figure 15.6 is a bar chart presentation of the distribution of scenic quality ratings by viewpoint and depicts the underlying scenic quality factors that contributed to the overall rating. The final map output from the model is the composite scoring map with distance zones applied, and the maximum value selected for any given area. Figure 15.7 is the adopted policy map for the TSP update and will be the starting point for future permit reviews when proposals are evaluated for potential visual impacts within areas of state jurisdiction.

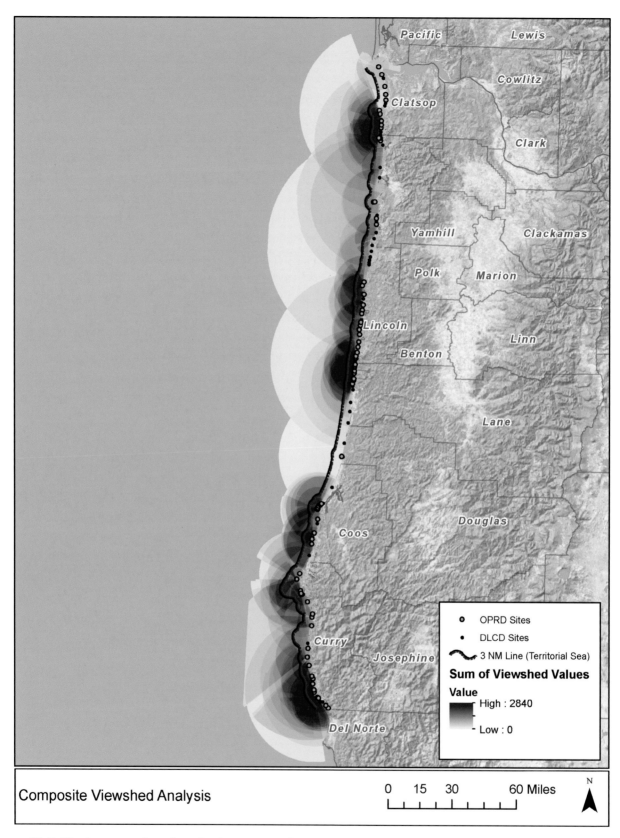

Composite Viewshed Analysis

0   15   30   60 Miles

N

Figure 15.5. **Final composite viewshed scores and extents, with the value depicted as the total score of all overlapping assessments, which illustrates the density and value of the views along the coast.**

Data sources: Esri, GEBCO, NOAA, National Geographic, DeLorme, HERE, Geonames.org, and other contributors.

Territorial Sea Scenic Quality Evaluation Scores*

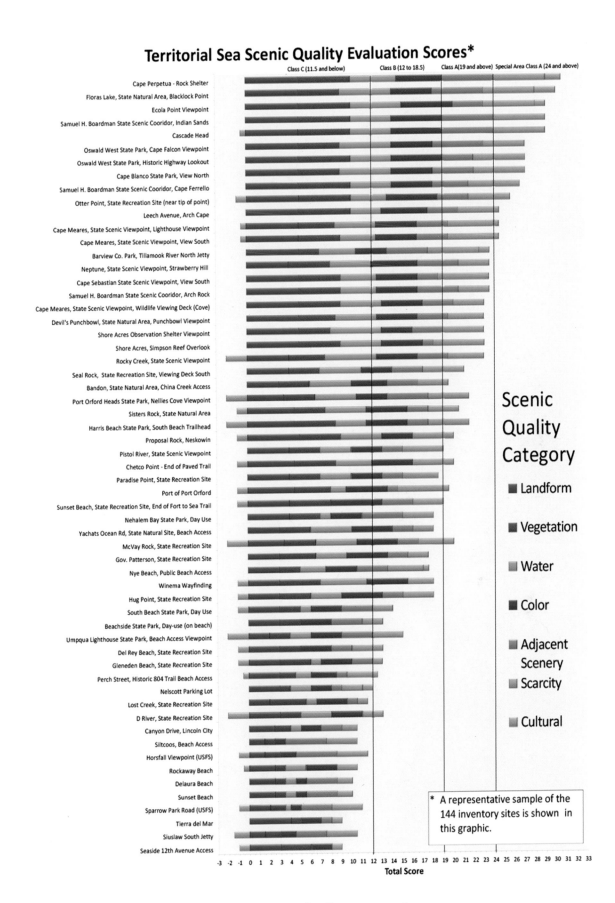

Figure 15.6. Results from the C-MAAR application on the Oregon coast.

Oregon Territorial Sea Plan Visual Resource Class Map

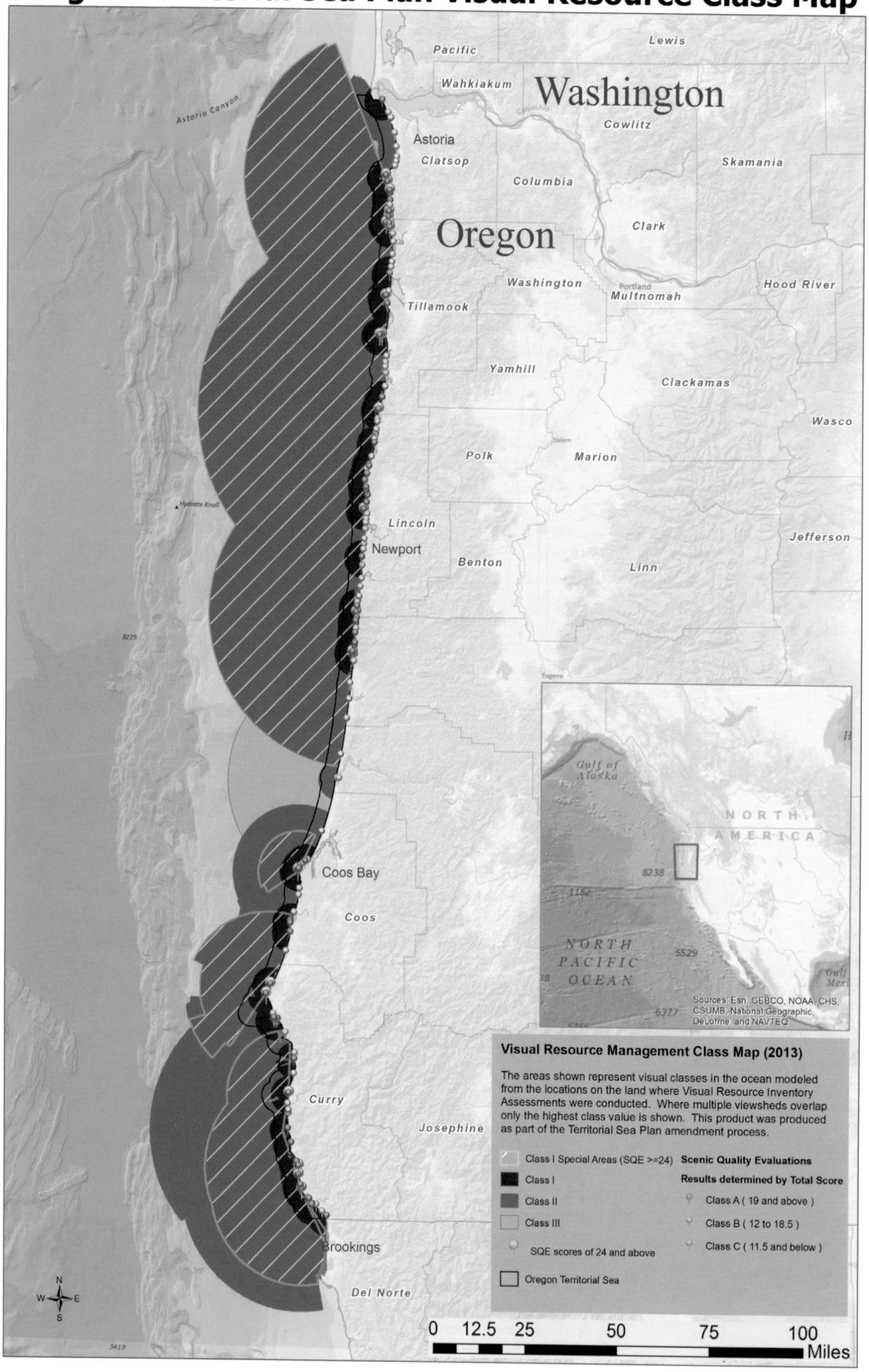

Figure 15.7. **Final visual resource management class map produced by C-MAAR.** Data sources: Esri, GEBCO, NOAA, National Geographic, DeLorme, HERE, Geonames.org, and other contributors. AND Inset map data sources: Esri, GEBCO, NOAA, CHS, CSUMB, National Geographic, DeLorme, and NAVTEQ.

The final viewshed map is the composite of the overlapping views, with the maximum class for any given area displayed. This resulted in a very large area of high-value classes. The prominent headlands and capes along the northern half of the coast all scored highly, and some were designated as special areas. Similar to the south, the steep headlands and cliffs produced large viewshed areas with high values. Upon initial review, this appears to create a uniform application of the highest-level view classification within Oregon's territorial sea. The committee charged with developing this model set policies that tended toward a more expansive definition of important views. This reflects a key value for the Oregon coast.

Several areas stand out as potential opportunities for marine renewable-energy development from a viewshed perspective. In the north, near the community of Astoria, a large section of territorial sea is in a class B ranking, permitting development as long as it does not dominate the view and only minimally attracts attention. This area is also a focus area for potential development because of its proximity to a large port, land-based infrastructure, and a potential power purchaser. Farther to the south, two more class B and one class C areas emerged from the analysis. These areas also meet other requirements for development of marine renewable energy. The viewshed classes are the starting point for reviewing permit applications. Applicants for marine renewable-energy projects will be required to complete visual impact analyses, including visual simulations of their intended development, and contrast evaluations from key viewing areas. This analysis will be reviewed by a joint agency review team (JART) and compared relative to the class standards included in TSP. If the applicants are able to demonstrate that their technology meets the intent of the visual resource standards for potentially affected viewpoints, the development will be allowed to proceed, at least on visual resource grounds. For example, a development is possible in the viewshed of a special area view as long as the applicant can show it will not change the view or attract attention, an easier task for distant developments offshore.

Since many types of technology are currently being developed, and the project reviews will be both technology and location specific, this model and associated standards allows the flexibility to pursue technology development in a precautionary fashion. The data from the C-MAAR application supported many of the initial concerns about the location of high-value visual resources. This was an important validation of the tool's use. The tool also identified areas for development that allow for a continued exploration of opportunities for marine renewable energy in Oregon.

Discussion and Conclusion

The C-MAAR application in Oregon presented a new opportunity to engage the public and stakeholders in a challenging resource evaluation. Visual resources are often defined complexly across communities, and coming to an agreed-on evaluation requires broad engagement and transparent tools. The Oregon experience suggests that there is a path forward, using an expert-based modeling system that engages a participatory field survey and codevelopment of methodology. In this final section of the chapter, we explore the implications of this tool for marine renewable-energy developers, the strengths and weaknesses of the model, and an assessment of the model's ability to face the challenges we identified in the literature on visual resources and decision-making.

The Oregon TSP update has resulted in a set of policies captured in spatial data for the use and management of the territorial sea. Moving forward, future energy developers will make their permit applications to the State of Oregon, and the data produced in the update will be part of the review. The visual resource inventory from the C-MAAR application is a starting point for developers. The permit process will ask for more detailed visual simulations based on the specifics of the devices proposed for use in the water. In this way, the model results from C-MAAR are general for steering interest and policy preferences. However, permitting may occur in areas with higher-value classes if the impacts can be demonstrated to meet the standards and objectives outlined in TSP.

One challenge in a standardized visual resource survey method is that the scoring criteria can focus values on particular areas based on the mix of attributes present. The C-MAAR application in Oregon focused on views that had complex colors, vegetation, water features, and steep relief or unique landforms. This resulted in high-value areas being the dramatic cliffs, headlands, and capes on the coast. Some areas along the Oregon coast, notably the Clatsop Plains to the north and the extensive sand dune country along the south central coast, tended to score lower because of the dominance of sand and beach grass views. These areas also tended to be lower in profile and have less diversity in colors, textures, and vegetation. They are beautiful places in their own right. Some public comments noted that these areas were uniquely special because of the "sea of beach grass" that stretches almost to the horizon. However, the committee and other reviewers felt that when the various components of a view were examined, and when results were compared, the more complex sites were truly thought of as more valuable by visitors and residents. However, this is a persistent problem in valuing natural resources in any setting. Experiences and values concerning the environment are difficult to quantify, and some trade-offs among sites are a necessary outcome of ranking views.

Similarly, field teams encountered challenges over agreeing on the impact of cultural resources and anthropogenic impacts to the view. In some cases, human activities contribute greatly to the view. Examples of this are the iconic lighthouses of the Oregon coast. These features are now just as defining of the Oregon coast view as natural components. Other human activities were more challenging. One portion of the coast has a unique dory fishing fleet that is launched using pickup trucks to back into the surf and release vessels from their trailers. This requires automobile traffic on the beach and includes other vehicles aside from the fishing fleet. Surfers, day-trip visitors, and others can drive on the beach and park vehicles. To some stakeholders and reviewers, this was a great detriment to the view. To other stakeholders, notably local residents, the vehicles pulling fishing vessels on the beach helped define the history of the area and did not detract from the view at all. They did not want to see their traditional uses as the basis for lowering the site's score. This was a challenge that was not easily solved. The views from this area scored well, but not at the very top of the scale. At the end of the process, some discomfort remained although these areas received the highest visual class ranking from overlapping views from adjacent viewpoints. This challenge highlights that these quantification exercises never fully resolve the politics over views, nor should they. The visual resource inventory was not an effort to "quantify out the politics," but rather an effort to craft a common language for the politics of views and create opportunities for the politics to be explored (Sarewitz 2004).

A challenge for this process was making a distinction with participants between what the model presents as visible versus what might actually be perceived by a viewer in any future development. Analyzing visibility differs from perception. Although much of the territorial sea is visible from shore, a question remains as to what viewers may or may not perceive. One part of this challenge is the variable ocean and sky conditions common for coastal areas, especially places like Oregon.

Sea state, fog, and precipitation can all reduce what is expected to be visible on the water at any given time. This model makes a conservative choice to use the best of sea and sky conditions as the baseline. From a policy perspective, this is justified as these types of conditions are also some of the most important to sightseers, tourists, and local residents. Another perception component is related to the viewer. We anticipate that local residents are far more likely to note changes on the ocean, whereas visitors are less likely to notice devices such as wave energy floats. If visitors note them, they may not know what they are without some interpretation or education. Further, there is also a distinct possibility that some viewers will find a new value in seeing the devices, just as marine traffic and navigational aids are desirable to some viewers.

Environmental controversies and the communities they impact are often framed as problems of missing science. In these situations, the response can be to turn to expert-driven solutions, but these can oversimplify problems and not improve the decision options. Local communities are often already actively producing their own information on options (Wynne 1992). Understanding this, communities are then important partners in developing a better understanding of the natural and social systems that policy choices impact. Measuring and valuing the natural systems society depends on calls for expanding expert-based systems to include stakeholder experiential values in research and practice (Daniel et al. 2012). The National Science Foundation (NSF) Advisory Committee for Environmental Research and Education claims that "place-based science" is at the heart of understanding "complex environmental systems, particularly in the 21st century" (Advisory Committee for Environmental Research and Education 2003, 63). The Oregon story affirms these findings. Striking a balance between the expert-driven model of assessing visual resources and the experiential model of participatory assessments was a goal of the Oregon C-MAAR application. This tool also wrestles with a tension between the academic (or research) interest to develop more certain or validated methods and the policy imperative to reach a decision, a tension that is apparent throughout this process. The model provides a quick and robust method to assess views while at the same time allowing for the coproduction of this data with communities that rely on these views.

The success of the application of C-MAAR is based on two observations from the process. The first is the experience in the field with the evaluation teams. The mix of local representatives and state planning staff allowed for evaluations to reflect on the importance of the site and its context in the state. The teams tended to converge on values for the sites and with only limited disagreement on the scoring. Only one site had a dispute over the final value, and ultimately it was a site with many disputes. The other basis for evaluating success was in sharing the findings with representatives and the public after fieldwork. In these settings, the process allowed for different elements of views to be understood in a dialog. As noted in the introduction, this effort relied on "hybrid forums," with a mix of actors, expertise, and concerns in the stakeholder committee (Callon et al. 2009). These forums are seen as opportunities to move from oppositional politics of for and against, and instead into an embrace of controversy where learning and exploration can occur. We believe that by coproducing the visual resource method, this was achieved and allowed for new understandings of the future of the Oregon territorial sea to be developed.

There are limits to this assessment procedure. It is worth noting that as a participatory approach, this tool does not claim to solicit feedback that is generalizable to a population. Surveys or assessments often sample a human population to solicit views that can then be accepted as representing the perspectives of that community. For this tool, we chose to engage actors with a representative role. This decision was made for two reasons: practicality and the policy setting that the assessments are a part of. Even with participatory GIS, there can be concerns that the expert role or technology limits true participation (Lejano 2008; Wright et al. 2009). A recognized barrier to the success of participatory approaches has been the institutional and resource constraints that limit complete participation in GIS projects (Brown 2012). Gaining complete participation by the desired representatives was also a challenge as some did not have the time or resources to participate. This may limit the success of assessments. Nevertheless, the application of C-MAAR in Oregon relied on reporting detailed results to commissions and other officials to allow for full-scale review to overcome this challenge. The Oregon planning experience has been highly collaborative, both by design and necessity. Oregon's land-use law that governs decision-making for the TSP update focuses on participatory approaches, and these have become a cultural norm in Oregon planning processes. Additionally, the TSP update was limited in funding, requiring partnerships with nongovernmental organizations, communities, and user groups. This resulted in a networked model of public process with many parties contributing to policy development.

This project extended many of the formal spatial tools used in CMSP into the social- and public-value arena by inventorying a key component of how most visitors and residents experience the Oregon coast. Views along the coastline are something that define the Oregon coast, supporting a thriving tourism industry, underpinning a deep sense of place, and building on a legacy of parks and recreation that is a strong part of the entire state's identity. These tools come with their own limitations and challenges. However, the C-MAAR experience in Oregon suggests that expert modeling tools can be paired with participatory approaches to coproduce data and findings that support durable policy agreements over the management of our coastal and marine resources.

Acknowledgments

We wish to acknowledge the stakeholder advisory committee members that were a part of the TSP update for their contributions in developing the method and participating in the field. Additionally, we thank the local city and county staff and representatives that joined field crews for assessments, as well as the staff from OPRD for their assistance in assessing all the sites. The comments and suggestions of two anonymous reviewers greatly improved this chapter.

References

Advisory Committee for Environmental Research and Education. 2003. *Complex Environmental Systems: Synthesis for Earth, Life, and Society in the 21st Century*. Washington, DC: National Science Foundation. http://www.nsf.gov/geo/ere/ereweb/ac-ere/acere_synthesis_rpt_full.pdf.

Apostol, D. 2009. *Prefiled Direct Testimony of Witness #21 Dean Apostol on Behalf of Intervenors Friends of the Columbia Gorge and Save Our Scenic Area*. Olympia, WA: Energy Facility Site Evaluation Council Testimony Document, Exhibit No. 21.01.

Basso, K. H. 1996. *Wisdom Sits in Places: Landscape and Language Among the Western Apache*. Albuquerque, NM: University of New Mexico Press.

BLM (Bureau of Land Management). 1980a. *Visual Contrast Rating*. Phoenix, AZ: Bureau of Land Management Handbook H-8431. http://on.doi.gov/1m4vEc1.

———. 1980b. *Visual Resource Inventory*. Phoenix, AZ: Bureau of Land Management Handbook H-8410-1. http://on.doi.gov/1m4vEc1.

Bonzon, K., R. Fujita, and P. Black. 2005. "Using GIS to Elicit and Apply Local Knowledge to Ocean Conservation." In *Place Matters: Geospatial Tools for Marine Science, Conservation, and Management in the Pacific Northwest*, edited by D. J. Wright and A. Scholz, 206–25. Corvallis, OR: Oregon State University Press.

Brown, G. 2012. "Public Participation GIS (PPGIS) for Regional and Environmental Planning: Reflections on a Decade of Empirical Research." *Journal of the Urban & Regional Information Systems Association* 24 (2): 7–18.

Callon, M., P. Lascoumes, and Y. Barthe. 2009. *Acting in an Uncertain World: An Essay on Technical Democracy*. Cambridge, MA: MIT Press.

CH2MHill. 1997. *Pacific Coast Scenic Byway Corridor Management Plan for US 101 in Oregon*. Portland, OR: Coastal Policy Advisory Committee on Transportation and the Oregon Department of Transportation.

Cheng, A. S., and S. E. Daniels. 2003. "Examining the Interaction between Geographic Scale and Ways of Knowing in Ecosystem Management: A Case Study of Place-Based Collaborative Planning." *Forest Science* 49 (6): 841–54.

Clark, J. K., and T. V. Stein. 2003. "Incorporating the Natural Landscape within an Assessment of Community Attachment." *Forest Science* 49 (6): 867–76.

Dakin, S. 2003. "There's More to Landscape than Meets the Eye: Towards Inclusive Landscape Assessment in Resource and Environmental Management." *Canadian Geographer/Le Géographe Canadien* 47 (2): 185–200.

Daniel, T. C. 2001. "Whither Scenic Beauty? Visual Landscape Quality Assessment in the 21st century." *Landscape and Urban Planning* 54 (1): 267–81.

Daniel, T. C., A. Muhar, A. Arnberger, O. Aznar, J. W. Boyd, K. M. A. Chan, R. Costanza, T. Elmqvist, C. G. Flint, P. H. Gobster, A. Grêt-Regamey, R. Lave, S. Muhar, M. Penker, R. G. Ribe, T. Schauppenlehner, T. Sikor, I. Soloviy, M. Spierenburg, K. Taczanowska, J. Tam, and A. von der Dunk. 2012. "Contributions of Cultural Services to the Ecosystem Services Agenda." In *Proceedings of the National Academy of Sciences* 109 (23): 8812–19.

De Groot, R. S., R. Alkemade, L. Braat, L. Hein, and L. Willemen. 2010. "Challenges in Integrating the Concept of Ecosystem Services and Values in Landscape Planning, Management and Decision Making." *Ecological Complexity* 7 (3): 260–72.

Duncan, S. 2006. "Mapping Whose Reality? Geographic Information Systems (GIS) and 'Wild Science.'" *Public Understanding of Science* 15 (4): 411–34.

Ehler, C., and F. Douvere. 2009. *Marine Spatial Planning: A Step-by-Step Approach toward Ecosystem-Based Management*. Paris: UNESCO Intergovernmental Oceanographic Commission Manual and Guide No. 53, ICAM Dossier No. 6.

Elwood, S. 2006. "Negotiating Knowledge Production: The Everyday Inclusions, Exclusions, and Contradictions of Participatory GIS Research." *The Professional Geographer* 58 (2): 197–208.

Frank, R. H. 2000. "Why Is Cost-Benefit Analysis So Controversial?" *The Journal of Legal Studies* 29 (2): 913–30.

Gobster, P. H. 1996. *Forest Aesthetics, Biodiversity, and the Perceived Appropriateness of Ecosystem Management Practices*. Washington, DC: United States Department of Agriculture Forest Service General Technical Report PNW 369:77–98.

Gross, M. 2010. *Ignorance and Surprise: Science, Society, and Ecological Design*. Cambridge, MA: MIT Press.

Hacking, I. 2000. *The Social Construction of What?* Cambridge, MA: Harvard University Press.

Kemmis, D. 1992. *Community and the Politics of Place*. Norman, OK: University of Oklahoma Press.

LaFranchi, C., and C. Daugherty. 2011. *Non-consumptive Ocean Recreation in Oregon: Human Uses, Economic Impacts and Spatial Data*. Santa Cruz, CA: Natural Equity, Surfrider Foundation, and Ecotrust Joint Technical Report.

Lejano, R. P. 2008. "Technology and institutions: A Critical Appraisal of GIS in the Planning Domain." *Science, Technology & Human Values* 33 (5): 653–78.

Lutzenhiser, M., and N. R. Netusil. 2001. "The Effect of Open Spaces on a Home's Sale Price." *Contemporary Economic Policy* 19 (3): 291–98.

Millennium Ecosystem Assessment. 2003. *Ecosystems and Human Well-Being: A Framework for Assessment.* Washington, DC: Island Press.

Norton, B. G. 2005. *Sustainability: A Philosophy of Adaptive Ecosystem Management.* Chicago, IL: University Of Chicago Press.

OPRD (Oregon Parks and Recreation Department). 2005. *Ocean Shore Management Plan.* Salem, OR: Oregon Parks and Recreation Department Technical Report. http://1.usa.gov/1vLApaZ.

Relph, E. 1976. *Place and Placelessness.* London: Pion.

Sabatier, P. A., and H. C. Jenkins-Smith. 1999. "The Advocacy Coalition Framework: An Assessment." In *Theories of the Policy Process*, edited by P. A. Sabatier, 117–66. Boulder, CO: Westview Press.

Sagoff, M. 2011. "The Quantification and Valuation of Ecosystem Services." *Ecological Economics* 70 (3): 497–502.

Sarewitz, D. 2004. "How Science Makes Environmental Controversies Worse." *Environmental Science & Policy* 7 (5): 385–403.

Shamai, S. 1991. "Sense of Place: An Empirical Measurement." *Geoforum* 22 (3): 347–58.

Shelby, B., and J. Tokarczyk. 2002. *Oregon Shore Recreational Use Study.* Salem, OR: Oregon Parks and Recreation Department Technical Report.

Smardon, R. C., J. F. Palmer, A. Knopf, K. Grinde, J. E. Henderson, and L. D. Peyman-Dove. 1988. *Visual Resources Assessment Procedure for US Army Corps of Engineers.* Vicksburg, MS: US Army Corps of Engineers Environmental Laboratory Instruction Report EL-88-1. http://bit.ly/1oxqbtW.

Stedman, R. 2003. "Is It Really Just a Social Construction? The Contribution of the Physical Environment to Sense of Place." *Society & Natural Resources* 16 (8): 671–85.

Steinback, C., S. Kruse, C. Chen, J. Bonkoski, T. Hesselgrave, N. Lyman, and E. Backus. 2010. *Supporting the Oregon Territorial Sea Plan Revision: Oregon Fishing Community Mapping Project.* Portland, OR: Ecotrust Technical Report.

Stone, D. 2001. *Policy Paradox: The Art of Political Decision Making*, rev. ed. New York: W. W. Norton.

Tuan, Y. 1977. *Space and Place: The Perspective of Experience.* Minneapolis: University of Minnesota Press.

US Department of Agriculture Forest Service. 1995. *Landscape Aesthetics: A Handbook for Scenery Management.* Washington, DC: US Department of Agriculture Forest Service Agriculture Handbook No. 701. http://1.usa.gov/1mVCQCE.

Williams, D. R., and S. I. Stewart. 1998. "Sense of Place: An Elusive Concept That Is Finding a Home in Ecosystem Management." *Journal of Forestry* 96 (5): 18–23.

Wright, D. J., S. L. Duncan, and D. H. Lach. 2009. Social Power and GIS Technology: A Review and Assessment of Approaches for Natural Resource Management." *Annals of the Association of American Geographers* 99 (2): 254–72.

Wynne, B. 1992. "Misunderstood Misunderstanding: Social Identities and Public Uptake of Science." Public Understanding of Science 1 (3): 281–304.

Supplemental Resources

Dawn J. Wright, ed.; 2015; *Ocean Solutions, Earth Solutions***; http://dx.doi.org/10.17128/9781589483651_d**

For the digital content for this chapter, explained in this section, go to the Esri Press "Book Resources" webpage at esripress.esri.com/bookresources. Then, in the list of Esri Press books, click

Ocean Solutions, Earth Solutions. On the *Ocean Solutions, Earth Solutions* resource page, click the chapter 15 link to access that webpage and the links to the digital content.

Territorial Sea Evaluation Tools

The book resource webpage for chapter 15 provides a link to download the following toolbox:

- Chapter15Tools.tbx: the ArcGIS toolbox developed for visual resource assessments, DLCD Viewshed Model

C-MAAR uses ModelBuilder to automate the steps required to translate field assessment results into spatial data. For the user, the model provides an interface to enter assumptions about the height of the viewer and the energy device at sea. The user is also asked to identify the point feature class to run the C-MAAR analysis on. The model steps through the viewshed method for each point in the assessment dataset.

The first step for the model is to create random points within a buffered distance of the viewpoint. This is to manage uncertainty in elevation data. The model then runs the ArcGIS Viewshed tool for each of these random points and assembles them into a single viewshed extent raster.

With the viewshed analysis completed, two map algebra steps are taken to apply the field data values and distance class values. The viewsheds come into this analysis with a basic value of 1. The field data results are multiplied against the viewshed. This results in each of the viewshed areas having the total score from the field assessment. The next step is to transform these values so the different distance classes are also in the score. The distance classes are foreground, midground, and background. Buffers around the viewpoint capture these distances and are converted into raster files. The raster values are then reclassified using 100, 10, and 1 for the three classes, where 100 is the foreground; 10, the midground; and 1, the background. These values are then multiplied by the viewshed values. Thus, a 27 value for the viewshed results in a 2700 value for the foreground, 270 for the midground, and 27 for the background. The logic of this system is to allow for analysis of overlapping values.

The final output is a viewshed for each point in the inventory, valued with the field score and distance classes applied. These are then ready to be assembled into various composite analyses based on policy rules for the planning process.

The book resource webpage for chapter 15 provides a link to view the following chart:

- ChartData.pdf: Territorial Sea Scenic Quality Evaluation Scores

This chart presents the assessment scores from the field for each site included in the Visual Resource Inventory Assessment (VRIA). For each site, the constituent attributes included in the total value are displayed in a stacked horizontal bar. The constituent scores in the stacked bar are (from left to right): Landform, Vegetation, Water, Color, Adjacent Scenery, Scarcity, and Cultural. The score threshold for classes A, B, and C is shown across the top of the chart.

Additional Oregon Territorial Sea Resources

URLs and QR codes are provided here for the following resources. Links are also available on the book resource webpage for chapter 15.

- Esri Story Map app: Oregon Territorial Sea Plan – Visual Resource Scenic Quality Class Map, at http://oregonocean.info/AGO_shortlist_StoryMap/index.html

This interactive map provides the locations of each of the locations assessed in VRIA. Each point is also a link to a pop-up window that includes a brief description of the site, a panoramic photo taken during the assessment, the final score, viewshed evaluation category, and a unique identifier code for the site.

- ArcGIS Online Map: *Visual Resource Management of the Oregon Territorial Sea*, at http://geo.maps.arcgis.com/home/item.html?id=bd285e90714544d5882a5a78dbf179ad

This online map displays the spatial outputs of VRIA. It is the composite viewshed map for use in managing the ocean off the Oregon coast. It includes the different distance-based viewshed scores, showing the extent of various views, how overlapping values are reconciled, and the final Viewshed Scenic Class Values. Map courtesy of the Oregon DLCD OPRD.

CHAPTER 16

Near Real-Time Oceanic Glider Mission Viewers

Shinichi Kobara, Christina Simoniello, Ruth Mullins-Perry, Ann Elizabeth Jochens, Matthew K. Howard, Stephanie M. Watson, and Stephan Howden

Abstract

The data portal of the Gulf of Mexico Coastal Ocean Observing System Regional Association was designed to aggregate and integrate data and model output from distributed providers and offer these, and derived products, through a single access point in standardized ways to diverse users. The portal evolved under funding from the US Integrated Ocean Observing System program led by the Interagency Ocean Observing Committee. In 2013, the Gulf of Mexico regional association participated in two pilot projects with two different glider platforms. The first project focused on the feasibility of using the Liquid Robotics Wave Glider to study ocean acidification. The second used the Teledyne Webb Research Slocum profile glider to study hypoxia over the Texas-Louisiana Shelf.

The first project, led by the University of Southern Mississippi, was a 36-day mission supporting the Ocean Acidification Program of the National Oceanic and Atmospheric Administration. The goals were to demonstrate that the wave gliders are suitable platforms to monitor ocean-atmosphere fluxes of carbon dioxide and give the regional association the opportunity to develop automated workflows for Liquid Robotics glider data. The second project involved two deployments of a subsurface Slocum G2 Glider (200 m) provided by Texas A&M University's Geochemical Environmental Research Group: the first during a portion of the August 2013 "Mechanisms Controlling Hypoxia" cruise conducted by Texas A&M's Department of Oceanography and the second for about 30 days in October. The goals of these two deployments were (1) to test the feasibility of collecting temperature, salinity, and dissolved oxygen data near the seabed in relatively shallow (20–40 m) water; and (2) to compare, validate, and merge cruise data with a glider's information.

One important outcome was the development of a web map application using the ArcGIS platform to show data acquired from the glider deployments. Because data from the projects was not directly reported to a Gulf of Mexico Coastal Ocean Observing System server, the Gulf of Mexico Coastal Ocean Observing System Regional Association had to download and reformat the data

Dawn J. Wright, ed.; 2015; *Ocean Solutions, Earth Solutions*; http://dx.doi.org/10.17128/9781589483651

to make it GIS-ready. Using ArcGIS for Server service and Python scripts, a time-aware glider-track layer was automatically generated in an enterprise database and updated in near real time. This process will be applied to visualize future near real-time observations from all surface and subsurface glider platforms.

Introduction

Stommel (1989) predicted that someday unmanned vessels would continuously survey the world's oceans while daily transmitting data to researchers. This prediction is well on its way to becoming a reality with the increasing number of glider vehicle technologies and international deployments (e.g., Glenn and Schofield 2009). Profiling gliders, also known as *long-range autonomous underwater gliding vehicles* (AUGVs), offer a means to collect in situ ocean data from a wide range of sensors at a relatively low cost compared with conventional methods, such as vessels and mooring arrays. Profiling gliders can be buoyancy driven or propeller driven and fly underwater in a sawtooth pattern, collecting horizontal and vertical profiles of hydrographic conditions (Rudnick et al. 2004; Alvarez et al. 2007) (figure 16.1).

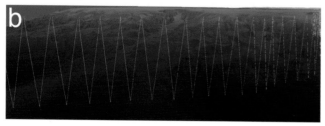

Figure 16.1. (a) Teledyne Webb Research Slocum G2 Glider. (b) The profile glider's sawtooth **movement underwater in the northern Gulf of Mexico.** Photo by Eddie Webb (Texas A&M University Geochemical and Environmental Research Group).

More recently, wave gliders, the first unmanned autonomous marine vehicles to generate propulsion using the ocean's endless supply of wave energy, have been deployed as platforms from which ocean data is gathered and communicated in near real time (figure 16.2). Together, these platforms offer unprecedented opportunities to observe and understand the world ocean.

Figure 16.2. Liquid Robotics Wave Gliders. Photo courtesy of Liquid Robotics.

Gliders complement traditional ocean sampling methods well. Foremost in a challenging economy is their cost-effectiveness. Operating gliders demands fewer man-hours relative to equivalent operations conducted using vessels, and gliders can be adaptively retasked and rerouted to fill data gaps, repeat a transect, or approach new targets of interest (Rudnick et al. 2004). Increasingly, glider development is moving toward "plug and play" capacity so that the platforms can carry varied and interchangeable instruments that measure and evaluate different oceanographic and air/sea interface attributes. In other words, sensors on the glider can be swapped out for different measuring devices, depending on the needs of each mission.

Recognizing the utility and cost-effectiveness of gliders, the US Integrated Ocean Observing System (IOOS) program (National Ocean Service 2014d) has embarked on an effort to outline a National Glider Network Plan (National Ocean Service 2014a). Data from these mobile platforms is expected to contribute significantly to the growing demand for sustained observations, which are needed to address issues ranging from climate change and severe weather forecasts to ecosystem management and water quality monitoring. For example, boundary currents such as the Gulf Stream along the Eastern Seaboard pass near the coastal ocean and are important drivers of climate change (Williams 2012). In the Gulf of Mexico, the Loop Current and its associated eddies dominate mesoscale circulation (Hurlburt and Thompson 1980; Gopalakrishnan et al. 2013) and the potential strength and track of hurricanes entering the Gulf. Gliders in the coastal and nearshore environment also serve many interests specific to understanding the exchange of water between the coast and continental shelf. For example, freshwater runoff and coastal currents can act as physical controls and drivers of the dispersion of pollutants (e.g., nonpoint source, oil), harmful algal blooms (HABs), and the extent and duration of hypoxia events resulting from low levels of dissolved oxygen on the ocean floor.

Expanding the use of gliders in the Gulf of Mexico is a necessary step in providing the geographical coverage, and long-term data, needed to support science-based decisions regarding the ecological health of the Gulf. Successful expansion can best be achieved through collaborative partnerships. It is critical that the Gulf glider community be prepared to develop the Gulf glider network according to the needs of Gulf stakeholders. The US IOOS Regional Associations and partners are heavily engaged in glider operations and, in many instances, are leading the way on their uses and mission applications. This chapter describes the current status of the Gulf of Mexico Coastal Ocean Observing System Regional Association (GCOOS-RA), with regard to acquiring, processing, and delivering ocean observing and monitoring data acquired from sensors on gliders, and frames this in the context of the National Glider Network Plan.

The Gulf of Mexico Coastal Ocean Observing System

GCOOS-RA is one of 11 regional components of the US IOOS, a cooperative effort of federal and nonfederal entities to provide new data, tools, and forecasts to improve marine safety, enhance the economy, and protect the US coastal and ocean environment (National Ocean Service 2014c) (figure 16.3). Because the Gulf of Mexico is a national ecological treasure and an economic driver for the country (Beck et al. 2000; Tunnell 2009), a delicate balance exists between environmental protection and economic development. All along the nearly 17,000 miles of shoreline from Florida to Texas (if bays and other inland waters are included), there are great demands from industries, including commercial and recreational fisheries, oil and gas extraction and exploration, ports, and tourism. The rich biodiversity of the region, combined with the dense human population, dictates that diligent monitoring take place to understand and protect the wealth of natural resources and protect human life and property.

Figure 16.3. The US IOOS program has 11 regional associations (RAs). These serve the nation's coastal communities, including the Great Lakes, Caribbean, and Pacific Islands and territories.

Since 2000, GCOOS has been growing toward becoming an integrated system of federal and nonfederal observing assets and data. Building GCOOS requires a partnership of many organizations, from governments to industry to academia to educators to the public, to integrate the measurements already being made and fill gaps where necessary to meet regional and national requirements. GCOOS priorities are organized around themes that illustrate the broad, beneficial uses of observing system activities. Rather than invest limited funds in its own instrumentation, GCOOS-RA has maximized funds by leveraging existing nonproprietary private, local, state, academic, and federal assets to build GCOOS. The program is in the early stages of development, with clearly articulated plans for growth (see more details in a Build-Out Plan, version 2.0 [GCOOS-RA 2014f]).

On April 20, 2010, the Deepwater Horizon (DWH) oil spill tragically began in the Gulf of Mexico. GCOOS contributions to the spill response effort were numerous and included the following: (1) providing ready access to near real-time and historical data; (2) creating new data layers for the Environmental Response Management Application (ERMA) tool used by the National Oceanic and Atmospheric Administration (NOAA) Office of Response and Restoration; (3) delivering model results from the Regional Oceans Model System and wind forecasts; (4) rapidly reinstalling high-frequency radar (HFR) along at-risk shorelines; (5) providing web access to compiled oil spill resources; and (6) conducting Gulf-wide education and outreach efforts to support effective communications about the spill (Office of Response and Restoration 2014).

In addition to GCOOS-RA, other IOOS regions contributed expertise during the DWH event, demonstrating the power of a coordinated network. GCOOS-RA, the Southeast Coastal Ocean Observing Regional Association (SECOORA), Mid-Atlantic Regional Association Coastal Ocean Observing System (MARACOOS), and Southern California Coastal Ocean Observing System (SCCOOS) all collaborated and coordinated glider missions to identify and track the subsurface oil plume from the Macondo Well, the well from which DWH was operating. A fleet of seven gliders with specialized sensors was deployed by IOOS partners to help identify the presence of oil in the water column. The gliders identified a search zone for the subsurface oil and helped define the likely movement of the oil. Temperature and salinity profiles were also acquired from sensors on the gliders. This was the first known spill response effort to execute a coordinated multiglider operation.

An important contribution of GCOOS-RA was to convert glider and float data from the Deepwater Gulf Response incident site (National Oceanographic Data Center 2012) into map service layers for display on-site, as well as for use in ERMA. These layers were two-dimensional, with metadata, including pointers to precomputed graphic displays of the associated profile data and the actual data. GCOOS-RA developed an interactive map showing the gliders' tracks, and temperature and salinity profiles at glider surfacing (GCOOS-RA 2014a) (figure 16.4).

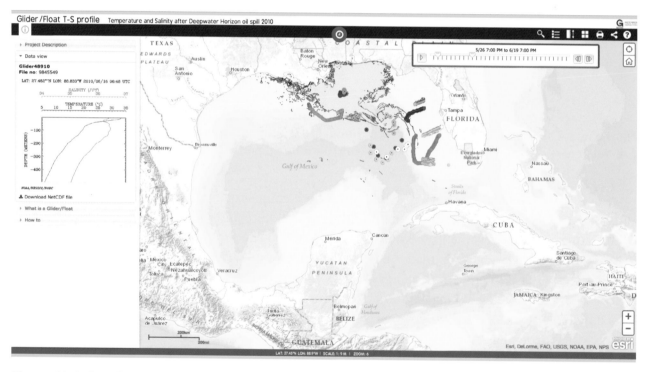

Figure 16.4. **A web map application with a time slider displays the paths of gliders and floats, and temperature and salinity profiles at their surfacing. This map product is part of the Gulf of Mexico Coastal Ocean Observing System Regional Association response to the Deepwater Horizon oil spill.** Image courtesy of Gulf of Mexico Coastal Ocean Observing System. Map data sources: Esri, DeLorme, FAO, USGS, NOAA, EPA, NPS.

This interactive map provided time-aware glider and float layers, which are stored information about the changing state of the glider and float over time. This function allowed users to track glider and float paths with a time slider. Despite important contributions to post-DWH monitoring, data processing, and information distribution, it was an ad hoc system, one that would be greatly improved with a thoughtful infrastructure in place. Developing monitoring plans need to include such infrastructure.

The National Context for Gliders: The National Glider Network Plan

As described, gliders are becoming increasingly common as effective ocean observing and monitoring tools. The US IOOS National Glider Network Plan (National Ocean Service 2014a) states that the ability to link coastal systems to the deep ocean requires sustained, cost-effective in situ sampling at appropriate spatial and temporal scales. Additionally, the ability to sample episodic events requires adaptive sampling capability combined with a rapid-response capability for deployment in specific regions.

A combination of ocean-observing platforms and assets, each with different advantages for providing spatial and temporal information, can be collectively used to meet these needs. For example, ships can be used to operate high-power, full-bandwidth high-resolution instruments to make observations and collect samples for various analyses, from the surface to great depths. However, because they are expensive to operate, most field campaigns using ships usually last for time scales of weeks to months. Also, the number of ships is limited, and they may not be available to respond quickly to events. Satellites make synoptic measurements over large areas of the ocean surface several times per day, but the measurements are largely limited to either a thin upper layer or information that is related to information integrated over the water column (e.g., sea surface topography). Moorings provide persistent time series at five- to 60-minute intervals over the water column at fixed locations and can acquire data for years. Gliders are becoming increasingly valuable complementary ocean-observing platforms because they are mobile and provide a persistent presence at a relatively low cost.

The IOOS National Glider Network is designed to function as a distributed system, with some centralized data management functions that apply consistent data standards and best practices to achieve integration among various glider-observing assets. As the lead federal agency for US IOOS, NOAA leads the overall plan and coordinates requirements and efforts to ensure consistency between national and regional needs and align the National Glider Network Plan with the national Data Management and Communication (DMAC) objectives.

To date, three versions of autonomous subsurface profiling gliders have demonstrated persistent observations in an operational capacity: (1) Scripps Institution of Oceanography Spray gliders, now marketed through Bluefin Robotics (Sherman et al. 2001; Instrument Development Group of the Scripps Institution of Oceanography 2014); (2) the University of Washington Seaglider, now marketed through Kongsberg Maritime (Eriksen et al. 2001; Applied Physics Lab of University of Washington 2014); and (3) Teledyne Webb Research Slocum Gliders (Webb et al. 2001; Teledyne Webb Research 2010). All three have displayed robust reliability and fulfilled the missions outlined here. Collectively, over five hundred of the three versions of profiling gliders have been manufactured and used. US IOOS is engaged in a collaborative effort to track where and when these missions occur.

The National Underwater Glider Network Map (National Ocean Service 2014b) includes current and historical glider missions dating back to 2005 and is a collaborative effort of GCOOS-RA,

SCCOOS, MARACOOS, the Northwest Association of Networked Ocean Observing System (NANOOS), and the Central and Northern California Ocean Observing System (CeNCOOS). US IOOS, NOAA, the Office of Naval Research (ONR), National Science Foundation (NSF), Environmental Protection Agency (EPA), and various universities, state agencies, and industries have funded the gliders displayed on the map (National Ocean Service 2014b).

The National Glider Network Plan provides several examples of specific issues that have benefited from subsurface data provided by profiling gliders and how these issues will further benefit from a national network. These include:

- emergency response (for example, the dispersion of oil spills by currents in the Gulf of Mexico);
- climate variability, including El Niño/La Niña in the California Current System, the Loop Current in the Gulf of Mexico, and the Gulf Stream along the US East Coast;
- hurricane intensity forecasting, relevant for the East Coast, Gulf of Mexico, and American territories;
- HABs throughout US coastal waters; and
- hypoxia (low dissolved oxygen in bottom water), particularly in the northern Gulf of Mexico.

The GCOOS Approach with Gliders

Clearly, glider observations for the Gulf of Mexico are becoming increasingly vital for responding to the issues identified in the National Glider Network Plan. In response to this need, GCOOS-RA is bringing together representatives from academic institutions, government organizations, and the private sector to plan and implement a glider network in the Gulf. Coordination began in 2011 through the development of the GCOOS Glider Plan, a component of the comprehensive GCOOS Build-Out Plan. In April 2013, GCOOS-RA assembled a team of Gulf glider experts and created the Gulf Glider Task Team (GGTT) to further develop the strategic and implementation plans for Gulf glider operations delineated in the GCOOS Build-Out Plan. Elements of the National Glider Strategy are also being incorporated.

The National Glider Plan calls for a set of two to three glider transects in each regional association domain. This will meet some, but not all, the needs of glider operations in the Gulf. However, the associated DMAC plan includes data ingested from all glider operations and can serve this data to any user. Glider pilot projects and partnerships between GCOOS-RA and stakeholders include collaborative strategies for detection of harmful algal blooms, monitoring ocean acidification (OA), monitoring hypoxia, routine measurements of water quality, and advancing a wide array of Gulf research.

In 2005, GCOOS-RA began developing the GCOOS Data Portal (GCOOS-RA 2008a), designed to aggregate observational data and model output from distributed providers and offer these, and derived products, through a single access point in standardized ways to a diverse set of users. GCOOS also has an associated Data Products page (GCOOS-RA 2008b), in which data is fused into products tailored for specific or general stakeholder groups. The goal of the GCOOS Data Portal and Data Products page is an automated, and largely unattended, data system that delivers high-quality data and products to Gulf stakeholders. GCOOS products are mostly related to satisfying the needs of consumers and are designed for quick online browsing for a particular purpose.

Because geospatial data already includes dozens of file formats and database structures, and continues to evolve and grow new types of data and standards, flexible and customizable automated data flow is required for real-time tracking systems. The data portal is custom built and includes web services based on Open Geospatial Consortium (OGC) standards-based Sensor Observation Service (SOS) with observations and measurements (O&M) encodings. Products appearing in the data portal are primarily map applications constructed using the ArcGIS platform.

Data access is through an online form and, for machines, is through a direct-access uniform resource locator (URL), Open Source Project for a Network Data Access Protocol (OPeNDAP), or an SOS access point. GCOOS designs web-based mission viewers for surface and subsurface (profiling) gliders using the ArcGIS platform. Web-based mission viewers run in near real time when stakeholders' gliders are operating in the Gulf and playback mode after a completed mission. Viewers show the glider trajectories and oceanographic data that is collected and are publicly available through the GCOOS website.

The mission viewers were developed for the following four glider projects:

1. Profiling gliders monitoring for subsurface oil following the DWH oil spill incident as described (GCOOS-RA 2010a)

2. Slocum profiling gliders monitoring HABs, particularly the dinoflagellate *Karenia brevis* in Southwest Florida (GCOOS-RA 2010b)

3. Pilot project 1: a surface wave glider monitoring OA in the northern Gulf of Mexico (GCOOS-RA 2014b)

4. Pilot project 2: Slocum profiling gliders being tested in the western Gulf of Mexico (GCOOS-RA 2013).

Profiling Gliders

GCOOS-RA partners at the University of South Florida (USF) College of Marine Science and Mote Marine Laboratory (MML) coordinated glider missions to gain a better understanding of the dominant Gulf of Mexico red-tide organism *Karenia brevis*. The science mission was to use tandem gliders to map the West Florida Shelf region with an emphasis on catching a red-tide bloom in progress to gain a better understanding of how to forecast and track harmful algal blooms. One specific project focus was to determine the abundance of *K. brevis* at the pycnocline because previous work had suggested the cells may be concentrated there.

In typical use, gliders profile from the surface to 500–1,000 m, taking three to six hours to complete a cycle from the surface to depth and back. During the cycle, the gliders travel 3–6 km horizontally for a speed of about 1 km/h. Deployments of three to six months are routine, during which the gliders' survey track extends well over 2,000 km. Sensors on gliders measure such physical variables as pressure, temperature, salinity, and currents; biological variables relevant to the abundance of phytoplankton and zooplankton; and ecologically important chemical variables such as dissolved oxygen and nitrate.

During short surface intervals on the order of a few minutes, gliders obtain location by GPS and communicate through the Iridium data telemetry satellite system. Glider data is transmitted to shore in near real time from that surface point, so an automated or semiautomated national distribution and archiving scheme is essential to making recently acquired and historical data readily available to a variety of users. At present, glider data is received by servers at local data nodes and distributed in a variety of formats and through several protocols to GCOOS and/or a national system such as the National Underwater Glider Network Map (National Ocean Service 2014b).

The joint USF/Mote Marine Laboratory coordinated HAB glider mission was the first opportunity for GCOOS-RA to experiment with automated workflow and learn how to display glider data in near real time via the GCOOS website. GCOOS-RA developed a mission viewer showing the gliders' tracks collected at locations where the gliders surfaced to obtain position fixes and transmitted data that had been continuously recorded in their data boxes (GCOOS-RA 2010b) (figure 16.5).

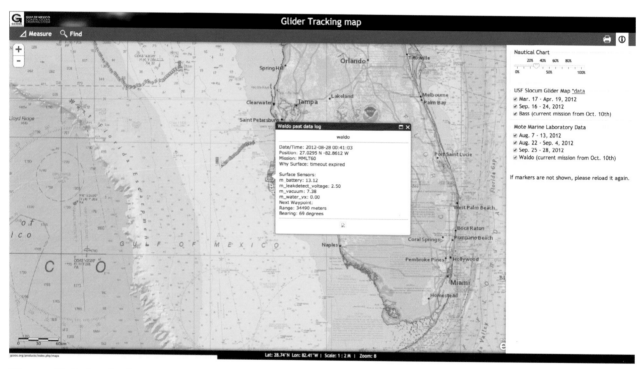

Figure 16.5. A mission viewer displays a glider's path and its system sensor information. Image courtesy of Gulf of Mexico Coastal Ocean Observing System (GCOOS). Data sources: Esri, GEBCO, NOAA, National Geographic, DeLorme, HERE, Geonames.org, and other contributors.

Pilot Project 1: Wave Glider Monitoring of Ocean Acidification in the Northern Gulf of Mexico

OA is the ongoing decrease in the pH of the oceans, caused by the uptake of carbon dioxide (CO_2) as the atmospheric CO_2 concentrations rise (Caldeira and Wickett 2003). It is estimated that approximately 30%–40% of the carbon dioxide released by humans into the atmosphere dissolves into the oceans, rivers, and lakes (Fabry et al. 2008; Doney et al. 2009). NOAA has initiated an active monitoring and research OA program (Ocean Acidification Steering Committee 2010). The principal goal of the program is to develop the monitoring capacity to quantify and track OA in open-ocean, coastal, and Great Lake systems (Ocean Acidification Steering Committee 2010).

In an effort to investigate the feasibility of new technologies to monitor changes in ocean acidity related to CO_2 fluxes between the atmosphere and ocean in the Gulf of Mexico, the University of Southern Mississippi (USM), NOAA, GCOOS, and Liquid Robotics collaborated to deploy a Liquid Robotics Wave Glider to measure CO_2, dissolved oxygen, pH, water temperature, conductivity, air temperature, barometric pressure, and wind speed and direction in the northern Gulf of Mexico. The CO_2 sensor package, which measures the mole fraction of CO_2 on either side of the air-sea interface, was developed by the Monterey Bay Aquarium Research Institute.

The team successfully deployed the glider in October 2012 from the USM R/V *Tom McIlwain* near the USM Central Gulf of Mexico Ocean Observing System (CenGOOS), a subset of GCOOS. The location was selected so that the glider traverse started and ended at the CenGOOS buoy in the northern Gulf, which has a similar CO_2 measuring system developed and built by the Pacific Marine and Environmental Laboratory. The glider was programmed to run a route around the Mississippi River Delta and traverse over stations where the NOAA R/V *Ronald H. Brown* sampled in July 2012 during the second Gulf of Mexico and East Coast Carbon program. These additional observations from the buoy and the ship-based sampling were used to help validate the measurements made from the wave glider platform.

Wave gliders offer many sampling benefits. Among these are greater spatial coverage compared with traditional buoys, collection of data at different depths near the surface, the ability to travel anywhere without need of refueling, remote navigation and mission control, and the ability to transmit data via satellite anywhere in the world. Unfortunately, like other autonomous platforms, wave gliders are subject to collisions at sea. Such was the case during this mission when the wave glider, at the southernmost part of its route, was struck by a vessel on the 27th night. The USM and Liquid Robotics team retrieved and relaunched the wave glider in early February 2013 (figure 16.6).

Figure 16.6. **The University of Southern Mississippi and Liquid Robotics team launch the wave glider in the Gulf of Mexico on February 12, 2013.** Photo by Stephan Howden (USM).

Similar to the viewer previously shown for the DWH glider mission (see figure 16.4), GCOOS-RA developed an interactive near real-time mission viewer for the OA mission (GCOOS-RA 2014b) (figure 16.7), showing the surface glider's location and data from the mission. The mission viewer also displays time-series charts by parameter.

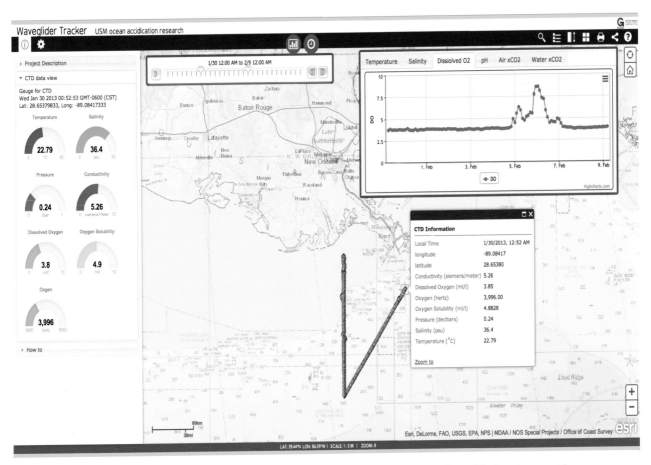

Figure 16.7. The wave glider mission viewer displays glider's path and multiple observation parameters in gauge and chart with a time slider. Image courtesy of Gulf of Mexico Coastal Ocean Observing System (GCOOS). Map data sources: University of Southern Mississippi (USM), Esri, DeLorme, FAO, USGS, EPA, NPS; NOAA NOS Special Projects Office of Coast Survey.

From aggregating data to showing updated sensor parameters every 10–20 minutes, several conditions needed to be considered during development of the near real-time wave glider mission viewer. Lessons learned from a previous mission by IOOS partners guided development. In an earlier demonstration project, two IOOS RAs, Northeastern Regional Association of Coastal and Ocean

Observing Systems (NERACOOS) and MARACOOS, had bandwidth problems for wave glider communications that led to failed initial attempts to establish SOS. Armed with this knowledge and knowing the OA wave glider mission was short, all the data from the wave glider was telemetered to a Liquid Robotics server, without automatic link to the GCOOS-RA server, thus bypassing the bandwidth issue.

As with any integrated observations, the heterogeneous devices and data formats create challenges for an organized data-storing process. In this case, there was no push system implemented, and a data aggregation system had to be built. Near real-time observations require a robust architecture from desktop to cloud, with a timely, extensible, and automated process. This poses similar problems that might arise in trying to add live weather or recent earthquake information to applications without writing long programs first. There are multiple approaches to tackle this challenge, but GCOOS chose ArcGIS 10.1 Tracking Server (predecessor of ArcGIS 10.2 GeoEvent Extension for Server) for this project. ArcGIS 10.1 Tracking Server is an independent server and the engine for processing and distributing real-time data to various clients via data links.

The process GCOOS-RA created to integrate the observations is as follows: First, GCOOS-RA aggregated information from Liquid Robotics Server. Liquid Robotics provided the specific URLs to retrieve CTD, weather, oxygen, and system data. Each file contained sensor information, location (latitude and longitude in World Geodetic System 1984 [WGS 84] datum), and time of collection, all in comma-separated value (CSV) file format. This process used simple Windows batch scripting and a handful of command-line utilities (cURL, wget) (Free Software Foundation 2012; Stenberg et al. 2014) to download automatically. Scripts were run every 10 minutes. For ArcGIS Tracking Server to operate, it must know how the data is formatted and how it is being received. Following the aggregation of Liquid Robotics data, GCOOS-RA created a new message definition and new generic input data link connection based on a received file (Esri 2012a and 2012b). Once the data was saved in a folder, another script ran to format the data for each database schema configured in the tracking server and pushed the data into the tracking server through a transmission control protocol (TCP) socket. Although ArcGIS Tracking Server is able to convert/format data, in this case it focuses on aggregating and logging data into an ArcSDE database. Esri provided C# and Java programs to send data through a TCP socket. GCOOS-RA created a simple Python script to do the same (TcpCommunication 2012). Four feature layers in ArcSDE database updated automatically every 10 minutes when new data was downloaded.

Although Liquid Robotics provided a web-based remote control system for the wave glider to give its current location, ArcGIS users can also see the glider location by connecting the tracking server in the ArcCatalog and ArcMap applications of ArcGIS. If users need real-time data viewing with analysis capabilities, the ArcGIS Tracking Analyst extension for ArcGIS for Desktop software allows users to display, analyze, and manipulate real-time and fixed-time data.

Once the data was logged into the ArcSDE database, the information was published through ArcGIS for Server as a mapping service. This whole process is shown in figure 16.8. Although ArcGIS Tracking Server is the engine for processing and distributing real-time data, a separate client application, such as ArcGIS for Desktop, is needed to view the data. GCOOS-RA developed a web map application, or glider mission viewer (GCOOS-RA 2014b), in the ArcGIS API for JavaScript application programming interface (API). The web-based application provides a common operating picture for monitoring and tracking an event. It has an interactive map, charts, gauges, and a time slider for scientific observing parameters (see figure 16.7). Basic functions such as a measurement tool, basemap catalog, print tool, and social sharing tool were also implemented.

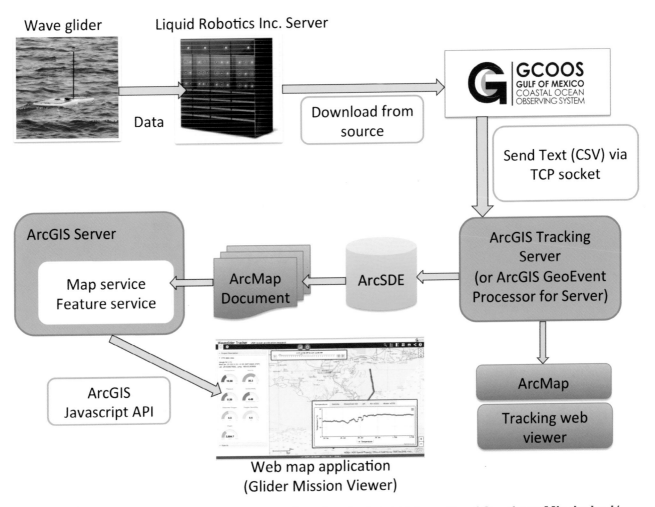

Figure 16.8. Diagram of the automated workflow for the joint University of Southern Mississippi/ Liquid Robotics/Gulf of Mexico Coastal Ocean Observing System Ocean Acidification wave glider mission viewer. Image courtesy of Gulf of Mexico Coastal Ocean Observing System (GCOOS). Map data sources: NOAA NOS Special Projects Office of Coast Survey, Esri, DeLorme, FAO. Photo courtesy of Liquid Robotics Inc.

This mission viewer uses gauges to show near real-time temperature, salinity, pressure, conductivity, and dissolved oxygen values. Other wave glider engineering and meteorological information, such as wind direction and speed, is provided with an overlaid nautical chart. The layers are also downloadable in Keyhole Markup Language (KML), an Extensible Markup Language (XML)–based file format used to display geographic data in an Earth browser such as Google Earth, and are accessible to the data through feature services for use with ArcGIS (GCOOS-RA 2014c).

Pilot Project 2: Development of a Profiling Glider Mission Viewer for Testing

The Texas A&M University (TAMU) Department of Oceanography and the Geochemical Environmental Research Group (GERG) deployed a Teledyne Webb Research Slocum G2 Glider (200 m) during the Mechanisms Controlling Hypoxia (MCH) cruise in August 2013 (DiMarco 2013), and again over the Texas-Louisiana Shelf in October during a Texas Automated Buoy System maintenance cruise (figure 16.9). The glider's mission was to test the capability of collecting temperature, salinity, and dissolved oxygen data over the shelf in relatively shallow waters of 20–40 m depth and collect hydrographic data near and around the Flower Garden Banks National Marine Sanctuary. In this pilot project, GCOOS tested a near real-time mission viewer compiled with a subsurface profiling glider's information.

Credit: E. Webb (GERG)

Figure 16.9. **Pilot test of a profiling glider for hypoxia studies in the northern Gulf of Mexico by the Texas A&M University Geochemical and Environmental Research Group.** Photo by Eddie Webb (Texas A&M University Geochemical and Environmental Research Group).

While at the surface, a glider obtains location by GPS and communicates through the Iridium data telemetry satellite system. Glider data is received by the GERG shore station in near real time and stored in a folder dedicated to that particular glider. There are two types of files created: (1) a glider system health and trajectory information (SBD) file produced while the glider was at the surface, and (2) a scientific data (TBD) file containing sensor measurements made while the glider was submerged. GCOOS-RA copied these files from the GERG web-accessible folder to a working folder, where they were processed into engineering units and basic products.

Trajectory files were processed using an Interactive Data Language (IDL) script to produce a cumulative file and daily file with the latest positional data. IDL software is often used in the analysis of one-, two-, and three-dimensional datasets. A cumulative file was also written in KML and made available to the IOOS map (National Ocean Service 2014b). Except for the processing of SBD and TBD files, the processes are almost identical to that of the wave glider. Files were processed four times per day in this project and sent to the ArcGIS 10.2 GeoEvent Extension for Server processor (upgraded from ArcGIS Tracking Server) and logged into an ArcSDE database.

Data files were processed, and graphic displays of measured parameters were produced and posted on the web. A mission viewer (GCOOS-RA 2013) showed the updated location of the profiling glider every six hours. Cumulative profile plots of temperature, salinity, and colored dissolved organic matter (CDOM) were also available on the mission viewer (figure 16.10). All glider tracks are available in a KML file as well. The glider tracking layers are also downloadable in KML and accessible to the data through feature services to enable copying into a user's local environment (GCOOS-RA 2014c).

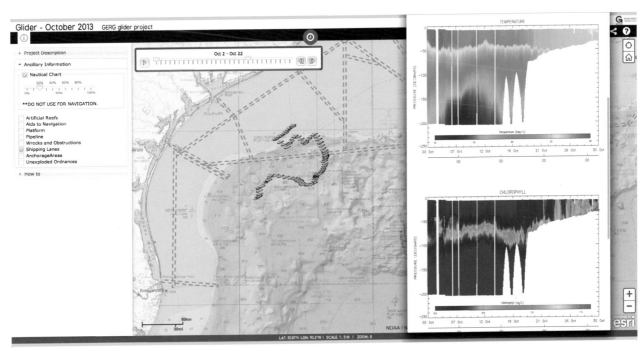

Figure 16.10. A mission viewer for Slocum profiling gliders being tested in the western Gulf of Mexico by the Texas A&M University Geochemical and Environmental Research Group. Image courtesy of Gulf of Mexico Coastal Ocean Observing System (GCOOS). Map data sources: NOAA NOS Special Projects Office of Coast Survey, Esri, DeLorme, FAO.

Lessons Learned and Future Plans

Developing mission viewers for the surface wave glider and profiling glider tests in the northern Gulf of Mexico pilot projects successfully demonstrated near real-time glider observations using ArcGIS web mapping APIs. These activities and resulting products were intended to validate data acquired from glider platforms, facilitate use of the information through user-friendly interfaces, and help establish a user-driven glider program that is responsive to Gulf of Mexico information needs.

GCOOS-RA will continue to broaden collaborations to advance the establishment of a comprehensive and continuous Gulf Glider Network and work with the US IOOS community on development of the National Glider Network Plan. In addition, GCOOS will continue working with operators and stakeholders to develop glider data visualization tools and products (e.g., 3D visualizers, education and outreach resources, and profiling operations on the GCOOS website). Establishing a glider server and asset map simplifies the distribution process for glider operators and data users alike. Archiving this data in a repository will ensure its continued availability to all users.

As one example of GCOOS-RA's future activity with gliders with a diversity of stakeholders, GCOOS is partnering with Shell Exploration and Production Co. and the NOAA National Data Buoy Center (NDBC) to improve hurricane prediction and forecast models with gliders. Details about the collaboration can be found at the GCOOS website (GCOOS-RA 2014d). Shell, NOAA NDBC, National Centers for Environmental Prediction (NCEP), IOOS, and USM are working together to operate profiling gliders (200 m and 1,000 m gliders) to collect real-time data to measure upper-ocean heat content in the Gulf of Mexico. The data collected from the gliders is being used to improve hurricane intensity predictions with the goal of improving real-time hurricane intensity forecasts. The project is currently in its third year of operation and coincides with additional federal projects to collect measurements necessary for hurricane prediction, such as the deployment of airborne expendable bathythermographs (AXBTs), which measure temperature as a function of depth, from the NOAA Hurricane Hunter aircraft that monitor hurricane development and progress. The GCOOS, Shell, and NOAA NDBC collaboration includes data sharing but also supports Gulf of Mexico stakeholder and community engagement, education, and outreach to promote hurricane research and monitoring.

As part of the collaboration, GCOOS has designed an interactive and real-time glider mission viewer and Web Graphics Library (webGL) visualizer in 3D (GCOOS-RA 2014e). Data shared in the viewer includes the Seaglider's surfacing and path location and hydrographic data collected during the mission. The viewer, when completed, will be published on the GCOOS Products page (GCOOS-RA 2008b).

The GCOOS/NOAA/Shell collaboration demonstrates the expansive versatility of gliders as platforms to collect data for projects ranging from applied research to real-time environmental monitoring. Data collected by gliders is valuable to address a number of conditions. Among these are high-resolution physical observations to better protect offshore oil and gas assets; improved tropical storm intensity and trajectory forecasts to better prepare Gulf Coast communities; and observations to address additional hazards, ranging from natural and anthropogenic pollution tracking to accurately estimating volume and areal extent of hypoxia events. For gliders to reach their full potential in the Gulf of Mexico, continued development of partnerships, such as those described here, between industry, government, military, and academia is critical. It is the only way to further research and expand ocean-observing capacity in the Gulf of Mexico.

Acknowledgments

We gratefully acknowledge the guidance, support, and funding of US IOOS and GCOOS-RA. The comments and suggestions of two anonymous reviewers greatly improved this chapter.

References

Alvarez, A., B. Garau, and A. Caiti. 2007. "Combining Networks of Drifting Profiling Floats and Gliders for Adaptive Sampling of the Ocean." In *Proceedings of the 2007 IEEE International Conference on Robotics and Automation*, 157–62, Rome, Italy, April 10–14.

Applied Physics Lab of University of Washington. 2014. *Seaglider—Summary.* http://www.apl.washington.edu/projects /seaglider/summary.html. Last accessed July 6, 2014.

Beck, M. W., M. Odaya, J. J. Bachant, J. Bergan, B. Keller, R. Martin, R. Mathews, C. Porter, and G. Ramseur. 2000. *Identification of Priority Sites for Conservation in the Northern Gulf of Mexico: An Ecoregional Plan.* Arlington, VA: The Nature Conservancy Technical Report.

Caldeira, K., and M. E. Wickett. 2003. "Anthropogenic Carbon and Ocean pH." *Nature* 425 (6956): 365.

DiMarco, S. 2013. Mechanisms Controlling Hypoxia. http://hypoxia.tamu.edu. Last accessed July 6, 2014.

Doney, S. C., V. J. Fabry, R. A. Feely, and J. A. Kleypas. 2009. "Ocean Acidification: The Other CO_2 Problem." *Annual Review of Marine Science* 1:169–92.

Eriksen, C. C., T. J. Osse, R. D. Light, T. Wen, T. W. Lehman, P. L. Sabin, J. W. Ballard, and A. M. Chiodi. 2001. "Seaglider: A Long Range Autonomous Underwater Vehicle for Oceanographic Research." *IEEE Journal of Oceanic Engineering* 26 (4): 424–36.

Esri. 2012a. Tracking Server: Creating Message Definitions. http://bit.ly/1ziqHl7. Last accessed July 16, 2014.

Esri. 2012b. Tracking Server: Using the Generic Input Data Link. http://bit.ly/1n62r0v. Last accessed July 16, 2014.

Fabry, V. J., B. A. Seibel, R. A. Feely, and J. C. Orr. 2008. "Impacts of Ocean Acidification on Marine Fauna and Ecosystem Processes." *ICES Journal of Marine Science* 65:414–32.

Free Software Foundation. 2012. GNU wget. https://www.gnu.org/software/wget. Last accessed July 6, 2014.

GCOOS-RA (Gulf of Mexico Coastal Ocean Observing System Regional Association). 2008a. GCOOS Data Portal. http://data.gcoos.org. Last accessed July 6, 2014.

———. 2008b. GCOOS Products. http://gcoos.org/products. Last accessed July 6, 2014.

———. 2010a. GCOOS Deepwater Horizon Glider TS Time Series Map. http://gcoos.org/products/maps /deepwaterhorizon/glider. Last accessed July 6, 2014.

———. 2010b. GCOOS Glider Tracker. http://gcoos.org/products/maps/glidertracker. Last accessed July 6, 2014.

———. 2013. GERG Glider. http://gcoos.org/products/maps/gerg/glider201310. Last accessed July 6, 2014.

———. 2014a. GCOOS Glider and Floats T-S Data after Deepwater Horizon Oil Spill. http://gcoos.org/products /maps/deepwaterhorizon/glider_float. Last accessed July 6, 2014.

———. 2014b. GCOOS Waveglider Tracker. http://gcoos.org/products/maps/waveglider/usm. Last accessed July 6, 2014.

———. 2014c. Glider/USM_Waveglider (Feature Server). http://gcoosmaps.tamu.edu/arcgis/rest/services/Glider /USM_Waveglider/FeatureServer. Last accessed July 6, 2014.

———. 2014d. Industry, Government, and Researchers Partner for Glider Operations in the Gulf of Mexico. http:// gcoos.org/?p=6070. Last accessed July 6, 2014.

———. 2014e. Shell Glider 553. http://bit.ly/1vOOuV8. Last accessed July 6, 2014.

———. 2014f. *A Sustained, Integrated Ocean Observing System for the Gulf of Mexico (GCOOS): Infrastructure for Decision-Making, Version 2.0.* College Station, TX: GCOOS Regional Association Technical Report. http://gcoos.org /BuildOut/BuildOutPlan-V2.pdf.

Glenn, S., and O. Schofield. 2009. "Growing a Distributed Ocean Observatory: Our View from the COOL Room." *Oceanography* 22 (2): 78–92.

Gopalakrishnan, G., B. D. Cornuelle, I. Hoteit, D. L. Rudnick, and W. B. Owens. 2013. "State Estimates and Forecasts of the Loop Current in the Gulf of Mexico Using the MITgcm and Its Adjoint." *Journal of Geophysical Research* 118 (7): 3292–3314.

Hurlburt, H. E., and J. D. Thompson. 1980. "A Numerical Study of Loop Current Intrusions and Eddy Shedding." *Journal of Physical Oceanography* 10 (10): 1611–51.

Instrument Development Group of the Scripps Institution of Oceanography. 2014. Underwater Glider Spray. http:// spray.ucsd.edu/pub/rel/index.php. Last accessed July 6, 2014.

National Ocean Service. 2014a. National Glider Strategy, US IOOS. http://www.ioos.noaa.gov/glider/strategy /welcome.html. Last updated July 6, 2014.

———. 2014b. National Underwater Glider Network Map. http://www.ioos.noaa.gov/observing/observing_assets /glider_asset_map.html. Last updated July 6, 2014.

———. 2014c. Regional US IOOS Partners. http://www.ioos.noaa.gov/regional.html. Last updated July 6, 2014.

———. 2014d. The US Integrated Ocean Observing Systems (IOOS). http://www.ioos.noaa.gov. Last updated July 6, 2014.

National Oceanographic Data Center. 2012. Ocean in situ Data: Deepwater Horizon Support. http://www.nodc.noaa .gov/deepwaterhorizon/insitu.html. Last accessed July 6, 2014.

Ocean Acidification Steering Committee. 2010. *NOAA Ocean and Great Lakes Acidification Research Plan*. Silver Spring, MD: National Oceanic and Atmospheric Administration Special Report.

Office of Response and Restoration. 2014. Maps and Spatial Data/Environmental Response Management Application (ERMA). http://response.restoration.noaa.gov/maps-and-spatial-data/environmental-response-management -application-erma. Last accessed July 6, 2014.

Rudnick, D. L., R. E. Davis, C. C. Eriksen, D. M. Fratantoni, and M. J. Perry. 2004. "Underwater Gliders for Ocean Research." *Marine Technology Society Journal* 38 (2): 73–84.

Sherman, J., R. E. Davis, W. B. Owens, and J. Valdes. 2001. "The Autonomous Underwater Glider 'Spray.'" *IEEE Journal of Oceanic Engineering* 26 (4): 437–46.

Stenberg, D., D. Fandrich, and Y. Tse. 2014. cURL and libcURL. http://curl.haxx.se. Last accessed July 6, 2014.

Stommel, H. 1989. "The SLOCUM Mission." *Oceanography* 2 (1): 22–25.

TcpCommunication. 2012. TcpCommunication: Python Wiki. https://wiki.python.org/moin/TcpCommunication. Last accessed July 6, 2014.

Teledyne Webb Research. 2010. Slocum Glider. http://www.webbresearch.com/slocumglider.aspx. Last accessed July 6, 2014.

Tunnell Jr., J. W. 2009. "Gulf of Mexico." In *Ocean: An Illustrated Atlas*, edited by S. A. Earle and L. K. Glover, 136–37. Washington, DC: National Geographic Society.

Webb, D. C., P. J. Simonetti, and C. P. Jones. 2001. "SLOCUM: An Underwater Glider Propelled by Environmental Energy." *IEEE Journal of Oceanic Engineering* 26 (4): 447–52.

Williams, R. G. 2012. "Oceanography: Centennial Warming of Ocean Jets." *Nature Climate Change* 2:149–50.

Supplemental Resources

Dawn J. Wright, ed.; 2015; *Ocean Solutions, Earth Solutions*; http://dx.doi.org/10.17128/9781589483651_d

For the digital content for this chapter, listed in this section, go to the Esri Press "Book Resources" webpage at esripress.esri.com/bookresources. Then, in the list of Esri Press books, click *Ocean Solutions, Earth Solutions*. On the *Ocean Solutions, Earth Solutions* resource page, click the chapter 16 link to access that webpage and the links to the digital content.

Python Script

The book resource webpage for chapter 16 provides a link to download the following script:
- Glider Time Track Python script to convert CSV file to KML

Additional ocean glider resources

URLs and QR codes are provided here for the following resources for near real-time oceanic glider mission viewers. Links are also available on the book resource webpage for chapter 16.

- GCOOS Data Portal, at http://data.gcoos.org

- GCOOS Products page, at http://gcoos.org/products

- Wave glider data via ArcGIS for Desktop, at http://gcoosmaps.tamu.edu/arcgis/rest /services/Glider/USM_Waveglider/FeatureServer

- Shell Glider in Esri CityEngine web viewer, at http://bit.ly/shell_sg553

- Tracking server reference: message definition, at http://resources.arcgis.com/en/help /tracking-server/10.1/index.html#/Creating_Message_Definitions/00r600000026000000

- Tracking server reference: generic input data link connection, at http://resources.arcgis.com/en/help/tracking-server/10.1/index.html# /Using_the_Generic_Input_Data_Link/00r60000000w000000

- Python reference: send a text via TCP socket, at https://wiki.python.org/moin /TcpCommunication

Contributors

The Editor

Dr. Dawn J. Wright

Environmental Systems Research Institute (Esri), 380 New York St., Redlands, CA 92373, USA;
dwright@esri.com; http://esriurl.com/dawn; +1-909-793-2853, ext. 2182 (phone)

Dawn Wright was appointed Chief Scientist of Esri in October 2011. In this role, she aids in formulating and advancing the intellectual agenda of the environmental, conservation, climate, and ocean sciences aspect of Esri's work, while also representing Esri to the national/international scientific community. Dr. Wright also maintains an affiliated faculty appointment as professor of geography and oceanography in the College of Earth, Ocean, and Atmospheric Sciences at Oregon State University. Her current research interests include data modeling, benthic terrain and habitat characterization, coastal/ocean informatics, and cyberinfrastructure. She serves on the US National Academy of Sciences Ocean Studies Board, US National Oceanic and Atmospheric Administration (NOAA) Science Advisory Board, the Blue Ribbon Panel of the World Bank's Global Partnership for Oceans, and many journal editorial boards. Wright is a fellow of the American Association for the Advancement of Science, as well as a fellow of Stanford University's Leopold Leadership Program. She holds an individual interdisciplinary PhD in physical geography and marine geology from the University of California, Santa Barbara (UCSB); an MS in oceanography from Texas A&M University (TAMU); and a BS cum laude in geology from Wheaton College, Illinois. Other interests include road cycling, 18th-century pirates, orange-flavored Gummi Bears, her dog Sally, and SpongeBob SquarePants. Follow her on Twitter at @deepseadawn.

The Authors

Jeff A. Ardron

Institute for Advanced Sustainability Studies (IASS), Berliner Straße 130, Potsdam 14467, Germany;
jeff.ardron@iass-potsdam.de; +49-331-288-22-385 (phone)

Jeff Ardron learned GIS in the late 1990s and applied spatial analyses to marine planning processes in British Columbia, Canada. After completing some early analyses (of which he was inordinately proud) and trying to get these accepted, he realized that not everyone shared his enthusiasm for spatial analysis tools. Through discussions with other conservation GIS users, he came to see there were many pitfalls and little guidance for bridging the science/public policy divide. With the Pacific Marine Analysis and Research Association (PacMARA) and University of Queensland, he led the development of the revised Marxan user manual, the *Marxan Good Practices Handbook*, and a set of training courses highlighting good practices in the use of such planning tools. He currently works for the Institute for Advanced Sustainability Studies (IASS), Germany, as a senior fellow, developing good practices in the transparency of marine resource planning, particularly deep-sea mining. He maintains his involvement with PacMARA as president of the board.

David Bader

Aquarium of the Pacific, Education Department, 100 Aquarium Way, Long Beach, CA 90802, USA; dbader@lbaop.org; +1-562-951-1633 (phone)

David Bader is the director of the Education Department at the Aquarium of the Pacific in Long Beach, California. He has played an integral role in developing and managing school and public programs since joining the Aquarium in 1998. He has established programming that utilizes the incredible living resources of the Aquarium to bring visitors the latest information in marine science. Bader has created a suite of capacity-building opportunities for Aquarium staff, university and agency scientists, and informal science educators throughout the Los Angeles area. David has a BS in marine biology from the University of California, Santa Cruz (UCSC), and an MS in biology, with an emphasis in educational media design, from the University of California, Irvine (UCI).

Cathryn R. Brandon

Coastal and Ocean Resource Analysis (CORAL) Group; Department of Geography, University of Victoria, PO Box 3060, Victoria, BC V8W 3R4, Canada; catyrose@uvic.ca; http://marinescapes.wordpress.com; +1-250-472-4624 (phone)

Cathryn Brandon is currently working on a PhD in marinescape geovisualization, building a coastal and underwater marine environment in a computer gaming engine using real-world spatial data and digital media to inform its creation. In addition to realism, ethics, and interactivity of geovirtual environments, Brandon is also interested in environmental psychology dimensions related to mental imagery of underwater marine environments, and its relationship to conservation and environmental attitudes, values, and behaviors and how virtual environments might affect this imagery. Her master's research involved a human–computer interaction study on the use of multiuser touch tables for collaborative geovisualization and decision-making in a marine spatial planning context, emphasizing how to measure collaboration in small groups. Brandon enjoys approaching her research from a psychology perspective, which usually underpins her research in some way. Research interests include marine spatial planning, participatory GIS and community mapping, 3D geovisualization, conservation psychology and behaviors, and marine protected areas.

Dr. Jennifer A. Brown

Monterey Bay National Marine Sanctuary, Monterey, CA 93940, USA; jennifer.brown@noaa.gov; montereybay.noaa.gov; +1-415-310-6764 (phone)

Jennifer Brown has been with the Monterey Bay National Marine Sanctuary (MBNMS) since 2004. As the ecosystem scientist for the Sanctuary Integrated Monitoring Network (SIMoN), she is responsible for developing and coordinating ecosystem assessments at both local and regional levels. These ecosystem models and condition reports help address resource management needs for integrated monitoring information. Dr. Brown supports ecosystem-based management of MBNMS by leading a collaboration with the California Current Integrated Ecosystem Assessment to develop key ecological indicators for sanctuary habitats, providing science support for the establishment of Sanctuary

Ecologically Significant Areas, and developing a collaborative groundfish essential fish habitat proposal to the Pacific Fishery Management Council. She has a BS in biology from the University of California, Los Angeles (UCLA), and a PhD in ecology and evolutionary biology from UCSC. Her doctoral research focused on evaluating the relative value of nearshore ecosystems at nursery grounds for juvenile flatfish. In addition to her work at the sanctuary, Brown is as an ecological consultant working on habitat assessments for fishes, especially flatfishes, in West Coast estuaries and evaluations of the nursery value of nearshore ecosystems to fish and invertebrate populations.

Chris Caldow

Channel Islands National Marine Sanctuary, University of California, Santa Barbara, Bldg 514, Santa Barbara, CA 93106-6115, USA; chris.caldow@noaa.gov

In 2014, Chris Caldow became research coordinator of the NOAA Channel Islands National Marine Sanctuary in Santa Barbara, California. His efforts are focused on establishing partnerships and promoting, conducting, and interpreting science aimed at addressing management priorities. Caldow has been with NOAA since 2000, when he became a John A. Knauss Marine Policy Fellow with the National Centers for Coastal Ocean Science (NCCOS) Biogeography Branch. The branch specializes in mapping the distributions of marine plants and animals to aid decision-makers faced with spatially explicit management decisions. In 2010, Caldow became the branch chief balancing the branch's efforts on core strengths such as seafloor characterization and science to support coastal zone management with emerging issues of national importance such as renewable-energy development and marine spatial planning. Caldow obtained a BS in aquatic biology from UCSB in 1996 and an MS in biology from the University of Houston in 2000.

Dr. Rosaline Canessa

Coastal and Ocean Resource Analysis Group, Department of Geography, University of Victoria, PO Box 3060, Victoria, BC V8W 3R4, Canada; Rosaline@uvic.ca; http://coral.geog.uvic.ca; +1-250-721-7339 (phone)

A BSc in marine biology at McGill University introduced Rosaline Canessa to the fascinating marine realm. But it was during her master's study in marine resource development and protection at Heriot-Watt University in the late 1980s that she discovered the curious intersection between GIS and the marine environment. That curiosity was nurtured during a PhD in the 1990s at the University of Victoria on coastal spatial decision support systems and her subsequent consulting work. Currently, Dr. Canessa is an associate professor in the Department of Geography at the University of Victoria and leads the CORAL Group. Her research focuses on place-based coastal management and the use of spatial technologies, such as GIS and seascape visualization, to support marine planning, particularly in interactive and collaborative settings.

Dr. Alexandra Carvalho

CMar Consulting LLC, Jacksonville, FL 32256, USA; alexandra@cmarconsulting.com; http://www.cmarconsulting.com; +1-904-993-4806 (phone)

Alexandra Carvalho first worked with GIS in the early 1990s while pursuing her degree in marine biology and fisheries at the Universidade do Algarve, Portugal. In 1994, she started her PhD in oceanography/coastal zone management at Florida Institute of Technology. She graduated in 2000. During her doctoral program, she continued working with coastal and marine spatial data management and analysis and started developing her own GIS applications. In 2001, Dr. Carvalho joined a coastal engineering consulting company, where she spent the next 10 years providing scientific planning, spatial data management and analysis, and project management support to the environmental, coastal, waterfront, water resources, and GIS groups. In 2011, she founded CMar Consulting, in which she is responsible for the company's environmental and GIS services. She specializes in coastal, marine, and water resource environmental and engineering spatial data management and analysis.

Alicia Clarke

NOAA National Centers for Coastal Ocean Science, 1305 East–West Hwy., Room 8110, Silver Spring, MD 20910, USA; Alicia.Clarke@noaa.gov; http://coastalscience.noaa.gov; +1-301-713-3028 (phone)

Alicia Clarke is a science communication specialist at NOAA NCCOS, a Consolidated Safety Services (CSS)-Dynamac contract position she has held since 2007. During that time, she has worked closely with scientists to develop products and tools that engage a broad audience. This includes leading the development of reports, story maps based on the Esri Story Map app, magazine articles, newsletters, educational videos, and other communication tools. Fusing her interests and background in marine science and journalism, Clarke has had the opportunity to cover a wide range of topics, such as the impacts of climate change in the Arctic, NOAA seafloor mapping efforts in the Caribbean, and the condition of coral reef ecosystems. Clarke also actively works to engage K-12 students and teachers in the world of science through hands-on experiences and bringing scientists into the classroom.

Dr. Heather M. Coleman

Pacific Marine Analysis and Research Association, PO Box 49002, Victoria, BC V8P5V8 Canada; hcoleman@pacmara.org; +1-778-300-1801 (phone)

Heather Coleman has acted as the science adviser for PacMARA since 2010. In this role, she provides advice on marine spatial planning and ecosystem-based management in an international and Canadian context. She coleads the development and delivery of training programs on systematic multiobjective planning, especially around the decision support tool Marxan, and communicating best practices for sound, integrated marine planning. She also facilitates working groups on topics such as ecosystem-based management implementation, science-policy integration, and stakeholder involvement in the planning process. Dr. Coleman has a PhD in marine ecology and an MA in environmental economics, both from UCSB.

Sophie de Beukelaer

Monterey Bay National Marine Sanctuary, Monterey, CA 93940, USA; sophie.debeukelaer@noaa.gov; http://montereybay.noaa.gov; +1-831-647-1286 (phone)

Sophie de Beukelaer has been with the Sanctuary since 2005 and is responsible for compiling, creating, and analyzing GIS data layers and related tools to augment efforts of MBNMS staff.

In addition, she is responsible for issuing and tracking all sanctuary research permits. De Beukelaer's GIS projects have focused on water quality issues, shoreline sensitivity issues, and most recently, developing Sanctuary Ecologically Significant Areas (SESAs). For this ecosystem-based management project, she developed biological, oceanographic, geologic, and other thematic layers to help inform future management decisions for the Sanctuary and contribute to improved characterizations of the region.

De Beukelaer holds a BA in natural sciences from New College and an MS in oceanography from TAMU. She focused her graduate research on the deep-sea ecology of the Gulf of Mexico and combined a variety of remote-sensing efforts, including submersibles and side-scan sonar, to identify the locations of chemosynthetic communities on the northern continental slope. De Beukelaer recently completed a two-year NOAA Coastal Management Fellowship with the Shoreline Management Program at the Washington State Department of Ecology to hone her GIS skills and learn more about coastal policy.

Daniel (Dan) S. Dorfman

Consolidated Safety Services Inc., 10301 Democracy Lane, Suite 300, Fairfax, VA 22030, USA, under contract for US Department of Commerce NOAA National Centers for Coastal Ocean Science, 1305 East–West Highway, SSMC4, Room 9216, Silver Spring, MD 20910, USA; dan.dorfman@noaa .gov; +1-301-713-3028, ext. 112 (phone)

Dan Dorfman is a marine conservation planner at NOAA NCCOS. His work is focused on applying scientific knowledge and information to support natural resource management decision-making. He specializes in the application of GIS and decision support tools. Dorfman served as the principal planner for intelligent marine planning. Past roles also include serving as senior marine planner, western regional GIS manager, and Hawaii GIS coordinator of The Nature Conservancy; working as the Hawaii Gap Analysis and Hawaii Marine Gap Analysis program manager of the Research Corporation of the University of Hawaii; and working with Greenpeace USA and the Rainforest Action Network. Dorfman holds a BA in ecology from the University of California, San Diego (UCSD), and an MA in biology from Boston University.

Dr. Dennis R. (Rob) Fatland

Microsoft Research, One Microsoft Way, Redmond, WA 98052, USA; rob.fatland@microsoft.com; http://research.microsoft.com; http://layerscape.org

Rob Fatland works collaboratively on applications of technology to data-driven problems in environmental science at Microsoft Research. His education includes a PhD in geophysics from the University of Alaska, Fairbanks, and a BS in physics from Caltech. He also worked in remote sensing at NASA/Jet Propulsion Laboratory (JPL) for six years. He joined Microsoft Research in 2010 and works with academics to build open solutions in the environmental data space, from sensor deployment to cloud processing with domain emphasis on oceanography. Dr. Fatland built a set of tools to enhance the Worldwide Telescope Virtual Globe, including an online learning/collaborating/ sharing website called Layerscape, at http://layerscape.org, and a set of developer tools called Narwhal that facilitate composition of complex data-driven visualizations from 30 million time-evolving geospatial data points.

Mark K. Finkbeiner

NOAA Office for Coastal Management, 2234 S. Hobson Ave., Charleston, SC 29405-2413, USA; mark.finkbeiner@noaa.gov; +1-843-740-1264 (phone)

Mark Finkbeiner is a physical scientist with the Office for Coastal Management and has over 30 years' experience in the remote sensing and GIS fields, with a BS in geography/remote sensing from Northern Arizona University. His work at NOAA has focused on mapping shallow intertidal and subtidal habitats with various aerial imagery and acoustic sensing technologies. He was also part of the development team of the Coastal and Marine Ecological Classification Standard (CMECS). His previous experience has included work on historical Superfund assessments, wetland jurisdictional determination, and satellite-based land-cover change detection. Currently, he is supporting the ocean planning community through the MarineCadastre.gov and Digital Coast efforts.

Kathy Fitzpatrick

Martin County Board of County Commissioners; Stuart, FL 34996, USA; kfitzpat@martin.fl.us; http://www.martin.fl.us; +1-772-288-5429 (phone)

Kathy Fitzpatrick is the coastal engineer of Martin County, Florida. She has worked in the coastal engineering field on the east coast of Florida since 1990. Her responsibilities have evolved from navigation maintenance for a small inlet district to countywide coastal management responsibilities in Martin County. Fitzpatrick's projects range from coastal storm protection, regional sand management, and marine and terrestrial environmental monitoring and management to large-scale estuarine, coastal, and marine environmental restoration. Fitzpatrick is a licensed professional engineer in Florida. She has a BA in social psychology and a BS in ocean engineering from Florida Atlantic University and an MS in engineering with a major in environmental engineering sciences from the University of Florida. Although she has not worked with GIS on a professional basis, she understands the benefits and uses of GIS for coastal programs and project planning and management.

Jessica Garland

Martin County Board of County Commissioners; Stuart, FL 34996, USA; jgarland@martin.fl.us; http://www.martin.fl.us; +1-772-288-5795 (phone)

Jessica Garland is associate project manager of the Coastal Engineering Division of Martin County. She has worked in the coastal engineering field since 2006. Garland's responsibilities have evolved from administrative to coastal project management, permit management, and grant management in Martin County. Garland is a self-trained GIS user. She started working with GIS in 2008 and appreciates the benefits and uses of GIS for project planning and management.

Kathleen L. Goodin

NatureServe; 4600 N. Fairfax Drive, 7th Floor, Arlington, VA 22203-1553, USA; kathy_goodin@natureserve.org; www.natureserve.org; +1-703-908-1883 (phone)

Kathy Goodin is deputy director of the Conservation Science Division and director of the marine program for NatureServe, a conservation science nonprofit organization. She is currently the lead marine scientist of NatureServe and is responsible for the development of CMECS, the

US national standard, and implementation of NatureServe's marine strategy. She has coordinated the development of national standards for terrestrial and marine ecological habitat classification, characterization, sampling, data management, and analysis. She has a BA in biology and environmental studies (1986) from Macalester College in St. Paul, Minnesota, and an MS in biology (1989) from the University of Miami, Miami, in Florida.

Karen F. Grimmer

Monterey Bay National Marine Sanctuary, Monterey, CA 93940, USA; karen.grimmer@noaa.gov; http://montereybay.noaa.gov; +1-831-647-4253 (phone)

Karen Grimmer has been with MBNMS since 1999 and is responsible for resource protection activities within the agency, leading a team of six. Prior to that, she served as deputy superintendent from 2007 to 2012 and has worked at the Sanctuary in a variety of positions, including Sanctuary Advisory Council coordinator and program coordinator of the Multicultural Education for Resource Issues Threatening Oceans (MERITO). She has over 22 years' experience in marine resource management and oversees the site's marine policy activities, including fisheries-related actions (i.e., essential fish habitat), vessel traffic compliance, lost fishing gear removal, and acoustic impacts, as well as some permitting actions (i.e., desalination). Most recently, Grimmer led a team to establish Sanctuary Ecologically Significant Areas within the Sanctuary and worked with fishermen and conservation groups to collaboratively propose boundary modifications for the trawl fishery.

Grimmer grew up in the San Francisco Bay Area and served as the director of education for the nonprofit Marine Science Institute (MSI) in Redwood City, California. Prior to MSI, she worked as an aquarist at Steinhart Aquarium in San Francisco. Karen received her BS in biology from San Francisco State University and her master's degree in education from California State University (CSU), Hayward.

Todd Hallenbeck

West Coast Ocean Data Portal, West Coast Governors Alliance on Ocean Health, Alameda, CA 94501, USA; Todd.r.hallenbeck@westcoastoceans.org; +1-408-482-6807 (phone)

Todd Hallenbeck is project coordinator of the West Coast Ocean Data Portal. He graduated with an MS in coastal watershed science and policy from CSU Monterey Bay, where he studied seafloor mapping, benthic ecology, and the application of GIS technology to ocean policy. Hallenbeck completed a two-year West Coast Governors Alliance and Sea Grant Fellowship, in which he worked directly with the California Governor's Office of Planning and Research Natural Resources Agency staff to provide GIS and staff support for the state's public process to develop a marine spatial plan for renewable ocean energy. During his fellowship, Hallenbeck also helped spearhead the creation and coordination of a West Coast–wide network of ocean and coastal data managers and users to help increase the discoverability and connectivity of ocean and coastal data and people to better inform regional resource management, policy development, and ocean planning.

Sarah D. Hile

NOAA National Centers for Coastal Ocean Science, 1305 East–West Hwy., Room 8110, Silver Spring, MD 20910, USA; Sarah.Hile@noaa.gov; http://coastalscience.noaa.gov; +1-301-713-3028 (phone)

Sarah Hile is a marine ecologist with NOAA NCCOS, a CSS-Dynamac contract position she has held since 2004. She has an MSc in coastal ocean management from the University of Ulster in Northern Ireland, with a current research focus on the condition of coral reef ecosystems, habitat utilization of reef fish, and management of marine protected areas. Hile is the comanager and quality control monitor of an extensive Caribbean-wide fish metrics and benthic composition database and a field coordinator of Caribbean field missions. This includes leading training seminars for field data collectors, ensuring the accuracy of data and metadata housed in a publicly served database, conducting data analyses, assisting in field mission planning and preparation, as well as generating products and publications. Hile is also involved in preschool and elementary science education and outreach efforts, with a focus on introducing science experiences at an early age.

Dr. Matthew K. Howard

Department of Oceanography, Texas A&M University, Oceanography and Meteorology Building, Room 614B, MS 3146, College Station, TX 77843, USA; mkhoward@tamu.edu, +1-979-862- 4169 (phone)

Matthew Howard is a physical oceanographer with 37 years' experience and training. His interests include data management, informatics, and interoperability. Dr. Howard has been involved with the US Integrated Ocean Observing System (IOOS) since 1998 and served on its Data Management and Communication Steering and Quality Assurance of Real-Time Oceanographic Data Program advisory teams. He is a member of the Marine Metadata Interoperability executive team. He is currently the data manager of the Gulf of Mexico Coastal Ocean Observing System (GCOOS) and a principal investigator in the Gulf of Mexico Research Initiative Information and Data Cooperative. He has been the principal data manager of some of the largest field experiments conducted in the Gulf of Mexico, including the Louisiana-Texas (LATEX), Northeastern Gulf of Mexico (NEGOM), and Mechanisms Controlling Hypoxia (MCH) surveys. Recently, he has become active in Gulf of Mexico glider deployments in support of hypoxia studies.

Dr. Stephan Howden

Department of Marine Science, University of Southern Mississippi, 1020 Balch Blvd., Stennis Space Center, MS 39529, USA; stephan.howden@usm.edu; +1-228-688-5284 (phone)

Stephan Howden is an associate professor in the Department of Marine Science (DMS) at the University of Southern Mississippi (USM). Dr. Howden teaches in the DMS Hydrographic Science academic program, which is one of two programs in the United States recognized by the International Federation of Surveyors/International Hydrographic Organization/International Cartographic Association International Board at the Category A level. He also operates the Central Gulf of Mexico Ocean Observing System (CenGOOS), at www.cengoos.org, and has been involved with GCOOS since 2003, for which he presently serves as a member of the board of directors.

Nuria Hermida Jiménez

Department of Marine Environment Protection/Marine Geology, Spanish Institute of Oceanography, 8 Corazón de María, 28002 Madrid, Spain; nuria.hermida@gmail.com; +34-91-510-75-19 (phone)

Nuria Jiménez is a technical surveying engineer and engineer in geodesy and cartography. She has over 14 years' experience in marine geomatics, including marine thematic mapping, designing, developing, and maintaining large spatial database systems, and applications and analytical issues in GIS for oceanographic data. She has participated in oceanographic research surveys, including the underwater volcanic eruption of El Hierro Island. Recently, she has focused on the development of a marine spatial data infrastructure composed of a metadata catalog, interactive web map viewers, standard map services, and search functions.

Dr. Christopher (Chris) F. G. Jeffrey

Consolidated Safety Services Inc., 10301 Democracy Lane, Suite 300, Fairfax, VA 22030, USA; under contract for US Department of Commerce NOAA National Centers for Coastal Ocean Science, 1305 East–West Highway, SSMC4, Room 9213, Silver Spring, MD 20910, USA; chris.jeffrey@noaa.gov; +1-301-713-3028, ext. 134 (phone)

Chris Jeffrey is a senior science professional and GIS practitioner with 20 years' experience in conducting and managing scientific research in coral reef ecosystems and communicating information to promote sustainable use of natural resources through minimal human impacts to the environment. He has designed, planned, and implemented field-based research programs to sample marine populations of interest, analyzed information to quantify spatial patterns in fauna and flora, and evaluated the status and condition of coastal marine ecosystems (e.g., coral reefs, sea grasses, and mangroves) in the US Caribbean. Dr. Jeffrey holds a PhD in ecology and an MS in conservation biology and sustainable development from the University of Georgia. In 1999, he was named a John A. Knauss Marine Policy Fellow with the NOAA NCCOS Biogeography Branch. He is employed by CSS-Dynamac and currently manages a team of NOAA-affiliated scientists that apply science and technology to create a healthier, safer, and environmentally sustainable future.

Dr. Ann Elizabeth Jochens

Department of Oceanography, Texas A&M University, College Station, TX 77843-3146; ajochens@tamu.edu; http://gcoos.org; +1-979-845-6714 (phone)

Ann Elizabeth Jochens is a research scientist in oceanography at TAMU. Her principal interest is development of integrated, comprehensive ocean observing systems through partnerships. She came to physical oceanography as a mathematician and attorney with extensive experience in environmental, safety, and permitting in the oil and gas and minerals industries. She was a principal investigator and project manager on 15 studies funded for over $35 million. These multi-institutional, interdisciplinary studies focused on the circulation, water properties, and habitat over the Texas, Louisiana, Mississippi, Alabama, and Florida shelf and slope and the deep waters of the Gulf of Mexico. She participated in an ocean observatory project off Oman and was the US World Ocean Circulation Experiment Project Scientist. Since 2005, Dr. Jochens has held leadership positions, serving as the first executive director in the Gulf of Mexico Coastal Ocean Observing System Regional Association (GCOOS-RA)—a multientity partnership of government agencies, academic and research institutions, and private industries that is working to develop a comprehensive ocean observing system for the Gulf.

Chad King

Monterey Bay National Marine Sanctuary, Monterey, CA 93940, USA; chad.king@noaa.gov; http://sanctuarysimon.org; +1-831-647-4248 (phone)

Chad King has been with the Sanctuary since 2002 and is responsible for the collection, analyses, and dissemination of spatial data for SIMoN and MBNMS. This data helps integrate past and present monitoring programs within the Sanctuary and is the foundation of decision-making tools such as interactive maps that are made available to the general public. He is a NOAA dive master and an active participant in subtidal research, including kelp forest monitoring and underwater photography and videography. Additionally, he produces short outreach films and significant content for the Sanctuary Exploration Center. King was instrumental in developing SeaPhoto, an iOS app that features imagery and life history content of MBNMS. King has an MS in marine science from Moss Landing Marine Laboratories, and his academic research focused on kelp forest ecology and subtropical ecological dynamics and genetics in the Gulf of California.

Melodi (Melo) King

Smallmelo: Geographic Information Services LLC, Albuquerque, NM 87114, USA; melo@smallmelo.com; http://smallmelo.com; +1-520-241-0314 (phone)

Melo King stumbled on GIS after abandoning a graduate program in chemical engineering and quickly recognized it as a natural framework for pursuing a career she was passionate about. After a year of side projects in mapping public art and water systems, she earned her MS in GIS from the University of Redlands in Redlands, California. There, she began building Whale mAPP, a GIS-based citizen science project that encourages participation in marine mammal research. With her company, Smallmelo, she combines her degrees in engineering and GIS with her research in user experience/user interfaces, design, and programming to help organizations collect and interact with spatial data.

Dr. Shinichi Kobara

Department of Oceanography, Texas A&M University, College Station, TX 77843-3146, USA; shinichi.kobara@gcoos.org; http://gcoos.org/products; +1-979-845-4089 (phone)

Shinichi Kobara is a marine geographer and certified GIS professional (GISP) with over 12 years' experience in geospatial analysis and the development of map applications. He has a BS and MS in bioengineering from Soka University in Japan and a PhD in geography from TAMU. Dr. Kobara has applied geospatial theory and GIS to studies in marine environment, specifically in prediction of reef fish spawning aggregations. His current research interests include the analysis of geospatial environmental information for natural resource management. He is currently involved in developing web map products for GCOOS.

Dr. Peter Kouwenhoven

CLIMsystems, PO Box 638, Waikato Mail Centre, Hamilton, 3204, New Zealand; pkouwen@climsystems.com; http://climsystems.com; +64-834-2999 (phone)

Peter Kouwenhoven is senior scientist with CLIMsystems. He started his career more than 20 years ago with Resource Analysis in Delft, the Netherlands, as an international consultant bridging the gap between science and policy. He worked on planning issues in the Netherlands and many other countries with an integrative approach that usually included work on environmental change and often climate change. Dr. Kouwenhoven brings a wide-ranging expertise and experience. He has worked on models on every scale (from microscopic organisms to global), many topics (e.g., ecology, hydrology, sedimentation, pollution, eutrophication, air traffic, economy), applications (e.g., risk assessment, adaptation options, planning), and time scales (current, through midterm and long term). He has developed modeling tools (science, technology, engineering, and math, or STEM, for differential equations and rapid assessment process, or RAP, for qualitative modeling) and applied these in his work with various stakeholder groups (local communities, experts, decision-makers). Internationally, Kouwenhoven is involved in GEO (Global Environmental Outlook) from the United Nations Environment Programme (UNEP) as well as CAN (Collaborative Adaptation Network). He has worked in more than 30 countries, including New Zealand, the Netherlands, Brazil, Nigeria, Canada, India, Bangladesh, Uzbekistan, China, Kazakhstan, Kyrgyzstan, the Philippines, Vietnam, Fiji, the Cook Islands, Samoa, Vanuatu, Nauru, Kiribati, Tonga, Tuvalu, and Mauritius. This gives him a unique perspective on global climate change, which he draws on for his consultancy and training work with CLIMsystems.

Kouwenhoven has a PhD in environmental management from the Technical Environmental Management Group in the Department of Chemical Engineering at the University of Twente in Enschede, Netherlands. He holds an MSc and BSc in biology from the University of Leiden, Netherlands. He also holds certificates in computer science, communication training, and small business management. Kouwenhoven is an associate member of the New Zealand Planning Institute and an adjunct associate professor in the Climate Change program at the University of the Sunshine Coast in Queensland, Australia.

Andy Lanier

Oregon Coastal Management Program, Department of Land Conservation and Development, Salem, OR 97301, USA; Andy.Lanier@state.or.us; +1-503-934-0072 (phone)

Andy Lanier is a coastal resource specialist with the Oregon Coastal Management Program, a division of the state Department of Land Conservation and Development. He is a graduate of the marine resource management master's degree program at Oregon State University, where he also earned his GISP certificate. Lanier joined the department after completion of a two-year NOAA Coastal Management Fellowship, in which he worked toward the creation of an ocean GIS database for use by state agencies and community interest groups. Lanier is currently participating in the West Coast Governors Alliance on Ocean Health as cochairman of the West Coast Ocean Data Portal Action Coordination Team, working to improve timely access to relevant coastal and ocean data on the West Coast. He has also been active within the Oregon coastal and marine spatial data networking community.

Alie LeBeau

Aquarium of the Pacific, Education Department, 100 Aquarium Way, Long Beach, CA 90802, USA; alebeau@lbaop.org; +1-562-951-1631 (phone)

Alie LeBeau is the education program manager at the Aquarium of the Pacific. She serves as primary project manager of many grant-funded projects within the Education Department, including the Institute of Museum and Library Services–funded Spherical Interpretation and Technology Integration staff professional development project and annual Boeing Teacher Institute. She has a background in elementary education, outdoor education, and community partnerships. Within the department, LeBeau facilitates teacher professional development offerings, manages program staff, develops strategic partnerships, and manages program budgets. She holds a bachelor's degree in biology and education from Earlham College in Richmond, IN.

Dr. Jennifer Lentz

Aquarium of the Pacific, Education Department, 100 Aquarium Way, Long Beach, CA 90802, USA; jlentz@lbaop.org; http://JenniferALentz.info; +1-562-951-1642 (phone)

Jennifer Lentz is an education specialist at the Aquarium of the Pacific. She joined the Aquarium's Education Department in April 2013 as a Marine Science Fellow. Dr. Lentz first encountered GIS in 2004 during a summer internship, studying the home range and habitat preferences of eastern box turtles. Her doctoral research focuses on developing geospatial analytical protocols that could be applied in the study of coral health. Her work at the Aquarium uses GIS to develop interactive ways to connect people of all ages to marine science and the ecology of the surrounding world. Lentz holds a BS in environmental science from Hamilton College in Clinton, NY, and a PhD in oceanography and coastal sciences from Louisiana State University, Baton Rouge.

Yinpeng Li

CLIMsystems, PO Box 638, Waikato Mail Centre, Hamilton, 3204, New Zealand; yinpeng@climsystems.com; http://climsystems.com; +64-834-2999 (phone)

Yinpeng Li is a senior climate scientist with CLIMsystems. He worked as a climate change science researcher for the International Global Change Institute (IGCI), New Zealand, and the Institute of Atmospheric Physics, Chinese Academy of Sciences. His main expertise is climate change integrated assessment model development. He developed a dynamic vegetation model (DVM) and food and water security models and tools for Food and Water Security Integrated Model System (FAWSIM) and SimCLIM. He is coordinating an Asia Pacific Research Network project on water and food security integrated assessment model development for northeast Asia, including China, Mongolia, Russia, and New Zealand. He was involved in the development of an Atmosphere Vegetation Interaction Model (AVIM). National and international peers have recognized AVIM, and a series of papers have been published. His research interests also extend to climate risk analysis, biocarbon sequestration assessment, data analysis, and vegetation ecology.

Paul Manson

Institute for Sustainable Solutions, Hatfield School of Government, Portland State University, PO Box 751-SUST, Portland, OR 97207, USA; mansonp@pdx.edu; http://www.naturalpraxis.com; +1-503-804-1645 (phone)

Paul Manson is a PhD student in the Public Affairs and Policy program of the Hatfield School of Government at Portland State University. He is also a fellow in the National Science Foundation–funded Integrated Graduate Education and Research Traineeship (IGERT) in Ecosystem Services for Urbanizing Regions (ESUR). His academic work is divided between scholarship on the role of models, spatial tools, and technology in participatory government efforts and applied practitioner support through participation in regional environmental planning processes. Recent efforts include development of spatially assisted public participation tools for ecosystem services planning in Nevada. This effort supports a local government policy for "no net loss" of ecosystem services. Other research includes Bayesian inferential modeling for marine ecosystem service functions through a National Oceanographic Partnership Program (NOPP)–funded project in partnership with Oregon State University. This project is developing a methodology to manage uncertainty in science and divergent stakeholder opinions. He also recently coauthored a Transportation Research Board report on the development of ecosystem service–based crediting methods to better inform transportation decisions and measure progress. Manson holds a BA in anthropology from Reed College and an MPA from Portland State.

Dr. Ruth Mullins-Perry

Shell Upstream Americas, PO Box 2463, Houston, TX 77252, USA; ruth.perry@shell.com;
+1-281-450-7148 (phone)

Ruth Mullins-Perry is a marine scientist specializing in physical and biological oceanography, ocean observing, and ocean management and policy. Dr. Perry earned her PhD in oceanography from TAMU. Her dissertation research focused on understanding the physical processes driving coastal hypoxia formation on the Texas coast and incorporated the use of real-time ocean observing systems, remote sensing, and GIS statistical modeling to quantify area and identify spatial and temporal trends of hypoxia in the western Gulf of Mexico. She joined GCOOS as a postdoctoral research associate. Her work focused on incorporating GIS into oceanographic applications, including the statistical mapping of coastal environmental hazards and ocean observing operations, such as profiling gliders and buoys, and outreach and stakeholder engagement efforts, such as K-12 curriculum. Perry recently joined Shell as a marine science and regulatory policy specialist, where she continues to use GIS as an adaptive tool.

Robert Newell

Coastal and Ocean Resource Analysis (CORAL) Group; Department of Geography, University of Victoria, PO Box 3060, Victoria, BC V8W 3R4, Canada; rgnewell@uvic.ca; +1-250- 472-4624 (phone)

Robert Newell has a BSc in biology and statistics, an MA in environment and management, and is currently a doctoral student of geography at the University of Victoria, British Columbia, Canada, researching the potential immersive geovisualization has for supporting coastal management objectives. Newell also works as a research associate of the Canadian Research Chair in Sustainable Community Development program, based in the School for Environment and Sustainability at Royal Roads University (RRU), and with RRU's School of Culture and Communications as a course developer and instructor. Newell has worked on research that includes developing creative and interactive ways of using digital media and online communication to disseminate sustainability research ideas, engage stakeholders and the public in sustainability, and facilitate knowledge sharing and collaborative strategies for sustainable development.

Dr. Simon J. Pittman

NOAA National Centers for Coastal Ocean Science Biogeography Branch, Silver Spring, MD 20910, USA; simon.pittman@noaa.gov; http://www.researchgate.net/profile/Simon_Pittman

Simon Pittman is an ecologist with a PhD in geographical sciences from the University of Queensland, Australia, and an MSc in environmental sciences from the University of

Wolverhampton, West Midlands, United Kingdom. He has applied GIS and remote-sensing data to address complex spatial problems in applied ecology since the early 1990s, when scripting for ArcInfo was the modus operandi. Dr. Pittman's primary area of research is the emerging field of seascape ecology, the application of landscape ecology concepts and tools to the marine environment. For the past decade, he has worked as a research scientist with the NOAA NCCOS Biogeography Branch, applying spatial technologies and predictive mapping to support marine spatial planning. He serves on the NOAA Caribbean Committee and Virgin Islands Reef Resilience Working Group and is editor in chief of the *NOAA in the Caribbean* newsletter. Pittman is director of Seascape Analytics, a spatial science consultancy.

Beatriz Ramos López

Department of Sciences, University of Alcala, Madrid, Spain; beatriz.ramosl@uah.es; http://www3.uah.es/marinebiodiversity/web; +34-60-067-41-21 (phone)

Beatriz Ramos López is a biologist with two master's degrees: a GIS emphasis, 2011–12, from Esri Spain, and a geographic information technologies emphasis, 2012–13, from the University of Alcala. She is currently a researcher in the EU–US Marine Biodiversity Research Group of the Department of Sciences at the University of Alcala, while also working toward her PhD at the Franklin Institute at the University of Alcala. Her PhD thesis focuses on the use and evolution of web GIS for marine spatial planning.

Rebecca A. Schaffner

Maine Department of Environmental Protection, 17 State House Station, Augusta, ME 04333-0017, USA; rschaffner@gmail.com; +1-310-795-6645 (phone)

Rebecca Schaffner began using GIS to support mapping of harmful algal blooms in New York and California, first while completing her MS at the State University of New York (SUNY), Stony Brook, and later as a research technician at the University of Southern California. From there, she moved on to the Southern California Coastal Water Research Project (SCCWRP), in Costa Mesa, California, where she collaborated as a GIS specialist on many different marine and aquatic research projects, including regional surveys of the Southern California Bight and mapping of riverine and wastewater discharge plumes. She has recently returned to the East Coast to work for the Maine Department of Environmental Protection as a GIS coordinator.

Kenneth (Ken) C. Schiff

Deputy director, Southern California Coastal Water Research Project, Costa Mesa, CA 92626-1437 USA; kens@sccwrp.org; http://www.sccwrp.org; +1-714-755-3202 (phone)

Ken Schiff is the deputy director of SCCWRP, an environmental research agency in California that provides unbiased technical information to environmental managers and policy makers so that they can make well-informed decisions about stewarding the region's rich natural resources. Schiff received a BS in biology from San Diego State University, an MS in biology from CSU Long Beach, and has been at SCCWRP for over 23 years. With nearly 60 peer-reviewed publications and book chapters to his credit, Schiff has become a national expert in monitoring design and interpretation. He currently leads several interdisciplinary, regionally based monitoring programs in California, including pollutant sources, fates, and effects across a multitude of habitats, including marine, estuarine, and freshwater ecosystems.

Lori Scott

NatureServe; 4600 N. Fairfax Drive, 7th Floor, Arlington, VA 22203-1553, USA; lori_scott@natureserve.org; www.natureserve.org; +1-703-908-1877 (phone)

Lori Scott is the chief information officer for NatureServe, a leading source of information about rare and endangered species and threatened ecosystems. She oversees software development, support, and IT infrastructure teams that produce GIS and web mapping applications for biodiversity data collection, management, and sharing. Scott holds a BS in mathematics from Bucknell University in Lewisburg, Pennsylvania, and has more than 20 years' experience in both corporate and nonprofit settings designing geospatial software solutions.

Dr. Christina Simoniello

Department of Oceanography, Texas A&M University, based at the University of South Florida College of Marine Science, 140 Seventh Avenue South, St. Petersburg, FL 33701, USA; simo@marine.usf.edu; +1-727-322-1318 (phone)

Christina Simoniello received a BS in biological sciences and a Certificate in Marine Science from Florida International University in 1988. She has conducted herpetological studies in Everglades National Park; worked in analytical chemistry for the Drinking Water Research Center, Miami; and researched bears, birds, seals, and otters for the Fish and Wildlife Service, Alaska, following the Exxon Valdez oil spill. Completing her PhD in biological oceanography at the University of South Florida in 2003, she has conducted research in the oceanic Gulf of Mexico, Florida Keys, Exumas, and southern ocean. Following five years of developing programs for the Southeast Atlantic Coastal Ocean Observing System as faculty in the University of Florida Sea Grant College Program, she presently directs outreach and education activities for GCOOS-RA. Current leadership roles include chairwoman, US Integrated Ocean Observing System Association Education and Outreach Committee, Steering Committee; past chairwoman, Gulf of Mexico Alliance Environmental Education Network; and Steering Committee member, NOAA Gulf of Mexico Regional Collaboration Team.

Stephen Sontag

Applied Science Associates Inc., 55 Village Square Drive, South Kingstown, RI 02879, USA; ssontag@asascience.com; +1-401-789-6224 (phone)

Stephen Sontag has a BS in geographic information management and a BS in remote sensing from Northern Arizona University. Sontag works on the technical team at RPS (Rural Planning Services) ASA, an environmental consulting company in Rhode Island. Sontag has a broad range of geographic data management experience and has most recently been active in the ocean sciences community. He is a web developer who incorporates his GIS data visualization skills to create resource management systems. Sontag has developed multiple rich web applications, leveraging standard and nonstandard geographic data formats for marine spatial planning and industry decision-makers.

Dr. Steven J. Steinberg

Department of Information Management and Analysis, Southern California Coastal Water Research Project, Costa Mesa, CA 92626-1437, USA; steves@sccwrp.org; http://www.sccwrp.org; +1-714-755-3260 (phone)

Steven Steinberg is a principal scientist at SCCWRP, an environmental research agency in California, with a focus on geospatial and data collection, management, analysis, and visualization systems. Dr. Steinberg first began working on natural resource applications of GIS in the early 1990s as part of his MS at the University of Michigan. From 1998 to 2011, Steinberg was a professor of geospatial science at Humboldt State University, California. He has been honored as a Fulbright Distinguished Chair (Simon Fraser University, Canada; 2004) and a Fulbright Senior Scholar (University of Helsinki, Finland; 2008). Steinberg is an active member and officer in several professional geospatial organizations, including the American Society for Photogrammetry and Remote Sensing (ASPRS), Urban and Regional Information Systems Association (URISA), Geographic Information Systems Certification Institute (GISCI), and California Geographic Information Association (CGIA). He also serves on multiple statewide and regional workgroups addressing data management and visualization, including the West Coast Governors Alliance on Ocean Health, California Coastal and Marine Geospatial Working Group, and California Water Quality Monitoring Council Data Management Workgroup. Steinberg has been a certified GISP since 2008.

Dr. Lei Lani Stelle

Department of Biology, University of Redlands, 1200 E. Colton Ave., Redlands, CA 92373, USA; leilani_stelle@redlands.edu; +1-909-748-8628, ext. 8628 (phone)

Lei Lani Stelle is an associate professor in the Department of Biology at the University of Redlands. She has studied marine mammals for over 20 years, starting when she earned her undergraduate degree in marine biology from UCSC. Dr. Stelle holds an MSC in zoology from the University of British Columbia and a PhD in organismic biology, ecology, and evolution from UCLA. Stelle's current research focuses on how humans impact marine mammal species in the busy waters off Southern California. She has supervised numerous graduate and undergraduate students in a range of projects, with an emphasis on using GIS to examine the relationships between marine mammal distributions and human activities. Her teaching has emphasized marine topics, including a course designed to use GIS to examine questions related to animal conservation. She has been involved in citizen science throughout her career and has led numerous expeditions that involve volunteer researchers.

Olvido Tello Antón

Department of Marine Environment Protection/Marine Geology, Spanish Institute of Oceanography, 8 Corazón de María, 28002 Madrid, Spain; olvido.tello@md.ieo.es; +34-91-510-75-35 (phone)

Antón earned a degree in geology from Complutense University, Madrid, Spain, in 1989. She has been on the research staff of the Spanish Institute of Oceanography since 1998, specializing in GIS and marine cartography. She also serves as Chief Scientist during oceanographic research cruises. She has participated in several national and international projects related to the exploration of the marine environment with high-resolution multibeam sonar, side-scan sonar, and seismics, as well as marine cartography and marine resource conservation. Antón has developed GIS projects for research institutions, companies, and the Spanish government, specializing in the interpretation of seabed geomorphology and sedimentology, the generation of 3D seabed models, marine data analysis and management, and the development of marine metadata with the International Organization for Standardization (ISO) 19115 standard.

Dr. Peter Urich

CLIMsystems, PO Box 638, Waikato Mail Centre, Hamilton, 3204, New Zealand; peter@climsystems.com; http://climsystems.com; +64-834-2999 (phone)

Peter Urich is the managing director of CLIMsystems. He has over 15 years' experience in community development, resource management, and climate change adaptation and risk assessment. Dr. Urich earned his PhD in human geography from the Research School of Pacific and Asian Studies at the Australian National University in Canberra and a BA and MA in geography from the University of Wisconsin, Madison. He also holds a diploma in agriculture from Guelph University in Canada. His work with rural economies and land managers across Asia and the Pacific and his publication record and affiliations across government and nongovernment organizations and private industry position him well to bring climate risk and impact assessment tools and technologies to a diverse array of end users. Urich has recently implemented the development and delivery of customized climate change risk assessment tools for the governments of Tonga and Vanuatu in the South Pacific. Urich is also the project leader for the Southeast Queensland Climate Change Mapping Project.

Dr. Tiffany C. Vance

NOAA National Marine Fisheries Service Alaska Fisheries Science Center, 7600 Sand Point Way NE, Seattle, WA 98115, USA; tiffany.c.vance@noaa.gov; +1-206-526-6767 (phone)

Tiffany Vance has an MS in marine geology and geophysics from the University of Washington and a PhD in geography and ecosystem informatics from Oregon State University. She works for the NOAA National Marine Fisheries Service and is an adjunct faculty member in the Department of Geography and Environment at San Francisco State University. Her research addresses the application of multidimensional GIS to both scientific and historical research, with an emphasis on the use and diffusion of techniques for representing 3D and 4D data. Ongoing projects include developing techniques to define and describe essential pelagic habitat; developing histories of environmental variables affecting larval pollock recruitment and survival in Shelikof Strait, Alaska; and the use of GIS and visualizations in the history of recent Arctic science.

Frank Veldhuis

NorthStar Geomatics Inc., Stuart, FL 34994, USA; frankv@nsgeo.com; http://www.nsgeo.com; +1-772-781-6400 (phone)

Frank Veldhuis is vice president and cofounder of NorthStar Geomatics, a GIS, surveying/mapping, and asset inventory firm in Stuart, Florida. A third-generation Florida native, Veldhuis obtained his BS in geomatics from the University of Florida in 1999. He then became licensed as a professional surveyor and mapper (PSM) in the State of Florida in 2007. His primary area of expertise is GIS, asset inventory, and IT support. Veldhuis's clients include county government, municipalities, water management districts, private drainage districts, and private engineering firms.

Stephanie M. Watson

Gulf of Mexico Coastal Ocean Observing System Communications lead; Stephanie.watson@gcoos.org; http://www.gcoos.org; +1-985-237-2727 (mobile)

Stephanie Watson is a consultant specializing in ocean observing, coastal and ocean management and policy, and data management. Watson has over 17 years' experience and is currently most active in the Gulf of Mexico region, where there is a great deal of cutting-edge ocean observing and monitoring technology, a diversity of stakeholder interests, many opportunities for private-public partnerships, and critical coastal and ocean resource management needs. Watson also has significant experience in the Gulf of Maine and the Central and Northern California coastal region. She has served as associate director of the Center for Gulf Studies at the University of Southern Mississippi and coordinator of the Central and Northern California Coastal Ocean Observing System. Watson has MS degrees in spatial information science and engineering and in environmental science and policy (marine) and has frequently used GIS as an adaptive tool.

Tim Welch

Technical consultant, Portland, OR, USA; tim.j.welch@gmail.com; +1-971-227-2357 (phone)

Tim Welch served as the technical lead for the West Coast Ocean Data Portal while working for Point 97, a technology company delivering easy-to-use technology solutions and program engagement strategies that improve marine and coastal management practices. For over eight years, he has helped develop award-winning web-based tools for spatial decision support and data collection in the United States and abroad. He is now an independent consultant based in Portland, Oregon.

Kyle Wilcox

Axiom Data Science, 1016 W. Sixth Ave., Suite 105, Anchorage, AK 99501, USA; kyle@axiomalaska.com; +1-360-450-3571 (phone)

Kyle Wilcox received his BS in computer science from the University of Rhode Island in 2006 and has since worked in the ocean sciences. He is a senior software engineer at Axiom Data Science of Anchorage, Alaska. Wilcox designed and developed the core of the LarvaMap applications and has also written and contributed to various Java and Python toolboxes for the ocean sciences. He currently resides in Rhode Island with his wife, Shannon, and their two children.

Emily Yam

Aquarium of the Pacific, Education Department, 100 Aquarium Way, Long Beach, CA 90802, USA; eyam@lbaop.org; +1-562-951-5378 (phone)

Emily Yam is the science interpretation supervisor at the Aquarium of the Pacific. In that role, she oversees grant-funded projects, strategic partnerships, and professional development of staff. Yam is especially interested in oceanography, climate change interpretation, data-driven interpretation, and spherical displays such as the Science on a Sphere. Yam holds a BS in biology and an MAT from the University of Virginia and an MS from the Virginia Institute of Marine Science, where she studied microbial and plankton ecology.

Index

The letter *f* following a page number denotes a figure.
The letter *t* following a page number denotes a table.

A

Aberrant Drift hypothesis, of fish stock variability, 5
Acidification, ocean, 92
    climate change affecting, 100–101, 101f
    in northern Gulf of Mexico, wave glider monitoring of,
        324–29, 326f, 327f, 329f
Acoustic Doppler current profiler, in marine drift
    experiment, 276, 277, 277f
Adaptation, definition of, 105
Adaptation costs, estimates of, for ocean changes, 105
Aesthetic resources. *See* Visual resources
Alaska, pollock fishery in, history and status of, 3–5
Alkalinity, ocean, total, climate change affecting,
    100–101, 101f
Android mobile phone, whale mAPP for, 155. *See also*
    Whale mAPP
Apache Solr, in West Coast Ocean Data Portal, 175, 175f
Aquarium of the Pacific (Long Beach, CA), 223–44
    description of, 224–25
    geospatial outreach by, through teacher workshops, 238–39
    geospatial programs of, 228–39
        ArcGIS Online website as, 232–35, 233f, 234f,
            235f, 236f
        floor-to-ceiling exhibit maps as, 228–29, 229f
        future, preview of, 239–40
        Google Liquid Galaxy exhibit as, 230–31, 230f
        new Story Map apps as, 240
        Science on a Sphere exhibit as, 231–32, 231f
            new content for, 239
        stand-alone web applications as, 236–38, 237f, 238f
    as Informal Science Education Institution, 223, 226
    interactive, web-based technologies in, 227
    learning theory employed by, 226
    next-generation science standards and, 226
ArcGIS
    applications in, LarvaMap output in, 19, 20f
    for Desktop
        applications of
            for Artificial Reef Program, 50f, 59–60
            for Beach Management Program, 62–64, 63f
        in Coastal GIS Program
            map optimization in, 56
            map organization in, 56

map solution in, 56
in developing collaborative proposal for essential
    fish habitat, 212, 212f, 218, 219f
in viewing wave glider acidification monitoring
    project data, 328
for Desktop Editor, in observation toolkit, 82
in GCOOS design of web-based mission viewers, 323
Online
    applications of, for Beach Management Program, 64
    in confederated visualization environments, 272, 273f
    web map applications
        of Aquarium of the Pacific, 232–35, 233f, 234f,
            235f, 236f
        for Artificial Reef Program, 60–61, 60f
        in developing essential fish habitat collaborative
            proposal, 213, 218
for Server
    in Cabrera map viewer development, 141–42
    in EcoDAAT, 7, 8f
    Ecosystems Fisheries Oceanography Coordinated
        Investigations Data Access and Analysis Tool
        use of, 1
    evolution of, 2
    in fisheries oceanography, 21
    in wave glider ocean acidification monitoring project
        mapping, 328, 329f
ArcGIS ModelBuilder, in Oregon territorial sea plan
    update, 299
ArcGIS Tracking Server, in wave glider ocean acidification
    monitoring project, 327–28, 329f
Archipelago de Cabrera National Park, Balearic Islands, Spain,
    map viewer for, 135–49. *See also* Map viewer, Cabrera
ArcMap, LarvaMap output in, 19, 20f
ArcPad, in observation toolkit, 82
ArcScene, LarvaMap output in, 20f
Artificial Reef Program (Martin County, FL), 58–61
    ArcGIS for Desktop applications for, 59–60, 59f
    ArcGIS Online web map applications for, 60–61, 60f
    data for, 58–59
Autonomous underwater vehicle (AUV), orbiting
    environmental sample processor in marine drift
    experiment, 274–77, 275f, 276f

Dawn J. Wright, ed.; 2015; *Ocean Solutions, Earth Solutions*; http://dx.doi.org/10.17128/9781589483651

B

C

overview of, 50–52
program history of, 51
Collaborative Ocean Visualization Environment (COVE)
3D marine geovisualization system developed by, 251
as monolithic visualization software, 272
Community-based Marine Assessment of Aesthetic Resources (C-MAAR)
challenges in, 308–9
limits to, 309–10
in Oregon territorial sea plan update
adoption and adaptation of established methods in, 293
conducting assessments for, 294–301, 297f, 298f, 300f, 302f
inventory planning in, 293–94
presenting results in, 302–3, 304f, 305f, 306f, 307
for visual resource assessment, 287–88
success of, bases for evaluating, 309
Computational fluid dynamics model, of Regional Ocean Modeling System, 279–81, 280f
Conductivity-temperature-depth (CTD) sensors, on autonomous underwater vehicle and environmental sample processor in marine drift experiment, 275, 276
Confederated visualization environments, monolithic *versus*, 272–74
Conservation planning processes
multiobjective, components of, 43
stakeholders in, working with, 36–39
Constructivism, 226
Copper, in pollutant exposure index for Southern California Bight, 112, 113
exposure to, by source, 120, 120t
Coral reefs
acidification-driven damage to, 100–101, 100f
of Martin County, Florida, 58–61 (*See also* Artificial Reef Program [Martin County, FL])
of Saint Croix East End Marine Park, ecosystem of, impacts of adjacent watersheds to, 185–200
background on, 186–87
results of, 196–97, 197f
survey sites for, 193, 193f
value and status of, 186
Corals, warm-water
sea surface temperatures and, 97, 97f
species of, sensitive to land-based pollution, 196–97
Costs
adaptation, estimates of, for ocean changes, 105
multiple, handling of, in Marxan, 36, 38
site selection results in Marxan and, 37, 38f
Critical Period hypothesis, of fish stock variability, 5
Crosswalk tool, in observation toolkit, 82–83, 88
Crosswalking, 82
Cruise Operations Database (MasterCOD), in EcoDAAT, 7
Curtain plot, in marine drift experiment, 276–77, 277f

Cyber-GIS
application of, to marine systems, 3
definition of, 1
description of, 3
in study of natural systems, 2

D
Darwin Core, 77, 78f, 79
Data
for essential fish habitat proposal, identifying, 208–9
multibeam, for bathymetric model of seamount, 257
ocean, complex, visualizing, using Worldwide Telescope, 269–86
oceanographic, relationship to research, framework for, 271–74
Data as a Service (DaaS)
attributes of, 272
in confederated visualization environment, 272–73, 273f
Data caching scheme, for LarvaMap circulation model, 16–17, 17f
Data sharing, regional, best practices for, 180–81
Data-to-visualization mapping, 278–79
Dead model, for LarvaMap, 16
Debris, marine, West Coast Ocean Data Portal expansion for, prioritizing, 177–80, 178f, 179f
Decision support tools (DSTs)
in marine conservation planning, 27–48
Marxan as, 29 (*See also* Marxan)
in systematic planning process, 28–29
use of, strengths and limitations in, 31
Deep Water Horizon (DWH) oil spill
gliders in response to, 320–21, 320f
response of Gulf of Mexico Coastal Ocean Observing System to, 319–20
Deoxygenation, ocean, 92
Department of Land Conservation and Development (DLCD), Oregon, 288
Diel model, for LarvaMap, 16
DIN. *See* Dissolved inorganic nitrogen (DIN)
Dirt roads, in terrestrial impacts to watershed impact zones, 191, 192f
DisMELS model, of fish larva distribution, 6
Dispersal Model for Early Life History Stages (DisMELS model), of fish larva distribution, 6
Dispersion model, for LarvaMap, 15, 15f
Dissolved inorganic nitrogen (DIN), in pollutant exposure index for Southern California Bight, 112, 116
exposure to, by source, 120, 120t, 121f
Dissolved organic matter (DOM), analysis of, using Fourier Transform Ion Cyclotron Resonance Mass Spectrometry, 281–84, 282f, 283f, 284f
Dissolved oxygen concentration, at ocean surface, climate change affecting, 99, 99f
DLCD. *See* Department of Land Conservation and Development (DLCD), Oregon
DOM. *See* Dissolved organic matter (DOM)

Geoform component feature layer, with CMECS, 76f, 87f

Geographic information system (GIS). *See also* ArcGIS
  in Informal Science Education Institutions, 227–28
  Martin County, Florida use of, in coastal programs, 49
    (*See also* Coastal GIS Program [Marlin County, FL])
  public participation (PPGIS), datasets of, in Oregon
    territorial sea plan, 289
  tools using
    to compute pollutant exposure index for Southern
      California Bight, 109–33
    benefits for using, 127–28
    to develop collaborative essential fish habitat
      proposal, 203–22 (*See also* Essential fish habitat
      [EFH], collaborative proposal for, GIS tools in
      developing)
  web, developments in, 136–37

Geography-specific species lists, for Whale mAPP, 157, 158f

GeoJSON track line from LarvaMap, 19

Geoportal Facets Customization (GFC), in West Coast
  Ocean Data Portal, 175, 175f

Geoportals, spatial data infrastructure and, 137–38, 139f

Geospatial approach, to Southern California Bight water
  quality threats assessment, 112–30

Geospatial programs, of Aquarium of the Pacific, 228–39.
  See also Aquarium of the Pacific (Long Beach, CA),
  geospatial programs of

Geovisualization(s)
  description of, 246
  immersive (*See also* Immersive geovisualizations)
    coastal and marine, 251–52, 252f
  marine, 3D, 249–50
  marinescape, 245–67 (*See also* Marinescape
    geovisualization)

GIS. *See* Geographic information system (GIS)

Gliders
  advantages of, 317–18
  in Deep Water Horizon oil spill response, 320–21, 320f
  future plans for, 331–32
  mission viewers for, web-based, 323, 324f
  national context for, 321–22
  National Glider Network Plan for, 321–22
  oceanic, near real-time, 315–35
  profiling, 316, 316f
    in harmful algal bloom monitoring, 323–24, 324f
    for hypoxia studies, mission viewer for, development
      of, 329–31, 331f
  wave, 316–17, 317f
    hazards encountered by, 325
    in monitoring of ocean acidification in northern
      Gulf of Mexico, 324–29, 326f, 327f, 329f

Global climate models (GCMs), 93
  in future projections, 94
  variables available per, 94, 95t

Global Position System (GPS) units, in observation
  toolkit, 82

Golfball coral *(Favia fragum),* in watershed impact zones
  of Saint Croix East End Marine Park, 197

Google
  Liquid Galaxy exhibit of, in Aquarium of the Pacific,
    230–31, 230f
  Ocean Street View project of, immersive visualizations
    in, 252

Google Earth, 2.5D format for, 249

Greenhouse gas emissions, ocean stressors driven by, 92

Groundfish
  essential fish habitat for, establishment for, 205
  habitat areas of particular concern for, 205
  species of, 207

Gulf of Alaska and Bering Sea pollock fishery, history
  of, 4–5

Gulf of Mexico, northern
  hypoxia studies in, using profiling glider, mission viewer
    development for, 329–31, 331f
  ocean acidification in, wave glider monitoring of,
    324–29, 326f, 327f, 329f

Gulf of Mexico Coastal Ocean Observing System
  (GCOOS), future glider projects planned by, 318–21,
  319f, 331–32
  Data Portal of, 322–23
  Data Products page of, 322
  in Deepwater Horizon oil spill response, 319–20
  gliders in approach of, 322–31

H

Habitat areas of particular concern (HAPCs), for
  groundfish, 205

HAGs. *See* Harmful algal blooms (HABs)

Harmful algal blooms (HABs), monitoring, profiling
  gliders in, 323–24, 324f

High-frequency radar, in riverine plume mapping, 114

Humans
  impact of, on marine mammals, 151–52, 152f, 153f
  influence of, on Southern California Bight, 111
  pollock uses by, 3
  use data and values of, in Marxan analyses, 37
  uses of Southern California Bight by, 110

Hutchinson Island Shore Protection Project (2013)
  in Beach Management Program, 62, 63f
  story map for, 64–68, 65f, 66f, 67f, 68f

Hydrodynamic caching strategy, for LarvaMap model
  runs, 16–17, 17f

I

IchPPSI, in EcoDAAT, 7

Immersive geovisualizations
  coastal and marine, 251–52, 252f
  computer game technology and, 253, 253f
  in exploring living sea, incorporating time and
    interactivity in, 252–53, 253f

Individual-based models, of fish larva distribution, 6

Infinite Scuba, immersive geovisualizations in, 253, 253f

Organic matter, dissolved, analysis of, using Fourier Transform Ion Cyclotron Resonance Mass Spectrometry, 281–84, 282f, 283f, 284f

Oxygen concentration, dissolved, at ocean surface, climate change affecting, 99, 99f

P

Pacific coast groundfish. *See also* Groundfish
    essential fish habitat for, establishment of, 205
    species of, 207

Pacific Fishery Management Council (PFMC), role of, 205

Pacific Marine Analysis and Research Association (PacMARA), Marxan experience and, 34

Paddle grass *(Halophila decipiens),* watershed impact zones of Saint Croix East End Marine Park and, 197

Parks Canada, observation toolkit implementation by, 88

Parque Nacional, Archipelago de Cabrera, Islas Baleares, España, map viewer for, 135–49. *See also* Map viewer, Cabrera

Particle tracking models, for LarvaMap, 14–19, 20f. *See* LarvaMap, in tracking transport and dispersion of pollock larvae, models for

PEI. *See* Pollutant exposure index (PEI)

Pelagic settlement, 16

Penguins, Magellanic, Story Map of, at Aquarium of the Pacific, 235, 236f

PFMC. *See* Pacific Fishery Management Council (PFMC)

pH, ocean, climate change affecting, 100–101, 101f

Phosphate, dissolved, concentration of, at ocean surface, climate change affecting, 102, 103f

Phytoplankton, net primary production of carbon by, climate change affecting, 98, 98f, 99f

Plumes in Southern California Bight
    extent of, pollutant exposures and, 128
    riverine
        mapping of, in pollutant exposure index, 114–16, 115f
        spatial extents of, 118, 119f
            precipitation variability and, 123, 124f
    sewage discharge
        mapping of, in pollutant exposure index, 116–17, 117f
        spatial extents of, 118, 119f

Pollock
    fishery for, in Alaska, history and status of, 3–5
    in food chain, 3, 4f
    human uses of, 3
    larvae of
        setting behaviors of, with Larval Behavior Library, 11–12, 11f, 12f
        tracking transport and dispersion of, 10–20, 11f
    life history of, 4
        early, understanding, using individual-based models, 5–6

Pollutant exposure index (PEI), for Southern California Bight
    computing, GIS tool for, 109–33

creation of
    merging pollutant exposure layers in, 117–18
    methods of, 113–18
    riverine plume mapping in, 114–16, 115f
    riverine pollutant loading in, 113–14
    sewage discharge plume mapping in, 116–17, 117f
    sewage discharge pollutant loading in, 116
    development of, benefits of, 112
    future directions for, 129–30
    pollutants in, 112–13
    regional, creation of, GIS in, benefits of, 127–28
    role of, in research and management, 129
    sensitivity analysis for, 118
    in state waters areas of interest, 122, 122t, 123f

Pollutant loading in Southern California Bight
    additional sources of, 129–30
    differential dispersal patterns in, 130
    exposure statistics on, 120, 120t
    riverine
        in pollutant exposure index, 113–14
        precipitation variability and, 123, 124f, 125t, 126f
    sewage discharge, in pollutant exposure index, 116

Pollution, land-based, impact on benthic habitats, 185–200

POTWs. *See* Publicly owned treatment works (POTWs)

Precipitation in Southern California Bight, interannual variability of
    riverine pollutant loading and, 123, 124f, 125t, 126f
    spatial extents of riverine plumes and, 123, 124f

Profiling gliders, 316, 316f
    in harmful algal blood monitoring, 323–24, 324f
    for hypoxia studies, mission viewer for, development of, 329–31, 331f

Project Elements web map, in Story Map app for 2013 Hutchinson Island Shore Protection Project, 65, 65f

Public Information web map, in Story Map app for 2013 Hutchinson Island Shore Protection Project, 66, 66f

Publicly owned treatment works (POTWs) polluting Southern California Bight, 110f, 111
    copper contribution by, 120, 120t
    dissolved inorganic nitrogen contribution by, 120, 120t, 121f
    plume mapping from, 116–17, 117f
    pollutant loading from, 116
    pollutants from, spatial statistics for, 120t
    relative risk from, 128–29
    spatial data and pollutant loading data sources for, 113t
    total suspended solids contribution by, 120t, 121–22, 121f

Puget Sound, coastal upwelling in computational fluid dynamics model of, 279–81, 280f

R

RCPs. *See* Representative concentration pathways (RCPs)

Reefs. *See* Coral reefs. *See also* Artificial Reef Program (Martin County, FL)

Regional data sharing, best practices for, 180–81